Catalysts for Fine
Chemical Synthesis

Volume 3

Catalysts for Fine Chemical Synthesis

Series Editors

Stanley M. Roberts and **Ivan V. Kozhevnikov**
University of Liverpool, UK

Eric Derouane
Universidade do Algarve, Faro, Portugal

Previously Published Books in this Series

Volume 1: Hydrolysis, Oxidation and Reduction
Edited by Stanley M. Roberts and Geraldine Poignant, *University of Liverpool, UK*
ISBN 0 471 98123 0

Volume 2: Catalysis by Polyoxometalates
Edited by Ivan. K. Kozhevnikov, *University of Liverpool*
ISBN 0 471 62381 4

Volume 3: Metal Catalysed Carbon-Carbon Bond-Forming Reactions
Edited by Stanley M. Roberts and Jianliang Xiao, *University of Liverpool, UK* and
John Whittall and Tom E. Pickett, *The Heath, Runcorn Stylacats Ltd, UK*

Forthcoming Books in this Series

Volume 4: Micro- and Mesoporous Catalysts
Edited by Eric Derouane, *Universidade do Algarve, Portugal*
ISBN 0 471 49054 7

Volume 5: Regio- and Stero-Controlled Oxidations and Reductions
Edited by Stanley M. Roberts, *University of Liverpool, UK*
ISBN 0 470 09022 7

Catalysts for Fine
Chemical Synthesis

Volume 3

Metal Catalysed Carbon-Carbon Bond-Forming Reactions

Edited by

Stanley M. Roberts and **Jianliang Xiao**
University of Liverpool, UK

John Whittall and **Tom E. Pickett**
The Heath, Runcorn Stylacats Ltd, UK

John Wiley & Sons, Ltd

Other Wiley Editorial Offices

John Wiley & Sons, Inc., 111 River Street, Hoboken, NJ 07030, USA

Jossey-Bass, 989 Market Street, San Francisco, CA 94103-1741, USA

Wiley-VCH Verlag GmbH, Pappellaee 3, D-69469 Weinheim, Germany

John Wiley & Sons Australia, Ltd, 33 Park Road, Milton, Queensland, 4064, Australia

John Wiley & Sons (Asia) Pte Ltd, 2 Clementi Loop #02-01, Jin Xing Distripark, Singapore 129809

John Wiley & Sons Canada Ltd, 22 Worcester Road, Etobicoke, Ontario, Canada, M9W 1L1

Wiley also publishes its books in a variety of electronic formats. Some content that appears
in print may not be available in electronic books.

Library of Congress Cataloging-in-Publication Data

TP248.65.E59 H98
660'.28443–dc21   2002072357

British Library Cataloguing in Publication Data

ISBN 0-470-86199-1

Typeset in 10/12pt Times by Thomson Press (India) Limited, New Delhi
Printed and bound in Great Britain by MPG Limited, Bodmin, Cornwall
This book is printed on acid-free paper responsibly manufactured from sustainable forestry
in which at least two trees are planted for each one used for paper production.

Contents

Catalysts for Fine Chemical Synthesis
Series Preface

During the early-to-mid 1990s we published a wide range of protocols, detailing the use of biotransformations in synthetic organic chemistry. The procedures were first published in the form of a loose-leaf laboratory manual and, recently, all the protocols have been collected together and published in book form (*Preparative Biotransformations*, Wiley, Chichester, 1999).

Over the past few years the employment of enzymes and whole cells to carry out selected organic reactions has become much more commonplace. Very few research groups would now have any reservations about using commercially available biocatalysts such as lipases. Biotransformations have become accepted as powerful methodologies in synthetic organic chemistry.

Perhaps less clear to a newcomer to a particular area of chemistry is *when* to use biocatalysis as a key step in a synthesis, and when it is better to use one of the alternative non-natural catalysts that may be available. Therefore we set out to extend the objective of *Preparative Biotransformations*, so as to cover the whole panoply of catalytic methods available to the synthetic chemist, incorporating biocatalytic procedures where appropriate.

In keeping with the earlier format we aim to provide the readership with sufficient practical details for the preparation and successful use of the relevant catalyst. Coupled with these specific examples, a selection of the products that may be obtained by a particular technology will be reviewed.

In the different volumes of this new series we will feature catalysts for oxidation and reduction reactions, hydrolysis protocols and catalytic systems for carbon–carbon bond formation *inter alia*. Many of the catalysts featured will be chiral, given the present day interest in the preparation of single-enantiomer fine chemicals. When appropriate, a catalyst type that is capable of a wide range of transformations will be featured. In these volumes the amount of practical data that is described will be proportionately less, and attention will be focused on the past uses of the system and its future potential.

Newcomers to a particular area of catalysis may use these volumes to validate their techniques, and, when a choice of methods is available, use the background information better to delineate the optimum strategy to try to accomplish a previously unknown conversion.

S. M. ROBERTS
I. KOZHEVNIKOV
E. DEROUANE
LIVERPOOL, 2002

Preface for Volume 3: Metal Catalysed Carbon-Carbon Bond-Forming Reactions

Volume 1 in this Series described practical tips for performing some topical oxidation and reduction reactions. This Volume features modern methods for carbon-carbon bond formation, which has always been at the heart of organic synthesis. In recent times, very efficient carbon-carbon bond-forming catalysts have been invented, to assist bench chemists in industry and academia to construct interesting target molecules.

The first Chapter presents a modern overview from an industrial perspective of the employment of catalysts in transformations of commercial importance.

Later Chapters in this Volume describe the use of some of the important catalysts for carbon-carbon bond-forming reactions in detail, often building very significantly on the information available in the primary Journals. Hints and tips on such things as crucial colour changes are included, as well as pointers to potential risks in the procedures. In the cases where the catalyst is not commercially available the preparation of the material is described, again in sufficient detail to make it accessible to the non-expert.

Thus, this Volume is divided into 12 Chapters encompassing different types of carbon-carbon bond forming reactions. The first two experimental Chapters cover alkylation reactions adjacent to carbonyl functionality (Chapter 2) and the asymmetric displacement of acetate groups situated in an allylic position with the resultant formation of optically active product possessing a new carbon-carbon bond (Chapter 3).

The procedures described in Chapters 4–7 all relate to a set of broadly similar transformations that are becoming exceptionally important and well used in laboratories worldwide. In Chapter 4, eight examples of the Suzuki coupling reaction are described; four accounts describe the use on activated alkene (vinyl bromide, triflate or tosylate) as the coupling partner for boronic acid derivatives. The other examples of Suzuki couplings involve aryl bromides and aryl chlorides. It is noteworthy that the methodology introduced by Nolan has been extended to include amination reactions.

Chapter 5 details conditions for performing the Heck reaction (including one asymmetric version) while Chapter 6 describes how Sonogashira reactions can be conducted successfully. (The conditions of Plenio and Luo can be used to perform Heck and/or Suzuki reactions). In Chapter 7 cross-coupling reactions involving

Grignard reagents and organoindium species are reported, together with a variety of cross-couplings invented by Lipshutz, Lautens and colleagues.

The regioselective allylation of selected aldehydes as well as asymmetric carbon-carbon bond-forming reactions using the same carbonyl species are featured in Chapter 8.

Three procedures have been gathered together in Chapter 9 in order to illustrate modern techniques for the use of olefin-methathesis reactions in organic chemistry (Chapter 9). Other cyclisation reactions form the basis of the content of Chapter 10; thus a procedure for the Pauson–Khand reaction is followed by two other methods for producing five-membered rings. In the same Chapter, two procedures leading to six-membered rings are documented, including a [2 + 2 + 2] reaction.

The final two Chapters contain descriptions of equally important reactions. Chapter 11 has experimental details for two asymmetric coupling reactions leading to aldol products as well as two protocols for performing asymmetric Michael reactions. Asymmetric and diastereoselective hydroformylation reactions form the bulk of Chapter 12; we have joined them together with a modern carbonylation procedure and an intriguing carboxylation reaction.

Last but certainly not least, the Editors wish to thank all the 40+ authors for providing their recipes without fuss and/or delay, according to our prescribed format. We hope the detailed descriptions will allow other scientists to have convenient access to a selection of reactions of ever-growing significance and importance in synthetic chemistry.

STANLEY M. ROBERTS
LIVERPOOL, 2004

Abbreviations

Ac	acetyl
acac	acetylacetone
Ar	aryl
b.p.	boiling point
BINAP	2,2′-Bis(diphenylphosphino)-1,1′-binaphthyl
BINOL	*to be confirmed*
BSA	*N,O*-bis-(trimethylsilyl)-acetamide
Bu	butyl
cat	catalyst
Chsalen	[N,N′-bis-(2′-hydroxybenzylidene)]-1,2-diaminocyclo-hexane
CLAMPS	cross-linked aminomethylpolystyrene
COD	*to be confirmed*
DBU	1,8-diazabicyclo[5.4.0]undec-7-ene
de	*to be confirmed*
DEPT	diethyl tartrate
DIPT	diisopropyl tartrate
DMAP	4-dimethylaminopyridine
DMF	Dimethyl formamide
DMM	dimethoxymethane
DMSO	dimethyl sulfoxide
dppf [or DPPF]	diphenylphosphinoferrocene
EDTA	ethylenediaminetetraacetic acid
ee	enantiomeric excess
eq	equivalent
Et	ethyl
EtOAc	Ethyl acetate
GC	gas chromatography
HPLC	high pressure liquid chromatography
ID	internal diameter
IR	infrared (spectroscopy)
L	ligand
lit.	literature
M	metal
m.p.	melting point

MCPBA ⎱	*meta*-chloroperbenzoic acid
m-CPBA ⎰	
Me	methyl
MTPA	methoxy-α-(trifluoromethyl)phenylacetyl
NaOMe	*to be confirmed*
NMR	nuclear magnetic resonance
Ph	phenyl
Pr	propyl
psi	pounds per square inch
PTC	Phase Transfer Catalyst
r.p.m.	rotation per minutes
R_f	retention factor
R_t	retention time
Salen	*to be confirmed*
TBHP	*tert*-butyl hydroperoxide
THF	tetrahydrofuran
TLC	thin layer chromatography
TMS	tetramethylsilane
UHP	urea–hydrogen peroxide
UV	ultraviolet
v:v	volume per unit volume

List of Chemical Names Used

CaH_2	Calcium hydride
$Ca(OH)_2$	Calcium hydroxide
CH_2Cl_2	Dichloromethane
CuCl	Copper (I) chloride
$CuCl_2$	Copper (II) chloride
DMF	Dimethyl formamide
EtOAc	Ethyl acetate
Et_2O	Diethyl ether
Et_2Zn	Diethyl zinc
HOAc	Acetic acid
$KHCO_3$	Potassium hydrogen carbonate
$KHSO_4$	Potassium hydrogen sulfate
$NaHCO_3$	Sodium hydrogen carbonate
Na_2SO_3	Sodium sulfite
$Pd(OAc)_2$	Palladium acetate

1 Considerations of Industrial Fine Chemical Synthesis

MARK W. HOOPER

Senior Chemist, Johnson Matthey, Orchard Road, Royston, Herts SG8 5HE, UK

CONTENTS

1.1 INTRODUCTION

When considering the best synthetic route to a fine chemical, it is very rare for there to be one definite answer. With the ever-expanding range of chemical transformations available, the modern chemist will often be faced with a choice of best options. At any stage there will be decisions made on which chemistry is 'best'. The decisions will include many factors. Some are general, such as reactivity, yield, selectivity, and are considered equally for academic and industrial use. There are, however, factors that are more relevant to industry than to academic research. These include cost obviously, but also less straight-forward factors such as availability of reagents, regulations on their use, health and safety implications on larger scale,

Catalysts for Fine Chemical Synthesis, Vol. 3, Metal Catalysed Carbon-Carbon Bond-Forming Reactions
Edited by S. M. Roberts, J. Xiao, J. Whittall, and T. Pickett
© 2004 John Wiley & Sons, Ltd ISBN: 0-470-86199-1

Intellectual Property Rights (IPR) considerations and practical scale-up issues (e.g. column chromatography is difficult on a tonne scale!). The intervention of these 'secondary' factors can lead to some superb academic synthetic chemistry being under-used (or impossible) on industrial scale.

This chapter explores some of the issues associated with commercialising catalytic synthesis and provides two examples of where metal catalysed carbon-carbon bond forming reactions are being used in industrial fine chemical synthesis. Other reviews detailing important industrial carbon-carbon bond forming reactions are available.[1]

1.2 TYPES OF PROCESSES – FLOW CHARTS

The following flow charts highlight differences between running a classical (non-catalytic) chemical process and running a catalytic process. There is also a specific example of a catalytic process where the catalyst contains a precious metal, e.g. palladium or rhodium (Pd or Rh).

1.2.1 CLASSICAL PROCESS

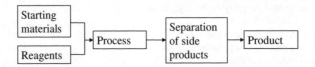

This is a traditional process, using stoichiometric reagents to convert the starting materials to the desired products. Often the processes are well established, e.g. Michael addition, Claisen condensation, and have little or no IPR issues associated with running them. The main costs are the starting materials, the one or more equivalents of reagents, possible multiple-steps needed to achieve the desired transformation and labour costs associated. Also, disposal of a stoichiometric equivalent of side products can involve cost and be detrimental to the environment. Contrary to some opinion, most chemists are strongly in favour of 'green' processes and will opt for the 'cleaner' option if the costs allow.

Another consideration is the increased cost or low availability of starting materials. Often non-catalytic processes require more expensive starting materials as the chemistry will not 'go' with less active, cheaper materials. An example of this is the use of aryl-bromides in place of cheaper aryl-chlorides owing to reactivity constraints. Also, if the desired product is homo-chiral, then the chirality must be introduced through a chiral starting material. The supply of these starting materials are often limited by what is naturally available, i.e. the chiral pool, and this can affect cost and quantity availability.

1.2.2 GENERAL CATALYTIC PROCESS

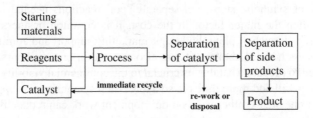

A catalytic process should have many advantages to outweigh the basic disadvantage of adding an extra component to the process. A catalytic system is intrinsically 'greener', i.e. environmentally friendly, as it involves a lower energy process with less waste (sub-stoichiometric reagents). A catalyst will often allow a reaction to occur that was previously impossible. In an ideal world, all reactions would be catalytic with no side product and the lowest possible energy utilisation for the process.

In the real, hard world of industry, however, catalysts bring with them various disadvantages that must be considered against the benefits. Firstly, there is the cost of the catalyst. This can be broken down to raw material cost plus the cost of fabricating the catalyst. Both are important. For example, when the raw material is a precious metal (e.g. palladium) which is not recovered, the cost can be significant. However, if a relatively cheap metal is used (e.g. iron) with a multi-step catalyst preparation, then the catalyst can still have significant cost.

The second major issue is the separation after reaction. In the simplest case, the catalyst is truly heterogeneous and can therefore be separated easily by physical means, e.g. filtration. Often, however, the catalyst is homogeneous or partially removed from the heterogeneous support (leaching). In this case a strategy for separation is needed. Various options are discussed later, including immobilising the catalyst and selective or non-selective scavenging. Ideally the catalyst could be re-cycled immediately without any re-work. However, in many (most) cases the active, turnover catalyst is a modified form of the original catalyst introduced, so that immediate re-use is impossible. Another issue is the IPR associated with use of catalysts. Many catalytic systems are relatively new and are covered by patents. This involves both direct costs (licences) and indirect costs (legal fees).

1.3 COSTS ASSOCIATED WITH USE OF CATALYSTS

1.3.1 CATALYST FABRICATION COSTS

These are costs associated with producing the catalyst or pre-catalyst that is to be added to the process. Various factors contribute to the cost. These include synthesis steps, handling/sensitivity, availability, preparation route and cost of materials. Each is discussed below with an example.

Synthesis steps

The number of synthetic steps, i.e. separate 'pot' reactions needed to prepare a catalyst, is often the major factor in the cost. It is common business practice to allocate overheads such as plant, buildings, marketing support and administration in terms of process hours. Labour costs are also counted in process hours. Therefore, the time taken to prepare a catalyst is crucial to the commercial viability. In general, each extra step will add more time on, so a three-step synthesis costs approximately three times a one-step synthesis. Good development work can reduce this cost, but in turn development is not free! So when considering a new catalyst it is worth bearing in mind how many reaction steps are involved in the preparation of the catalyst.

Handling/sensitivity

The very nature of a catalyst implies that it is reactive or promotes reactivity. This is advantageous to the chemical process, but can be less helpful in handling, storing and preparing catalysts. A recent example is tri-tert-butylphosphine. This is a remarkably good ligand for palladium-catalysed coupling reactions. In the pure state, however, it is a low-melting, pyrophoric, air-sensitive solid. This makes it difficult, and therefore costly, to handle. There are various alternatives to handling the pure solid include pre-forming into complexes, such as $Pd(P^tBu_3)_2$ (**1**) or $[Pd(P^tBu_3)Br]_2$ (**2**), or forming a salt, e.g. $HP^tBu_3BF_4$. These are a lot more stable and can be handled in air.

Pd(P^{t}Bu$_3$)$_2$ **(1)**   [Pd(P^{t}Bu$_3$)Br]$_2$ **(2)**

It is important to remember that, in general, organic reactions are run under an inert atmosphere for safety reasons (i.e. don't heat flammable solvents in the presence of air!). It is normally possible to handle liquids and solutions under anaerobic conditions, e.g. nitrogen atmosphere. Most industrial plants, however, are not set up to handle and transfer solids under anaerobic conditions. It is normal practice to 'load' the solids under air and then put under nitrogen by vacuum cycling or purging. This means that for most uses, solid catalysts must have stability for up to 30 minutes in air. Catalysts can be stored and supplied under nitrogen, so long term air-stability is less important. Therefore, an industrial definition of critical air-sensitivity of solids is: "can the activity be significantly affected by short-term (up to 30 minutes) exposure to air?"

Availability

Another issue that affects cost is the availability of certain chemicals on an industrial scale. It is often the case that the cost and availability of a chemical in a research catalogue do not reflect the situation on a larger scale. This can be illustrated in the following two examples.

Rhodium-catalysed hydroboration is a powerful tool for introducing chirality into a styrene-derivative. (Figure 1.1)[2] This was in competition to the established route based on chiral resolution using separation of diastereomers formed from reaction of the racemic amines with homo-chiral acids (natural pool). However, although the process appeared favourable from the chemical synthetic route, the process was practically impossible owing to there being no supplier of catecholborane on large scale at the time.

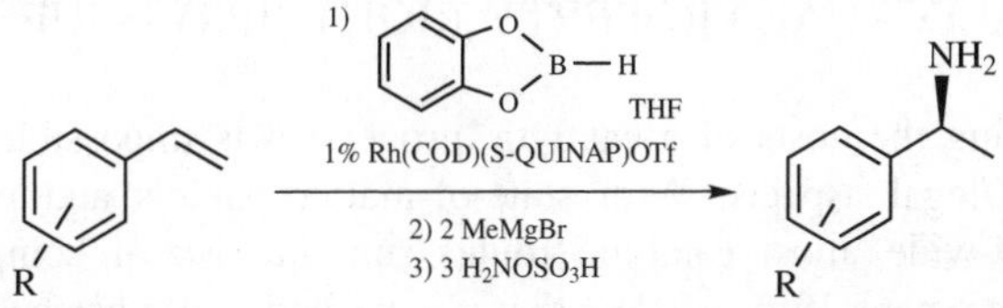

Figure 1.1 Rhodium-catalysed hydroboration of a styrene derivative.

A second example is found in ruthenium-catalysed hydrogenation studies. A lot of the academic work is carried out using a catalyst generated *in-situ* from Ru(benzene)Cl$_2$ plus a ligand. This leads to industrial research starting from Ru(benzene)Cl$_2$. However, most of the major suppliers of transition metal catalysts do not list the benzene complex as a sales product but do list the *p*-cymene complex, Ru(*p*-cymene)Cl$_2$. This is not due to any major chemical difference in the use or preparation of the two complexes. The complexes are prepared from the corresponding cyclohexadienes, and the *p*-cymene pre-cursor is approximately 10 times cheaper and available on larger scale. So unless there is a specific advantage of the benzene complex, the *p*-cymene complex is used in industrial applications.

Preparation route

Another factor in the cost is the route of preparation. An example of this can be illustrated using two common palladium pre-cursors, Pd(MeCN)$_2$Cl$_2$ and Pd(PhCN)$_2$Cl$_2$. Again, in academic work it is common to see the benzonitrile complex used, presumably due to its slightly better solubility in a range of organic solvents. However, the acetonitrile complex has better crystallisation characteristics, and is therefore much cheaper to produce. So Pd(MeCN)$_2$Cl$_2$ is more common in industrial applications.

Other considerations are environmental impact and health and safety. Chlorinated solvents have a bad reputation, with carbon tetrachloride mostly banned owing to its carcinogenic properties, chloroform being phased out and questions over the use of dichloromethane. Therefore, complexes that require the use of chlorinated solvents in their preparation will entail either more care in the synthesis, or more development work towards alternative solvents. Either way more cost is involved.

Cost of materials

Finally, more complex ligands will generally add extra cost. This can be illustrated in the field of palladium-catalysed coupling chemistry.[3] Within the literature two palladium-precursors dominate; $Pd(OAc)_2$ and $Pd(dba)_2$ (or $Pd_2(dba)_3$) (dba = dibenzylideneacetone). These are often interchangeable, with no simple pattern to which is more efficient. However, in an industrial setting, the cost of the ligands will affect the overall cost of the catalyst pre-cursor. The acetate ligand is derived from acetic acid, which is substantially cheaper than dba.

1.3.2   INTELLECTUAL PROPERTY RIGHT (IPR) ISSUES

When considering the costs of a catalytic process, it is impossible these days to ignore the IPR/legal aspects. With state-of-matter patents and process patents available world-wide, most catalytic routes run the risk of being 'covered' to some extent in a patent. If the process that it is wished to use has been claimed in a patent then there are two basic choices, after determining that it is not possible to avoid the patent by 'designing around'. Firstly some sort of license can be arranged. This commonly involves either paying a straight fee to gain access to use of the technology, payment of an amount linked to the volume/revenue generated from use of the technology, or negotiating some form of cross-licence. These can be considered as direct costs attributable to the process.

Secondly, the patent can be challenged. This obviously involves a direct cost of legal fees, which can be very substantial, and indirect costs of time delay and executive time involved. At all stages of the patent process, legal costs are incurred, both in writing and submitting a patent, and in assessing and applying for a licence or for patent revocation. The outflow of cash (ultimately of profit) from the chemical industry to the legal industry is evident at each point. This must be justified by the advantages of the patented system, and adds an extra burden on the choice of route.

Another possible 'indirect' cost of patented technology is the potential loss of control of the process. For example, if a pharmaceutical product of company A has just one synthetic route approved which involves a patented step controlled by company B, then there is potential for company B to exploit the position. This often leads to company A either paying extra legal fees to protect itself, or abandoning the technology. A real example of the control of patented technology is the industrial use of BINAP, which is one of the most versatile and active chiral ligands ever discovered. However, its industrial use has been limited as the patent has been strongly defended. Therefore the potential of the technology has been never fully realised. Often better chemical routes are abandoned due to the costs (direct and indirect) associated with the patents.

Patenting is a strong way of protecting a real or potential source of income. If a process is being run and revenue is being generated, then patenting can defend that income from other entrants. There is a growing tendency of patenting in academic institutions to try and generate future revenues for the researcher/department/institution.

1.3.3 SEPARATION COSTS

On using a catalyst in a process, by definition an extra component is being added to the reaction mixture. This means that to get a pure product, the catalyst must be separated at the end of the reaction. There are various strategies for this operation, which can be divided into pre-reaction strategies, such as immobilisation, and post-reaction strategies, including classical routes and selective or non-selective scavenging. Separation can be critical in making a process commercially viable.[4] This can be a major difference between academic work and industrial work. For example, many academic reactions end with a purification step *via* flash column chromatography. However, on larger scale chromatography becomes very expensive or impossible, with the volumes of solvents required being costly and certainly not environmentally efficient. Therefore the development of alternative separation techniques can often be the major hurdle in commercialisation of academic technology.

1.3.4 PRE-REACTION/IMMOBILISATION

Many catalysts are homogeneous, i.e. they react in the same phase as the substrates. One strategy to aid the separation is to immobilise the homogeneous catalyst onto a support that is heterogeneous to the reaction phase. The catalyst can then be physically separated easily. The important factors for an immobilised catalyst are:

1. The catalytic activity should be as close to the original homogeneous catalyst as possible.
2. The catalyst should not 'leach', i.e. it should stay attached to the support.
3. The support should add as little cost as possible to the catalyst

The issue of added cost can be reduced if the immobilisation allows the catalyst to be re-cycled. However, this is often complicated by the application of Good Manufacturing Practice (GMP) discouraging re-use of a catalyst in different batches of product (possible cross-batch contamination). Also, the catalyst is often more sensitive after the reaction, and filtration/storage can lead to catalyst de-activation.

Strategies for immobilisation include supporting the catalysts *via* a ligand, anchoring or absorbing onto a solid surface, and encapsulating the catalyst in a heterogeneous media.[5] An example that illustrates some of these concepts is detailed below.

FibreCat®

FibreCat® is a new generation of catalysts, which combine the selectivity of homogeneous catalysts with the ease of handling and separation of heterogeneous catalysts.[6] The catalysts are anchored to a series of functionalised fibres; the composition of the fibres can be modified to ensure compatibility with a wide range of solvent and reaction systems. FibreCat® catalysts demonstrate activities that are comparable to their homogeneous analogues.

FibreCat® catalysts consist of a polymer fibre that is inert and insoluble in all solvents to which functional groups, ligands and the precious metal can be added. The fibres are functionalised *via* graft co-polymerisation, which results in a high density of active functional sites being generated on the polymer. Further modification of the functional groups on the fibre is then possible to meet any linking requirements. It is also possible to have the reagents directly associated with the catalyst so that they do not need to be added to the process. The functionalised polymer can then be used as a support for particulate metal catalysts or for metal complexes that are active homogeneous catalysts. The range of products is suitable for many catalytic reactions, for example the FibreCat® 1000 series for carbon-carbon coupling reactions. These include: FibreCat® 1001, FibreCat® 1007 and FibreCat® 1026.

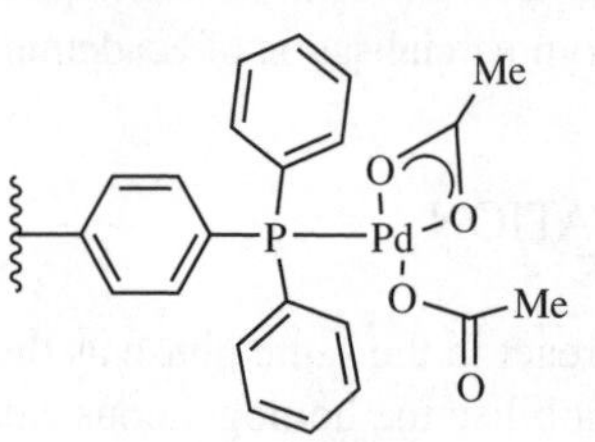

FibreCat® 1001

PPh₂-fibre/Pd(OAc)₂

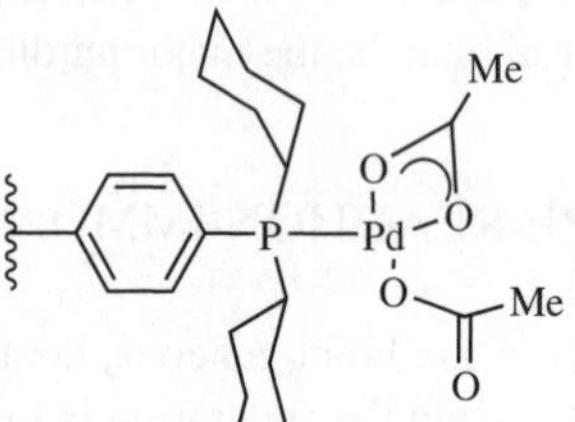

FibreCat® 1007

PCy₂-fibre/Pd(OAc)₂

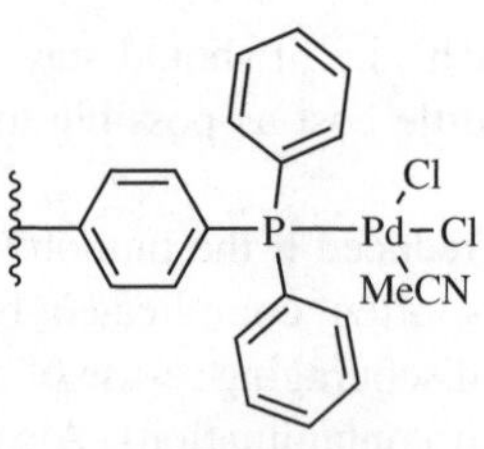

FibreCat® 1026

PPh₂-fibre/PdCl₂/MeCN

1.3.5 POST REACTION – SEPARATION

There are many 'classical' separation techniques used in chemical processes. These are equally applicable to separating out catalysts.

Common techniques include:

- crystallisation of product or catalyst,
- distillation of the product,

- phase-separation using an added non-miscible component, e.g. solvent extraction,
- chemical treatment to destroy or precipitate the catalyst or product.

It is also common to use scavengers to remove catalysts from the product. These can be non-selective, i.e. absorbing both catalyst and product, or selective, i.e. targeting just the desired component/catalyst. Examples of non-selective scavengers include activated charcoal/carbon, silica, alumina and Keiselgel. Generally, column chromatography is not favoured in industrial applications.

There has recently been extensive development of selective scavenging systems. These include ion-exchange resins (beads), functionalised polymers[7] and functionalised silica. The scavengers can target the whole active catalyst, or just the metal or ligand components if desired. It is possible in some cases to remove the catalyst from the scavenger and recycle it directly.

1.3.6   INDUSTRIAL EXAMPLES

Two examples are highlighted below where precious metal catalysts are used to produce fine chemicals on an industrial scale via carbon-carbon bond forming reactions. The first (a) is rhodium-catalysed hydroformylation in the 'oxo-process', which is a well established industrial process. The second (b) highlights a new process developed by Lucite involving a palladium-catalysed methoxy-carbonylation. Many of the points mentioned above in this article are illustrated in the examples, with efficient recycle of catalyst (precious metal) and the extra cost of ligands being justified by the costs savings of the novel chemistry.

(a) Rh Oxo process (developed by ICI)

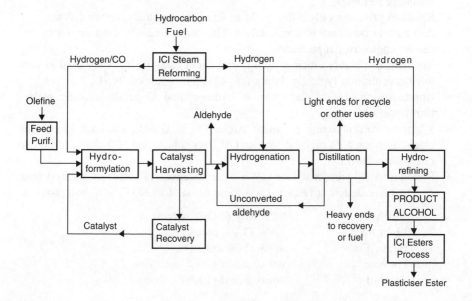

The Rh Oxo process comprises two main steps, hydroformylation followed by hydrogenation. The hydroformylation is carried out in the liquid/gas phase with a homogeneous rhodium catalyst. This is the carbon-carbon bond forming step.

$$R-CH=CH-R' \quad + \quad CO + H_2 \quad \rightarrow \quad R-CH_2-CH(CHO)-R'$$
$$\text{olefin} \qquad\qquad \text{syngas} \qquad\qquad \text{aldehyde}$$

This is followed by a hydrogenation reaction, which is carried out in the vapour or liquid phase over heterogeneous catalyst

$$R-CH_2-CH(CHO)-R' \quad + \quad H_2 \quad \rightarrow \quad R-CH_2-CH(CH_2OH)-R'$$
$$\text{aldehyde} \qquad\qquad \text{hydrogen} \qquad\qquad \text{alcohol}$$

Characteristics of unligated rhodium (I)

In general, a rhodium catalyst is used as it efficiently converts olefin to alcohol:

e.g. Heptenes $\rightarrow$ iso-Octanol, Octenes $\rightarrow$ iso-Nonanol, Nonenes $\rightarrow$ iso-Decanol

Unligated rhodium has the ability to hydroformylate a wide range of different olefins, both branched internal forms as well as linear. Virtually no water is used or created in the process (unlike a cobalt-catalysed system which needs water for catalyst recovery). Other key advantages of a rhodium (I) system are:

- Rhodium has very little hydrogenation capability, which minimises by-product formation. (Cobalt is an alternative hydrogenation catalyst, which leads to significant alcohol formation and causes the formation of acids and esters including formates).
- Rhodium produces only a low yield of heavy by-products (unlike cobalt).
- Rhodium is an isomerisation catalyst. This feature can be used or not by the correct choice of temperature.
- Rhodium is a highly efficient hydroformylation catalyst. It is only used in very low concentration typically 4 ppm (cf. 400 ppm cobalt catalyst).
- Approximately 1% of the olefin is hydrogenated to paraffin during hydro-formylation.
- When hydroformylating a Linear Alpha Olefin (LAO), rhodium inserts the aldehyde group 52% onto carbon one (70% for cobalt) and 48% onto carbon two (30% for cobalt).
- Rhodium will hydroformylate LAOs to an aldehyde of sufficient purity to use directly to make derivatives, e.g. hydroformylation of a LAO C_{12}-C_{14} will produce:

Aldehyde	94% w/w (44.6% cobalt)
Light Ends	1.4% (3.7% cobalt)
Heavy Ends	3.9% (37% cobalt)
Formate	not detected (4.3% cobalt)
Alcohol	not detected (10.4% cobalt)

- The catalytic entity is a rhodium carbonyl species. To be present, this requires high partial pressures of carbon monoxide and hydrogen. Hence, hydroformylation with unliganded rhodium requires the use of high pressure (in excess of 200 bar).

As it is of significant value, the rhodium is recovered and recycled very efficiently and it has little environmental impact.

Characteristics of 'liganded' rhodium (I)

By attaching a ligand containing a Group V element to rhodium, the stability of the Rh:CO linkage can be changed. Phosphorus was the element of choice. By attaching three phenyl groups to the phosphorus (triphenyl phosphine, TPP), the selectivity to linear aldehydes is enhanced. CO and hydrogen will attach to rhodium liganded with TPP and the complex is stable at low pressures unlike unliganded rhodium. If the pressure is increased, the CO displaces the TPP. This problem was solved by operating with a large excess of TPP to maintain the stability of the rhodium:TPP

The presence of the large TPP group attached to the rhodium makes the rhodium less active catalytically. Hence, a higher concentration of rhodium is required and the process is used for hydroformylating the most active olefins. The large size of the rhodium:TPP species brings with it steric effects: for an alpha olefin, insertion of the aldehyde group is almost entirely on carbon atom one ($> 90\%$); for branched, internal olefins, little or no insertion takes place. The liganded rhodium catalyst is retained within the plant, but at long intervals it is taken out for rhodium recovery and replacement with a new charge of catalyst.

(b) Methylmethacrylate process (Lucite's new homogeneous palladium-catalysed process)

Methylmethacrylate (MMA) is the building block for a wide range of products. It polymerises easily to form transparent resins and polymers, e.g. kitchen and bathroom surfaces, co-polymerises with other monomers, and is used in paints and coatings.

Lucite's 'alpha process'

This is a new technology developed by Lucite (formerly Ineous, formerly ICI).[8] It involves a two-step process: a liquid phase methoxy-carbonylation followed by a gas phase condensation.

$$CO + C_2H_4 + MeOH \rightarrow CH_3CH_2CO_2CH_3$$
$$CH_3CH_2CO_2CH_3 + CH_2O \rightarrow CH_2C(CH_3)CO_2CH_3$$

The first step involves a homogeneous palladium-catalyst based on a chelating bis-phosphine (3). Some general features of the process are:

- 1 kg of catalyst will produce 200 tonnes of methyl propionate.
- Catalyst allows mild conditions: 10 bar, 100 °C.
- Virtually no by-products. Pilot plant in operation at Lucite, Teeside.
- 100,000 tonne/year plant planned for 2006.
- 20% cost advantage over conventional cyanohydrin technology.

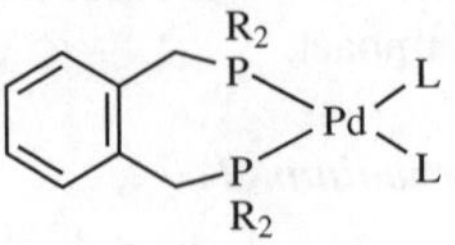

R = ^{i}Pr, Cy, ^{t}Bu, Ph

Lucite's methylpropionate catalyst (3)

Key issues associated with the process are the efficient use of catalyst, high turn-over numbers, high selectivity of catalyst to prevent by-product formation, effective separation of metal from product in high yield and quick recovery of metal, refining and recycle as fresh catalyst.

REFERENCES

1. *Asymmetric Catalysis on Industrial Scale* (ed. H.U. Blaser and E. Schmidt) Wiley-VCH, Weinheim, **2004**; *Metal-Catalyzed Cross Coupling Reactions* (eds. F. Diederich and P. J. Stang) Wiley-VCH, Weinham, **1998**; *Transition Metals for Organic Synthesis* (eds. M. Beller and C. Bolm) Wiley-VCH, Weinheim, **1998**.
2. C. M. Crudden and D. Edwards, *Eur. J. Org. Chem.*, **2003**, 4695–4712: J. M. Brown et al. *Chem. Commun.* **1997**, 173–4.
3. For advances in fine chemical synthesis using homogeneous palladium catalysts see: A. Zapf and M. Beller, *Topics in Catalysis*, **2002**, *19*, 101.
4. For example the recycle of palladium catalysts is very important in commercially-viable Heck-type reactions see: J. G. de Vries, *Can. J. Chem.*, **2001**, *79*, 1086–92.
5. For a recent review on asymmetric C-C and C-heteroatom bond-forming reactions using polymer-bound catalysts see: S. Bräse, F. Lauterwasser and R. E. Ziegert, *Adv. Synth. Catal.*, **2003**, *345*, 869–929.
6. S. Collard in *Catalysis of Organic Reactions* (ed. D. G. Morrell) Marcel Dekker, New York, **2002**, 49–60.
7. For example Smopex®, S. Buckley, *Speciality Chemicals*, October **2002**, 12–13.
8. (a) *European Chemical News*, 23–29 September **2002**, 29. (b) Clegg, W.; Eastham, G. R.; Elsegood, M. R. J.; Tooze, R. P.; Wang, X. L.; Whiston, K. *Chem. Commun.*, **1999**, 1877.

2 Alkylation and Allylation Adjacent to a Carbonyl Group

CONTENTS

Catalysts for Fine Chemical Synthesis, Vol. 3, Metal Catalysed Carbon-Carbon Bond-Forming Reactions
Edited by S. M. Roberts, J. Xiao, J. Whittall, and T. Pickett
© 2004 John Wiley & Sons, Ltd   ISBN: 0-470-86199-1

2.1   THE RuH$_2$(CO)(PPh$_3$)$_3$-CATALYSED ALKYLATION, ALKENYLATION, AND ARYLATION OF AROMATIC KETONES VIA CARBON-HYDROGEN BOND CLEAVAGE

FUMITOSHI KAKIUCHI,* SATOSHI UENO and NAOTO CHATANI

Department of Applied Chemistry, Faculty of Engineering, Osaka University, Suita, Osaka 565-0871, Japan

Catalytic functionalisation of otherwise unreactive carbon-hydrogen bonds has been one of the most attractive research subjects during this decade.[1] One of the most important factors in the success of these reactions involves chelation-assistance by a heteroatom. Thus, the co-ordination of the heteroatom to the metal, brings the metal closer to the carbon-hydrogen bond and stabilises the thermally unstable C-M-H species formed by the oxidative addition of a carbon-hydrogen bond to a low valent transition metal complex. In addition, the use of the chelation-assistance leads to a high regioselectivity, an essential factor in organic synthesis. To date, several types of catalytic reactions have been developed by means of chelation-assistance, such as alkylation with olefins,[2,3] alkenylation with acetylenes,[4] arylation with organometallic reagents,[5] acylation with olefins and carbon monoxide,[1] and silylation with hydrosilanes.[1]

Described in this section is a convenient synthetic procedure for preparation of RuH$_2$(CO)(PPh$_3$)$_3$,[3,6] which is a highly effective catalyst for the conversion of carbon-hydrogen bonds to carbon-carbon bonds, and typical procedures for RuH$_2$-(CO)(PPh$_3$)$_3$-catalysed reactions of aromatic ketones with olefins,[2,3,7] acetylenes,[4] and arylboronates,[5] giving *ortho* alkylation, alkenylation, and arylation products, respectively.

2.1.1   PREPARATION OF CARBONYLDIHYDROTRIS (TRIPHENYLPHOSPHINE)RUTHENIUM[3,6]

$$RuCl_3\text{-}3H_2O + PPh_3 + HCHOaq + KOH \xrightarrow[\text{reflux, overnight}]{\text{EtOH 180 mL}} RuH_2(CO)(PPh_3)_3$$

    2 mmol   12 mmol   20 mL   10 mmol   55–65%

- Hydrated ruthenium trichloride (RuCl$_3$-3H$_2$O), 0.52 g, 2 mmol
- Triphenylphosphine (>99%), 3.14 g, 12 mmol
- Formalin (40 wt%) 20 mL
- Potassium hydroxide (85%), 0.71 g, 10 mmol
- Ethanol (>99%), 240 mL (commercial grade)
- Deionized water, 50 mL
- Hexane, 50 mL
- Benzene, 430–630 mL

- Alumina (Merck aluminium oxide 90 active, neutral activity I), *ca* 80 g
- Methanol, 150 mL

- One 300-mL two-necked, round-bottomed flask
- One magnetic stirring bar
- One rubber septum
- One reflux condenser
- One funnel
- Three 50-mL Erlenmeyer flasks
- One Büchner funnel, diameter 4 cm
- One glass sintered funnel, diameter 3 cm
- One glass column for alumina column chromatography, 30-mm inner diameter

Procedure

In a 300-mL two-necked, round-bottomed flask equipped with a magnetic stirring bar, a rubber septum, and a reflux condenser are placed triphenylphosphine (3.14 g, 12 mmol) and 140 mL of ethanol. The flask is carefully evacuated using a 5 mmHg vacuum line at room temperature until the solution just begins to boil, and then the flask is rapidly refilled with nitrogen at atmospheric pressure. This procedure is repeated three times. The solution is refluxed. To the boiling ethanol solution of triphenylphosphine, an ethanol (20 mL) solution of hydrated ruthenium trichloride (0.52 g, 2 mmol), formalin (40 wt%, 20 mL, *ca* 267 mmol), and ethanol (20 mL) solution of potassium hydroxide (85% content, 0.71 g, 10 mmol) are added successively *via* a syringe. The solution is heated under reflux overnight (for 12–15 h) and then cooled in an ice bath. The resulting light brown precipitate is filtered using Büchner funnel (diameter 4 cm), washed with 30 mL of ethanol, 50 mL of deionised water, 30 mL of ethanol, and 50 mL of hexane, and then dried in *vacuo*. The crude product is dissolved in 30 mL of warm benzene (*ca* 50–60 °C) in a 50-mL Erlenmeyer flask; the solution is filtered to remove insoluble solids using a funnel. The benzene solution is passed through a neutral alumina column (packing: Merck aluminum oxide 90 active, neutral activity I; 30 mm inner diameter × 120 mm length; eluent: benzene (*ca* 400–600 mL)). The elution of $RuH_2(CO)$ $(PPh_3)_3$ is monitored by UV_{254}. The benzene eluent is concentrated to *ca* 100 mL by rotary evaporation. To the concentrated solution is added 150 mL of methanol. At this point a white solid is precipitated. The resulting liquid is concentrated to *ca* 50 mL by rotary evaporation. The concentrated liquid is cooled in an ice bath to enforce precipitation of the solids. A colorless solid of pure $RuH_2(CO)(PPh_3)_3$ is filtered using a glass sintered funnel (diameter 40 mm), washed with hexane, and dried in *vacuo*. Yield 1.0–1.2 g (55–65% yield).

Note: This reaction can be carried out under air. When all manipulations are performed under air, the ruthenium complex is obtained in ca 50% yield (0.9 g).

^{1}H NMR (270 MHz, C_6D_6) –8.29 (dtd, $^2J_{PH}$ = 72.3 Hz, $^2J_{PH}$ = 28.4 Hz, $^2J_{HH}$ = 5.9 Hz, 1H, Ru-H), –6.43 (dtd, $^2J_{PH}$ = 30.2 Hz, $^2J_{HH}$ = 14.6 Hz, $^2J_{HH}$ = 5.9 Hz, 1H, Ru-H), 6.8–8.0 (m, 45H, Ph).
^{31}P NMR (109 Hz, C_6D_6) 45.86 (t, J_{PP} = 17.6 Hz), 57.99 (d, J_{PP} = 17.6 Hz).

2.1.2   SYNTHESIS OF 8-(2-TRIETHOXYSILANYLETHYL)-3,4-DIHYDRO-2H-NAPHTHALEN-1-ONE[2,3,7]

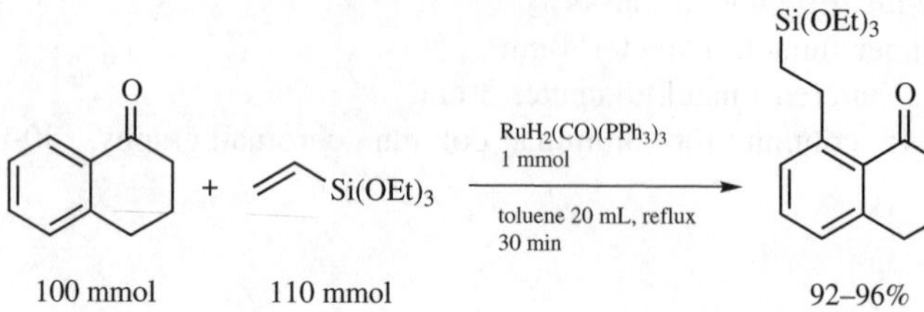

100 mmol   110 mmol   92–96%

Materials and equipment

- α-Tetralone, 14.62 g, 100 mmol
- $RuH_2(CO)(PPh_3)_3$, 0.918 g, 1 mmol
- Triethoxyvinylsilane, 20.93 g, 110 mmol
- Toluene, 30 mL

- One 100-mL two-necked, round-bottomed flask
- One magnetic stirring bar
- One reflux condenser
- One nitrogen inlet tube sealed with a rubber septum
- One 50-mL round-bottomed flask
- One distillation apparatus

Procedure

An apparatus consisting of a 100-mL two-necked round-bottomed flask, a reflux condenser connected to a vacuum/nitrogen line, an inlet tube sealed with a rubber septum, and a magnetic stirring bar is evacuated, then flushed with nitrogen. This cycle is repeated four times. The apparatus is flame-dried under a flow of nitrogen, and then cooled to room temperature under a nitrogen atmosphere. Carbonyl-dihydridotris(triphenylphosphine)ruthenium(II), $RuH_2(CO)(PPh_3)_3$, (0.918 g, 1.00 mmol) is placed in the flask under a slow flow of nitrogen. Addition of 20 mL of toluene to the flask gives a suspension of white solids. To the suspension are added triethoxyvinylsilane (20.93 g, 110 mmol) and α-tetralone (14.62 g, 100 mmol), in this order, through the rubber septum using a syringe. The resulting mixture containing the white solids is refluxed in an oil bath (temperature 135 °C). After heating for 30 minutes, the reaction mixture is cooled to room temperature. About half of the reaction mixture is transferred to a 50-mL round-bottomed flask. Toluene and triethoxyvinylsilane are removed under reduced pressure. After almost all volatile materials are removed, the other half of the reaction mixture is transferred to the flask. The reaction vessel is rinsed with two 5 mL portions of toluene; the rinses are also transferred to the flask. The volatile materials of the combined reaction mixture and rinses are evaporated. Distillation of the residue under reduced pressure gives 31–32.5 g (92–96% yield) of 8-(2-trimethylsilanylethyl)-3,4-dihydro-2H-naphthalen-1-one as a pale yellow liquid, bp 133–135 °C/0.2 mmHg.

^{1}H NMR (270 MHz, CDCl$_3$) 0.96–1.02 (c, 2H, SiCH$_2$), 1.26 (t, $J = 7.02$ Hz, 9H, CH$_3$), 2.07 (quint, $J = 6.48$ Hz, 2H, CH$_2$CH$_2$CH$_2$), 2.64 (t, $J = 6.48$ Hz, 2H, ArCH$_2$), 2.95 (t, $J = 6.48$ Hz, 2H, C(O)CH$_2$), 3.10–3.16 (c, 2H, ArCH$_2$CH$_2$Si), 3.88 (q, $J = 7.02$ Hz, 6H, OCH$_2$), 7.09 (d, $J = 7.56$ Hz, 1H, ArH), 7.14 (d, $J = 7.56$ Hz, 1H, ArH), 7.33 (t, $J = 7.56$ Hz, 1H, ArH).

^{13}C NMR (67.5 Hz, CDCl$_3$) 12.11, 18.06, 22.68, 28.41, 30.82, 40.79, 58.01, 126.56, 128.99, 130.17, 132.15, 145.48, 147.78, 199.28.

2.1.3 SYNTHESIS OF 8-(1-METHYL-2-TRIMETHYLSILANYLVINYL)-3,4-DIHYDRO-2H-NAPHTHALEN-1-ONE[4]

Materials and equipment

- α-Tetralone, 0.292 g, 2 mmol
- RuH$_2$(CO)(PPh$_3$)$_3$, 0.110 g, 0.12 mmol
- 1-trimethylsilylpropyne, 0.448 g, 4 mmol
- Toluene, mL

- One 10-mL two-necked, round-bottomed flask
- One magnetic stirring bar
- One reflux condenser
- One nitrogen inlet tube sealed with a rubber septum
- One 10-mL round-bottomed flask

Procedure

The apparatus, consisting of 10-mL two-necked flask equipped with a reflux condenser connected to a nitrogen line, a rubber septum, and a magnetic stirring bar, is flame dried under a flow of nitrogen. In the flask are placed the ruthenium complex (0.110 g, 0.12 mmol), 3 mL of toluene, α-tetralone (0.292 g, 2 mmol), and 1-trimethylsilylpropyne (0.448g, 4 mmol). The resulting mixture is refluxed under a nitrogen atmosphere. After refluxing for 3 hours, the reaction mixture is transferred into a 10-mL round-bottomed flask, and the reaction vessel is rinsed with two 2-mL portions of toluene. The reaction mixture and rinses are evaporated under reduced pressure to remove the volatile materials (toluene and trimethylsilylpropyne). Bulb-to-bulb distillation under reduced pressure gives 0.428 g (83% yield) of 8-(1-methyl-2-trimethylsilanylvinyl)-3,4-dihydro-2H-naphthalen-1-one as a pale yellow liquid, 110 °C/2 mmHg.

^{1}H NMR (CDCl$_3$) 0.02 (s, 9H, CH$_3$), 2.04 (s, 3H, CH$_3$), 2.10 (quint, J = 6.3 Hz, 2H, CH$_2$), 2.63 (t, J = 6.3 Hz, 2H, C(O)CH$_2$), 2.96 (t, J = 6.3 Hz, 2H, ArCH$_2$), 5.24 (s, 1H, CH =), 7.01 (d, J = 7.6 Hz, 1H, ArH), 7.14 (d, J = 7.6 Hz, 1H, ArH), 7.35 (t, J = 7.6 Hz, 1H, ArH).

^{13}C NMR (CDCl$_3$)–0.29, 22.81, 23.02, 30.51, 40.20, 124.42, 127.33, 128.01, 129.26, 132.26, 144.87, 149.76, 157.65, 197.68.

2.1.4   SYNTHESIS OF 1-BIPHENYL-2-YL-2,2-DIMETHYLPROPAN-1-ONE[5]

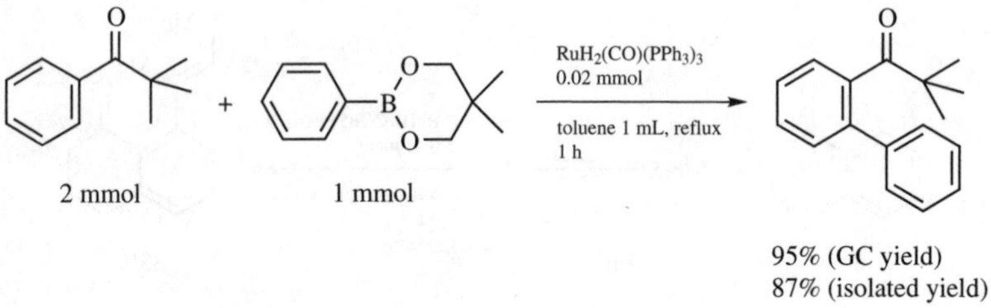

2 mmol   1 mmol

95% (GC yield)
87% (isolated yield)

Materials and equipment

- 1-Phenyl-2,2-dimethylpropan-1-one, 0.324 g, 2 mmol
- RuH$_2$(CO)(PPh$_3$)$_3$, 0.018 g, 0.02 mmol
- 5,5-Dimethyl-2-phenyl-[1,3,2]dioxaborinane, 0.190 g, 1 mmol
- Toluene, 1 mL

- One 10-mL two-necked, round-bottomed flask
- One magnetic stirring bar
- One reflux condenser
- One nitrogen inlet tube sealed with a rubber septum
- One 10-mL round-bottomed flask
- One glass column for silica gel column chromatography, 30-mm inner diameter

Procedure

The apparatus, consisting of a 10-mL two-necked flask equipped with a reflux condenser connected to a nitrogen line, a rubber septum, and a magnetic stirring bar, is flame dried under a flow of nitrogen. In the flask are placed RuH_2 $(CO)(PPh_3)_3$ (0.02 mmol), 1 mL of toluene, 1-phenyl-2,2-dimethylpropan-1-one (2 mmol), and 5,5-dimethyl-2-phenyl-[1,3,2]dioxaborinane (1 mmol). The resulting mixture is refluxed for 1 hour under nitrogen atmosphere. The reaction mixture is placed in a 10-mL round-bottomed flask and is concentrated. Silica gel column chromatography (30-mm inner diameter × 150 mm length; silica gel: Merck 60, 230–400 mesh, *ca* 50 g; eluent: hexane/EtOAc = 50/1) of the residue gives 0.207 g (87% isolated yield; 95% GC yield) of 1-biphenyl-2-yl-2,2-dimethylpropan-1-one as a colorless solid, m.p. 47 °C.

Note: In some cases, the arylation product cannot be isolated in a pure form by a silica gel column chromatography. In this case, a further purification of the crude product by bulb-to-bulb distillation or by GPC (gel permeation chromatography) affords the pure product.

^1H NMR $(CDCl_3)$ 0.87 (s, 9 H, CH_3), 7.14 (d, $J = 3.8Hz$, 1 H, ArH), 7.3–7.5 (m, 8 H, ArH).
^{13}C NMR $(CDCl_3)$ 27.29, 45.00, 125.59, 126.65, 127.48, 128.33, 128.60, 129.48, 129.63, 137.92, 140.82, 140.95, 216.38.

2.1.5   CONCLUSION

The catalytic methods for conversion of carbon-hydrogen bonds to carbon-carbon bonds have already become a useful tool in organic synthesis. The protocol is simple. A carbon-hydrogen bond is cleaved by transition metal complexes, and then reacts with olefins, acetylenes, and arylboronates resulting in a new carbon-carbon bond. For these reactions, the co-ordination of heteroatom to the metal center is important to attain high regioselectivity and high efficiency. In the case of alkylation, aromatic ketones show high reactivity (Table 2.1). In addition to a

Table 2.1   Alkykation of aromatic ketones with triethoxyvinylsilane.

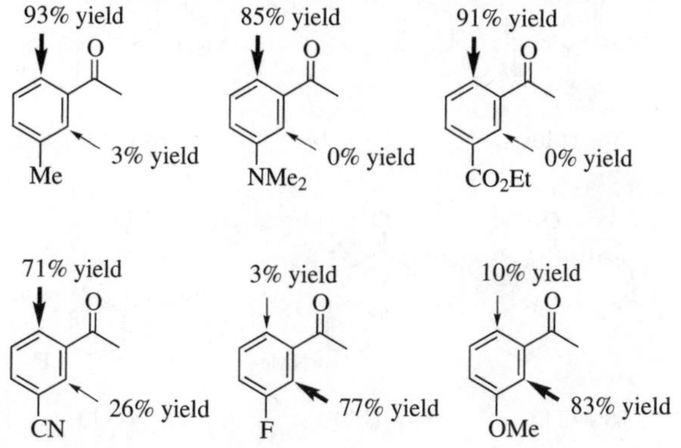

Table 2.2   Alkykation of several aromatic compounds.

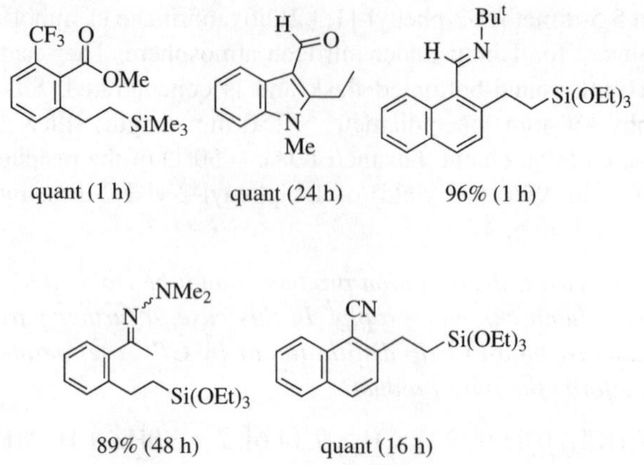

| quant (1 h) | quant (24 h) | 96% (1 h) |

| 89% (48 h) | quant (16 h) |

quant = quantitative yield

ketone carbonyl group, other functional groups, such as carbonyl groups in esters and aldehydes, imino groups in imines and hydrazones, and nitriles, can be used for this reaction (Table 2.2).

Arylation of carbon-hydrogen bonds with organoboron compounds are a new synthetic protocol for obtaining biaryl compounds. The functional group compatibility of this reaction is wide. This coupling reaction is tolerant of a variety of functional groups (Table 2.3).

Table 2.3   Arylation of aromatic ketones with arylboronates.

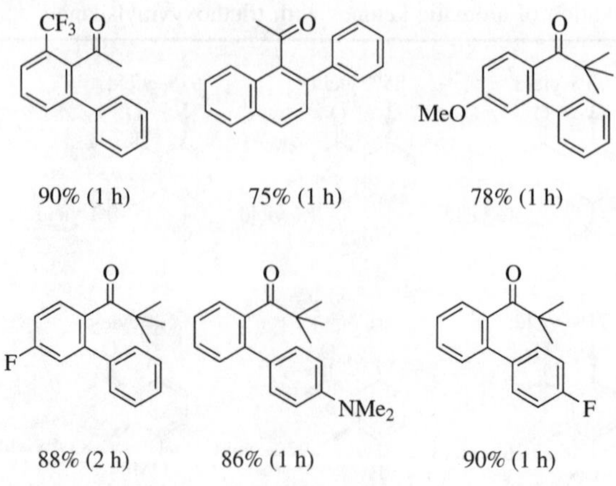

| 90% (1 h) | 75% (1 h) | 78% (1 h) |

| 88% (2 h) | 86% (1 h) | 90% (1 h) |

As described above, $RuH_2(CO)(PPh_3)_3$-catalysed functionalisation of aromatic carbon-hydrogen bonds have already become a reliable synthetic protocol for making carbon-carbon bonds. Thus, the first stage of these types of catalytic reactions has been clarified. The second stage of this area will include applications of this methodology in the synthesis and preparation of highly valuable chemicals. Recently, some promising results have been reported by Woodgate[8] and by Sames.[9] They utilised the carbon-hydrogen bond-activation for the synthesis of natural products. These works strongly imply that carbon-hydrogen bond-activation can now be used as a synthetic tool in the synthesis of complex molecules.

In the past several years, the fundamental features of the catalytic functionalism of carbon-hydrogen bond in organic synthesis have been defined. In the coming decade, it is likely that developments of fascinating synthetic protocols involving unreactive carbon-hydrogen bond cleavage will be reported in organic synthesis.

REFERENCES

1. Kakiuchi, F. and Chatani, N. *Adv. Synth. Catal.* **2003**, *345*, 1077.
2. Murai, S., Kakiuchi, F., Sekine, S., Tanaka, Y., Kamatani, A., Sonoda, M. and Chatani, N. *Nature* **1993**, *366*, 529.
3. Kakiuchi, F., Sekine, S., Tanaka, Y., Kamatani, A., Sonoda, M., Chatani, N. and Murai, S. *Bull. Chem. Soc. Jpn.* **1995**, *68*, 62.
4. Kakiuchi, F., Yamamoto, Y., Chatani, N. and Murai, S. *Chem. Lett.* **1995**, 681.
5. Kakiucih, F., Kan, S., Igi, K., Chatani, N. and Murai, S. *J. Am. Chem. Soc.* **2003**, *125*, 1698.
6. Ahmad, N., Levison, J. J., Robinson, S. D. and Uttley, M. F. *Inorg. Synth.* **1974**, *15*, 50.
7. Kakiuchi, F. and Murai, S. *Org. Synth.* **2003**, *80*, 104.
8. Harris, P. W. R., Rickard, C. E. F. and Woodgate, P. D. *J. Organomet. Chem.* **2000**, *601*, 172.
9. Johnson, J. A., Li, N. and Sames, D. *J. Am. Chem. Soc.* **2002**, *124*, 6900.

2.2 CATALYTIC, ASYMMETRIC SYNTHESIS OF α,α-DISUBSTITUTED AMINO ACIDS USING A CHIRAL COPPER-SALEN COMPLEX AS A PHASE TRANSFER CATALYST

MICHAEL NORTH* and JOSE A. FUENTES

Department of Chemistry, King's College London, Strand, London WC2R 2LS, UK

α,α-Disubstituted amino acids are components of a number of potential pharmaceuticals. Whilst a number of chiral auxiliary based methods are available for the asymmetric synthesis of α,α-disubstituted amino acids, a catalytic method would have significant financial and practical advantages. Recently, we have shown that a copper-salen complex (Figure 2.1) is capable of catalysing the asymmetric alkylation of alanine enolates under very mild reaction conditions[1]. This reaction leads

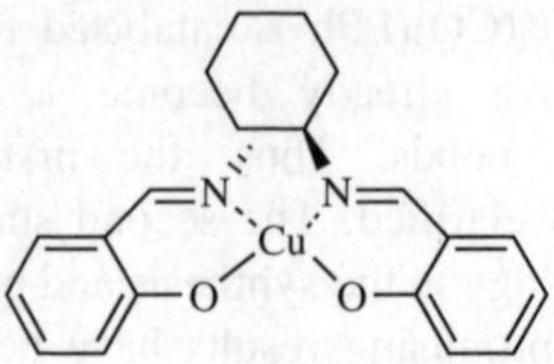

Figure 2.1 Copper-salen complex.

to α-methyl-α-amino acids with high enantiomeric excesses (up to 92%). Optimal results are obtained using a *para*-chlorobenzylidene imine[2], and the chemistry is compatible with the use of methyl ester substrates[3].

2.2.1 SYNTHESIS OF (CHSALEN)

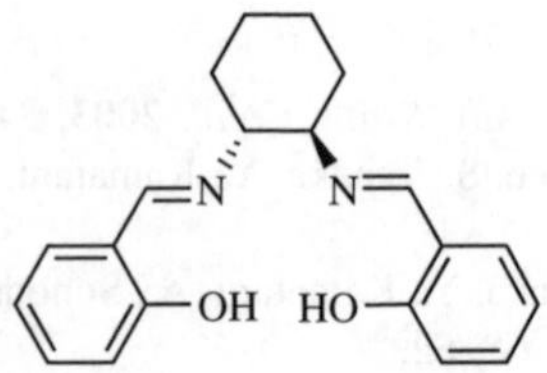

Materials and equipment

- Salicylaldehyde, 2.44 g, 20.0 mmol
- (1*R*, 2*R*)-diaminocyclohexane dihydrochloride salt, 1.87g, 10.0 mmol
- NaOMe, 1.46 g, 27 mmol
- Methanol, 47 mL
- Ethanol, 37 mL
- Dichloromethane, 110 mL
- Water, 60 mL
- Saturated solution of sodium chloride
- Anhydrous magnesium sulfate

- 100 mL and 250-mL round-bottomed flasks
- One magnetic stirrer bar
- Magnetic stirrer hot plate
- One Liebig condenser
- One Büchner funnel, diameter 65 mm
- One Büchner flask, 250 mL
- One 250-mL separating funnel
- Filter paper
- Rotary evaporator

Procedure

1. Salicylaldehyde (2.44 g, 20.0 mmol) was placed in a 250-mL round-bottomed flask equipped with a magnetic stirrer bar. Methanol (37 mL) and ethanol (37 mL) were then added. To this solution, (1R, 2R)-diaminocyclohexane dihydrochloride salt (1.87g, 10.0 mmol) was added and the resulting mixture was added to a solution of NaOMe (1.46 g, 27 mmol) in methanol (10 mL). The resulting bright yellow solution was stirred under reflux for 18 hours.

2. The solution was cooled to room temperature and then filtered into a Büchner funnel with the aid of a water aspirator and the filtrate evaporated in a rotary evaporator. The yellow residue was taken up in dichloromethane (80 mL), filtered again and then the organic layers were washed with water (2 × 30 mL) and brine (1 × 30 mL) in a separatory funnel. The combined aqueous layers were back extracted with dichloromethane (30 mL). The combined organic layers were dried over anhydrous magnesium sulfate and evaporated to dryness to leave the desired compound as a yellow gel (2.96 g, 92%). The purity of the crude product is high enough to be used in the copper complex synthesis, but can be further purified by vacuum distillation of the unreacted salicylaldehyde.

$[\alpha]_D^{20} = -386$ ($c = 1.000$, $CHCl_3$). ^{1}H-NMR (360 MHz, $CDCl_3$): 1.32–1.44 (m, 2H, cyclohexyl-H), 1.59–1.67 (m, 2H, cyclohexyl-H), 1.77–1.86 (m, 4H, cyclohexyl-H), 3.19–3.26 (m, 2H, CH-N × 2), 6.68–6.81 (m, 4H, ArCH), 7.04–7.17 (m, 4H, ArCH), 8.17 (s, 2H, H-CN × 2).

^{13}C-NMR (90 MHz, $CDCl_3$): 24.6 (CH_2-cyclohexyl), 33.5 (CH_2-cyclohexyl), 73.0 (CH), 117.1 and 119.0 (ArCH), 119.0 (ArC), 131.9 and 132.6 (ArCH), 161.3 (ArC), 165.1 (CN).

2.2.2   SYNTHESIS OF COPPER(II)(CHSALEN)

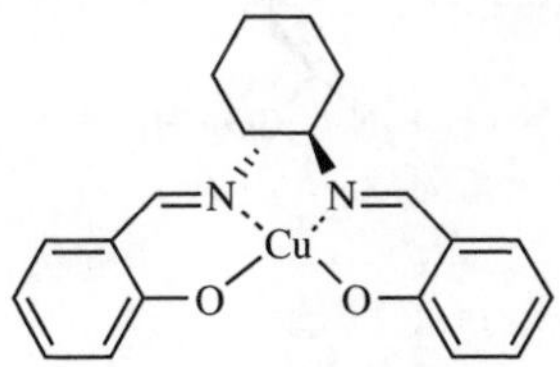

Materials and equipment

- (1R,2R)-[N,N'-bis-(2'-hydroxybenzylidene)]-1,2-diaminocyclohexane, 2.96 g, 9.2 mmol
- Copper(II) bromide, 2.05 g, 9.2 mmol
- NaOMe, 0.50 g, 9.2 mmol
- Methanol, 51 mL

- Two 100-mL round-bottomed flasks
- One magnetic stirrer bar

- Magnetic stirrer plate
- One glass column, diameter 35 mm
- Sephadex LH-20 (25–100 μm)
- Ethanol
- Toluene
- Rotary evaporator

Procedure

1. (1*R*,2*R*)-[*N*,*N*′-bis-(2′-hydroxybenzylidene)]-1,2-diaminocyclohexane (2.96 g, 9.2 mmol) was placed in a 100-mL round-bottomed flask equipped with a magnetic stirrer bar. Methanol (46 mL) was then added. A solution of NaOMe (0.50 g, 9.2 mmol) in methanol (5 mL) was then added followed by copper(II) bromide (2.05 g, 9.2 mmol). The resulting mixture was stirred for 3 hours at room temperature and then the solvent was removed in a rotary evaporator.
2. The copper complex was isolated from the crude residue by column chromatography on Sephadex LH-20 (EtOH/toluene 25:75) (2.2 g, 63% yield); m.p. >270 °C.

   $[\alpha]_D^{20} = -930$ ($c = 0.010$, CHCl$_3$). IR (KBr, cm^{-1}): 3021 (w), 2931 (m), 2856 (w), 1632 (s), 1602 (s), 1537 (s).

2.2.3   ALKYLATION OF ALANINE METHYL ESTER SCHIFF BASE BY CHIRAL SALEN-METAL CATALYSTS, α-BENZYL-ALANINE METHYL ESTER

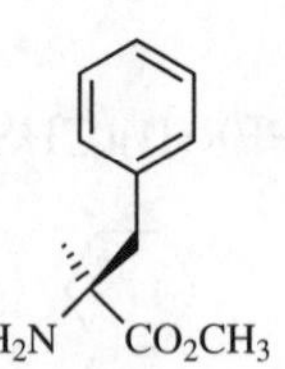

Materials and equipment

- *para*-Chlorobenzylidene alanine methyl ester, 0.237 g, 1.05 mmol
- Copper(II)(chsalen), 8 mg, 0.02 mmol
- Sodium hydroxide, 0.146 g, 3.66 mmol
- Dry toluene, 2.5 mL
- Benzyl bromide, (98%), 0.15 ml, 1.26 mmol

- 25-mL and 50-mL round-bottomed flasks
- One magnetic stirrer bar
- Magnetic stirrer plate
- Methanol, 2 ml
- Acetyl chloride, (98%), 0.5 mL

- Ethanol
- Ethyl acetate
- Ninhydrin (99%)
- Silica gel (60, 40–63 μm)
- TLC plates
- S-(−)-1-phenylethyl isocyanate (99% ee/GC)
- Rotary evaporator

Procedure

1. *para*-Chlorobenzylidene alanine methyl ester (0.237 g, 1.05 mmol) was placed in a 25-mL round-bottomed flask equipped with a magnetic stirrer bar. Dry toluene was then added. To this solution, 2 mol% of copper(II)(chsalen) (8 mg, 0.02 mmol) and powdered sodium hydroxide (0.146 g, 3.66 mmol) were added. An argon atmosphere was then generated by purging the system with a balloon filled with argon. Finally, benzyl bromide (0.15 mL, 1.26 mmol) was added using a syringe.

2. The mixture was stirred for 19 hours under an argon atmosphere and at room temperature. After that time, methanol (2 mL) and then acetyl chloride (0.5 mL) were added to the mixture, the latter added dropwise.

Attention: very exothermic reaction. If the reaction is scaled up, an ice-water bath should be used for the acetyl chloride addition.

3. The mixture was stirred for a further four hours and then the solvents were removed in a rotary evaporator. The crude residue was purified by column chromatography (silica gel, ethyl acetate first and then EtOH/EtOAc 20:80). The amino ester was visualised with ninhydrin solution in ethanol (0.14 g, 71% yield).

   ^{1}H-NMR (360 MHz, CDCl$_3$): 1.35 (s, 3H, CH_3), 2.12 (br s, 2H, NH_2), 2.76 (d, 1H, $J = 13.2$ Hz, CH_2), 3.07 (d, 1H, $J = 13.2$ Hz, CH_2), 3.63 (s, 3H, CH_3-O), 7.07–7.24 (m, 5H, ArCH).

   ^{13}C-NMR (90 MHz, CDCl$_3$): 28.3 (CH$_3$), 48.6 (CH$_2$), 53.7 (CH$_3$-O), 60.5 (C), 128.6, 130.0 and 131.6 (ArCH), 138.2 (ArC), 179.2 (CO$_2$).

   Note: The free amino acid can be obtained by an alternative aqueous work up. Instead of adding methanol and acetyl chloride (part 2 above), the reaction was filtered and the toluene removed in a rotary evaporator. Dilute hydrochloric acid was added to the residue and the mixture was refluxed overnight. Extraction of by-products with dichloromethane and evaporation to dryness of the aqueous phase gave the desired free amino acid.

4. S-(−)-1-Phenylethyl isocyanate (1 or 2 drops) were added to an NMR sample of α-benzyl-alanine methyl ester and allowed to react overnight until there was no unreacted amino ester. The de and therefore ee (92%) was determined from integration of the α-methylene proton region of the resulting diastereomers.

2.2.4   CONCLUSION

The synthesis of the chiral copper catalyst is very easy to reproduce. The complex catalyses the asymmetric alkylation of enolates of a range of amino acids, thus allowing the synthesis of enantiomerically enriched α,α disubstituted amino acids with up to 92% ee. The procedure combines the synthetic simplicity of the Phase Transfer Catalyst (PTC) approach, with the advantages of catalysis by metal complexes. The chemistry is compatible with the use of methyl ester substrates, thus avoiding the use of *iso*-propyl or *tert*-butyl esters which are needed for cinchona-alkaloid catalyzed reactions[4], where the steric bulk of the ester is important for efficient asymmetric induction. Another advantage compared with cinchona-alkaloid systems is that copper(II)(chsalen) catalyses the alkylation of substrates derived from a range of amino acids, not just glycine and alanine (Table 2.4).

Table 2.4   Alkylation of amino acid derivatives.

Entry	X	R	Alkylating agent	Catalyst (mol%)	Time (days)	Yield (%)	ee (%)
1	H	CH_3	Benzyl bromide	2	1	91	81
2	Cl	CH_3	Benzyl bromide	2	1	71	92
3	H	CH_3	*para*-Nitrobenzyl bromide	2	1	78	85
4	H	CH_3	1-bromomethyl-naphthalene	2	1	85	86
5	H	CH_3	Allyl bromide	2	1	75	72
6	H	CH_3	Propargyl bromide	2	1	70	43
7	Cl	CH_3	Propargyl bromide	2	1	82	58
8	Cl	CH_2CH_3	Benzyl bromide	2	2	91	82
9	Cl	CH_2CH_3	Allyl bromide	2	1	46	80
10	Cl	CH_2CH_3	*para*-Nitrobenzyl bromide	2	2	83	79
11	Cl	$CH_2CH(CH_3)_2$	Benzyl bromide	10	7	54	55
12	Cl	$CH_2CH(CH_3)_2$	*para*-Nitrobenzyl bromide	10	7	63	56
13	Cl	$CH_2CH(CH_3)_2$	Allyl bromide	10	1	46	22
14	Cl	CH_2Ph	Allyl bromide	10	1	40	31
15	Cl	CH_2Ph	*para*-Nitrobenzyl bromide	10	1	67	34
16	H	$CH_2CH=CH_2$	Benzyl bromide	2	2	69	42
17	H	$CH_2CH=CH_2$	*para*-Nitrobenzyl bromide	2	2	45	47
18	H	$CH_2CH=CH_2$	1-Bromobut-2-yne	2	2	46	25

REFERENCES

1. (a) Belokon', Y. N., North, M., Kublitski, V. S., Ikonnikov, N. S., Krasik, P. E. and Maleev, V. I. *Tetrahedron Lett.* **1999**, *40*, 6105; (b) Belokon', Y. N., North, M., Churkina, T. D., Ikonnikov, N. S., Maleev, V. I. *Tetrahedron* **2001**, *57*, 2491.
2. Belokon', Y. N., Davies, R. G., Fuentes, J. A., North, M. and Parsons, T. *Tetrahedron Lett.* **2001**, *42*, 8093.
3. Belokon', Y. N., Davies, R. G. and North, M. *Tetrahedron Lett.* **2000**, *41*, 7245.
4. Lygo, B., Crosby, J. and Peterson, J. A. *Tetrahedron Lett.* **1999**, *40*, 8671.

2.3 ASYMMETRIC PHASE-TRANSFER CATALYSED ALKYLATION OF GLYCINE IMINES USING CINCHONA ALKALOID DERIVED QUATERNARY AMMONIUM SALTS

BARRY LYGO* and BENJAMIN I. ANDREWS

School of Chemistry, Nottingham University, Nottingham, NG7 2RD, UK

The asymmetric alkylation of glycine derivatives constitutes a general means of accessing a wide range of natural and unnatural α-amino acids.[1] Recently it has been established that the quaternary ammonium salt, (1*S*,2*S*,4*S*,5*R*,1′*R*)-1-anthracen-9-yl)methyl-2-[benzyloxy(quinolin-4-yl)methyl]-5-ethyl-1-azoniabicyclo [2.2.2]octane bromide, is a highly effective catalyst in the asymmetric liquid-liquid phase-transfer alkylation of *tert*-butyl *N*-(diphenylmethylene)glycinate. Subsequent hydrolysis of the imine provides access to a wide range of α-amino acid *tert*-butyl esters in high yield and with high enantiomeric excess.[2,3]

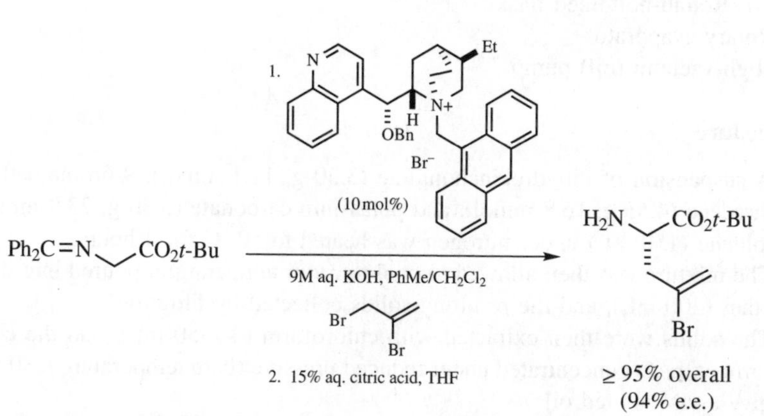

2.3.1   SYNTHESIS OF (1S,2S,4S,5R,1′R)-1-(ANTHRACEN-9-YLMETHYL)-5-ETHYL-2-[HYDROXY(QUINOLIN-4-YL)METHYL]-1-AZONIABICYCLO[2.2.2]OCTANE BROMIDE

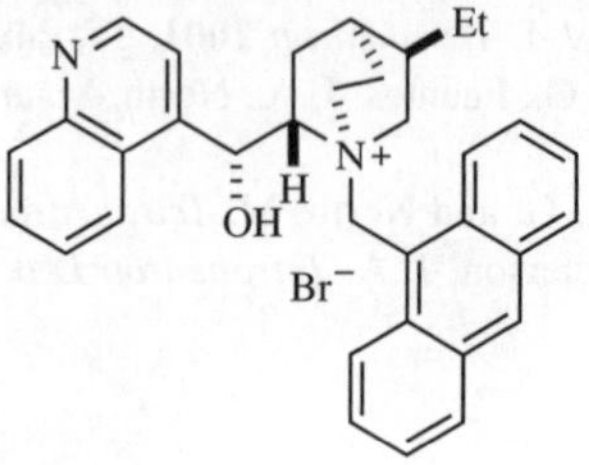

Materials and equipment

- Dihydrocinchonidine, 3.30 g
- 9-Bromomethylanthracene, 4.55 g
- Potassium carbonate, 3.30 g
- Toluene, 150 mL
- Diethyl ether, 400 mL
- Chloroform, 200 mL

- 250-mL Round-bottomed flask with magnetic stirrer bar
- Liebig condenser
- Nitrogen supply
- Magnetic stirrer hotplate
- Oil bath
- 500-mL Erlenmeyer flask
- Büchner funnel, diameter 7.5 cm
- Büchner flask, 1 L
- Büchner funnel, diameter 5 cm
- Büchner flask, 500 mL
- Two filter papers
- 1-L Round-bottomed flask
- Rotary evaporator
- High-vacuum (oil) pump

Procedure

1. A suspension of dihydrocinchonidine (3.30 g, 11.1 mmol), 9-bromomethylanthracene (4.55 g, 16.8 mmol), and potassium carbonate (3.30 g, 23.9 mmol) in toluene (150 mL) under nitrogen was heated to 50 °C for 4 hours.
2. The mixture was then allowed to cool to room temperature, poured into diethyl ether (300 mL), and the resulting solids collected by filtration.
3. The solids were then extracted with chloroform (4 × 50 mL), and the chloroform extracts concentrated under reduced pressure (bath temperature <30 °C) to give a viscous red oil.

4. Diethyl ether (100 mL) was then added, and the mixture stirred vigorously at room temperature for 2 hours. The resulting solid was collected by filtration and dried *in vacuo* to give (1S,2S,4S,5R,1′R)-1-(anthracen-9-ylmethyl)-5-ethyl-2-[hydroxy(quinolin-4-yl)methyl]-1-azoniabicyclo[2.2.2]octane bromide as a pale yellow solid (5.97 g, 94%); m.p. 173–175 °C, [α]$_D$–220 (c 0.7, CHCl$_3$).

^{1}H NMR (500 MHz, CDCl$_3$) δ 8.91–8.87 (1H, m), 8.84 (1H, d, J 4.4 Hz), 8.77 (2H, app. t, J 9.4 Hz), 8.05 (1H, d, J 4.3 Hz), 8.01 (1H, s), 7.69 (1H, d, J 7.7 Hz), 7.64–7.60 (2H, m), 7.54 (1H, dd, J 7.8, 7.3 Hz), 7.30–7.22 (2H, m), 7.21–7.12 (4H, m), 7.05–7.02 (1H, m), 6.58 (1H, d, J 13.5 Hz), 6.35 (1H, d, J 13.5 Hz), 4.85–4.78 (1H, m), 4.71–4.61 (1H, m), 3.80 (1H, d, J 12.5 Hz), 2.63 (1H, dd, J 12.5, 11.0 Hz), 2.33 (1H, dd, J 11.5, 9.5 Hz), 1.86–1.78 (2H, m), 1.65 (1H, br. s), 1.32–1.20 (2H, m), 1.09–0.99 (3H, m), 0.56 (3H, t, J 7.0 Hz).

^{13}C NMR (125 MHz, CDCl$_3$) δ 149.6 (CH), 147.3 (C), 145.5 (C), 133.3 (C), 132.6 (C), 131.3 (CH), 130.4 (2 × C), 129.6 (CH), 128.9 (CH), 128.6 (CH), 128.4 (CH), 128.0 (CH), 127.4 (2 × CH), 126.0 (CH), 125.9 (CH), 125.0 (CH), 124.9 (CH), 124.3 (C), 124.2 (CH), 120.3 (CH), 117.9 (C), 67.1 (CH), 67.0 (CH), 63.9 (CH$_2$), 55.0 (CH$_2$), 50.9 (CH$_2$), 37.2 (CH), 26.6 (CH$_2$), 26.3 (CH$_2$), 23.4 (CH), 23.1 (CH$_2$), 11.6 (CH$_3$).

Notes: The ^{1}H nmr is concentration and purity dependant and may vary significantly from that reported above.

This salt can be used directly for most PTC alkylation reactions.

2.3.2 SYNTHESIS OF (1S,2S,4S,5R,1′R)-1-(ANTHRACEN-9-YLMETHYL)-5-ETHYL-2-[BENZYLOXY(QUINOLIN-4-YL)METHYL]-1-AZONIABICYCLO[2.2.2]OCTANE BROMIDE

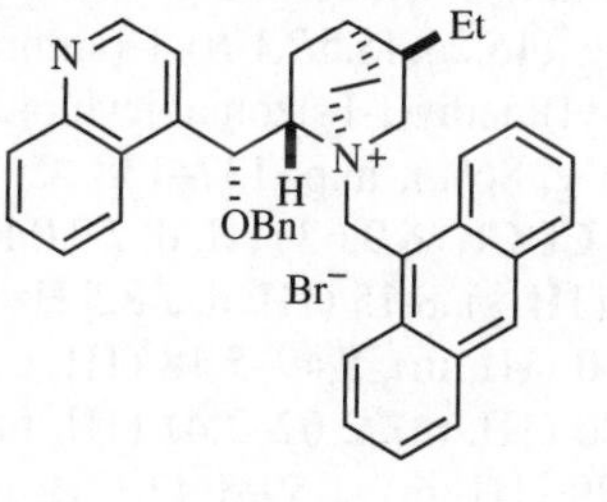

Materials and equipment

- (1S,2S,4S,5R,1′R)-1-(Anthracen-9-ylmethyl)-5-ethyl-2-[hydroxy(quinolin-4-yl)methyl]-1-azoniabicyclo[2.2.2] octane bromide, 5.97 g
- Benzyl bromide, 1.38 mL
- Dichloromethane, 160 mL
- 9M Aqueous sodium hydroxide, 3 mL
- Water, 80 mL
- Diethyl ether, 100 mL
- Magnesium sulfate (anhydrous)

- 250-mL Round-bottomed flask with magnetic stirrer bar
- Magnetic stirrer plate
- 250-mL Separating funnel
- 500-mL Erlenmeyer flask
- Büchner funnel, diameter 7.5 cm
- Büchner flask, 500 mL
- 500-mL Round-bottomed flask
- Büchner funnel, diameter 5 cm
- Büchner flask, 250 mL
- Two filter papers
- Rotary evaporator
- High-vacuum (oil) pump

Procedure

1. Benzyl bromide (1.38 mL, 11.6 mmol) was added to a solution of (1S,2S,4S, 5R,1$'R$)-1-(anthracen-9-ylmethyl)-5-ethyl-2-[hydroxy(quinolin-4-yl)methyl]-1-azoniabicyclo[2.2.2]octane bromide (5.97 g, 10.52 mmol) in dichloromethane (80 mL). The resulting solution was then treated with 9M aqueous sodium hydroxide (3 mL) and the mixture stirred vigorously for 1.5 hours at room temperature.
2. Water (80 mL) was then added and the aqueous layer extracted with dichloromethane (2 × 40 mL). The combined organic fractions were dried (magnesium sulfate) and concentrated under reduced pressure (bath temperature <30 °C).

Note: Excessive heating during the concentration of the solution can lead to decomposition of the salt.

3. Diethyl ether (100 mL) was then added, and the mixture stirred vigorously at room temperature for 2 hours. The resulting solid was collected by filtration and dried *in vacuo* to give (1S,2S,4S,5R,1$'R$)-1-(anthracen-9-ylmethyl)-5-ethyl-2-[benzyloxy(quinolin-4-yl)methyl]-1-azoniabicyclo[2.2.2]octane bromide as a pale yellow solid (5.93 g, 86%), m.p. 137–138 °C, [α]$_D$–215 (c 1.1, CHCl$_3$).

^{1}H NMR (500 MHz, CDCl$_3$) δ 9.62 (1H, d, J 9.0 Hz), 9.32–9.08 (1H, m), 9.00 (1H, d, J 4.0 Hz), 8.49 (1H, s), 8.15 (1H, d, J 8.5 Hz), 8.03–7.26 (14H, m), 6.95–6.57 (2H, m), 5.95–5.60 (1H, m), 5.49–5.18 (1H, m), 4.95–4.78 (2H, m), 4.65–4.29 (2H, m), 2.96–2.70 (1H, m), 2.62–2.41 (1H, m), 2.39–2.11 (2H, m), 2.04–1.90 (1H, m), 1.89–1.79 (1H, m), 1.59–1.47 (2H, m), 1.46–1.13 (2H, m), 0.66 (3H, t, J 7.0 Hz).

^{13}C NMR (125 MHz, CDCl$_3$) δ 148.9 (C), 136.5 (C), 134.1 (C), 133.3 (C), 132.3 (CH), 131.5 (C), 131.0 (C), 130.1 (CH), 129.3 (2 × CH), 129.0 (CH), 128.7 (CH), 128.6 (CH), 127.6 (CH), 127.2 (CH), 126.0 (CH), 125.0 (CH), 123.4 (CH), 118.2 (C), 71.6 (CH$_2$), 66.4 (2 × CH), 63.0 (CH$_2$), 55.2 (CH$_2$), 51.3 (CH$_2$), 37.1 (CH), 26.5 (CH$_2$), 26.4 (CH$_2$), 23.5 (CH, CH$_2$), 11.7 (CH$_3$).

Note: This material is usually of sufficient purity to be used in subsequent reactions, however if required further purification can be achieved by flash chromatography on silica gel (dichloromethane/methanol, 19:1).

2.3.3   SYNTHESIS OF (2S)-TERT-BUTYL 2-AMINO-4-BROMOPENT-4-ENOATE

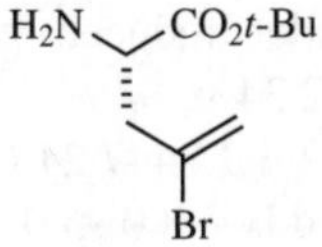

$$H_2N \diagdown \diagup CO_2t\text{-}Bu$$

Br

Materials and equipment

- (1S,2S,4S,5R,1′R)-1-(Anthracen-9-ylmethyl)-5-ethyl-2-[benzyloxy(quinolin-4-yl)methyl]-1-azoniabicyclo[2.2.2]octane bromide, 0.33 g
- *tert*-Butyl N-(diphenylmethylene)glycinate, 1.50 g
- 2,3-Dibromopropene, 0.59 mL
- Toluene, 40 mL
- Dichloromethane, 25 mL
- 9M Aqueous potassium hydroxide, 10 mL
- Ethyl acetate, 240 mL
- Diethyl ether, 150 mL
- Sodium sulfate (anhydrous)
- Tetrahydrofuran, 60 mL
- 15% Aqueous citric acid, 30 mL
- Saturated aqueous potassium carbonate

- Two 250-mL Round-bottomed flasks with magnetic stirrer bar
- Magnetic stirrer plate
- Ice-water bath
- Nitrogen line
- 100-mL Separating funnel
- 250-mL Separating funnel
- Three 250-mL Erlenmeyer flasks
- Two Büchner funnels, diameter 7.5 cm
- Two Büchner flasks, 250 mL
- Two filter papers
- Rotary evaporator

Procedure

1. (1S,2S,4S,5R,1′R)-1-(Anthracen-9-ylmethyl)-5-ethyl-2-[benzyloxy(quinolin-4-yl)-methyl]-1-azoniabicyclo[2.2.2]octane bromide (0.33 g, 0.51 mmol) was added to a solution of *tert*-butyl N-(diphenylmethylene)glycinate (1.50 g, 5.08 mmol) and 2,3-dibromopropene (0.59 mL, 6.09 mmol) in toluene/dichloromethane (40 mL/25 mL). The solution was then placed under a nitrogen atmosphere, and cooled in an ice-water bath. 9M aqueous potassium hydroxide (10 mL) was added dropwise, the ice-water bath removed, and the resulting two-phase mixture was stirred vigorously at room temperature for 16 hours.

Note: The toluene alone can be used as the organic phase if required.

2. The mixture was then separated and the aqueous phase extracted with ethyl acetate (3×30 mL). The combined organic fractions were dried (sodium sulfate) and concentrated under reduced pressure to give crude alkylated imine as a pale yellow solid (2.34 g, 94% ee).

^{1}H NMR (400 MHz, CDCl$_3$) δ 7.84–7.24 (10H, m), 5.72–5.70 (1H, m), 5.45 (1H, d, J 1.5 Hz), 4.28 (1H, dd, J 8.0, 5.0 Hz,), 3.09 (1H, ddd, J 14.5, 8.0, 1.0 Hz), 3.04 (1H, ddd, J 14.5, 5.0, 1.0 Hz), 1.47 (9H, s).

^{13}C NMR (100 MHz, CDCl$_3$) δ 171.5 (C), 170.2 (C), 139.8 (C), 136.3 (C), 130.4 ($2 \times$ CH), 130.3 (C), 129.0 ($2 \times$ CH), 128.7 (CH), 128.5 (CH), 128.4 ($4 \times$ CH), 119.7 (CH$_2$), 81.7 (C), 64.0 (CH), 45.7 (CH$_2$), 28.2 ($3 \times$ CH$_3$).

Chiral HPLC: Chiralcel OD-H; hexane/isopropanol 99.5:0.5; 0.5 ml/min; $R_T = 10.3$ min (S)-isomer, 12.3 min (R)-isomer; observing at 250 nm.

This material was used in the following hydrolysis step without further purification.

Note: If dichloromethane is used in the extraction, the catalyst can usually be recovered in good yield (80–90%) by precipitation from the crude product using diethyl ether.

3. The crude imine (2.34 g) was dissolved in tetrahydrofuran (60 mL), and 15% aqueous citric acid (30 mL) added. The mixture was then stirred for 18 hours at room temperature.

4. Diethyl ether (50 mL) was then added, and the aqueous phase washed with diethyl ether (2×50 mL). The aqueous phase was then basified with saturated aqueous potassium carbonate, and extracted with ethyl acetate (3×50 mL). The combined ethyl acetate fractions were dried (sodium sulfate) and then concentrated under reduced pressure to give ($2S$)-*tert*-butyl 2-amino-4-bromopent-4-enoate as a very pale yellow oil (1.28 g, *ca* 100%).

^{1}H NMR (400 MHz, CDCl$_3$) δ 5.71–5.69 (1H, m), 5.56–5.54 (1H, m), 3.69 (1H, dd, J 8.5, 5.0 Hz), 2.90–2.83 (1H, m), 2.57 (1H, ddd, J 14.5, 8.5, 1.0 Hz), 1.58 (2H, br. s), 1.48 (9H, s).

^{13}C NMR (100 MHz, CDCl$_3$) δ 173.6 (C), 130.2 (C), 120.1 (CH$_2$), 81.8 (C), 53.3 (CH), 46.8 (CH$_2$), 28.2 (CH$_3$).

Notes: Benzophenone can be recovered from the diethyl ether extracts if required.

This material is usually of sufficient purity to be used in subsequent reactions, however if required further purification can be achieved either by flash chromatography on silica gel (dichloromethane/methanol, 9:1) or by conversion to the corresponding hydrochloride salt (HCl, Et$_2$O) and recrystallisation.

2.3.4 CONCLUSION

The asymmetric alkylation of *tert*-butyl N-(diphenylmethylene)glycinate as described above works well with a range of primary alkyl halides. Enantioselectivities at room temperature are typically in the range 89–94% ee (Table 2.5), and

$$Ph_2C=N\diagdown CO_2t\text{-}Bu \xrightarrow[\text{2. 15\% aq. citric acid, THF}]{\text{9M aq. KOH, PhMe, R-X}} H_2N\diagdown CO_2t\text{-}Bu$$

Table 2.5   Room temperature alkylation of *tert*-butyl *N*-(diphenylmethylene)-glycinate using (1*S*,2*S*,4*S*,5*R*,1′*R*)-1-(anthracen-9-ylmethyl)-5-ethyl-2-[benzyloxy(quinolin-4-yl)methyl]-1-azoniabicyclo[2.2.2]octane bromide as catalyst.

Entry	R-X	ee (%)
1	$CH_2=C(Me)CH_2Br$	92 (*S*)
2	$CH_2=C(Cl)CH_2I$	94 (*S*)
3	$CH_2=CHCH_2Br$	92 (*S*)
4	$CH\equiv CCH_2Br$	89 (*S*)
5	$PhCH_2Br$	93 (*S*)
6	$4\text{-Br-}C_6H_4CH_2Br$	93 (*S*)
7	$2\text{-Naphthyl}CH_2Br$	93 (*S*)
8	n-BuI	90 (*S*)

can be increased to 96–99% ee by running the reactions at lower temperatures.[3] The corresponding (*R*)-amino acid derivatives can be prepared with similar levels of selectivity simply by employing the cinchonine derived salt, (1*S*,2*R*,4*S*,5*R*,1′*S*)-1-(anthracen-9-ylmethyl)-5-vinyl-2-[benzyloxy-(quinolin-4-yl)methyl]-1-azoniabicyclo[2.2.2]octane bromide.[2,4]

REFERENCES

1. O'Donnell, M. J. *Aldrichimica Acta* **2001**, *34*, 3.
2. Lygo, B. and Wainwright, P. G. *Tetrahedron Lett.* **1997**, *38*, 8595; Lygo, B., Crosby, J., Lowdon, T., Peterson, J. A. and Wainwright, P. G. *Tetrahedron* **2001**, *57*, 2403; Lygo B. and Humphreys, L. D. *Tetrahedron Lett.* **2002**, *43*, 6677; Lygo, B. and Andrews, B. I. *Tetrahedron Lett.* **2003**, *44*, 4499.
3. Corey, E. J., Xu, F. and Noe, M. C. *J. Am. Chem. Soc.* **1997**, *119*, 12414; O'Donnell, M. J., Delgado, F., Hostettler, C. and Schwesinger, R. *Tetrahedron Lett.* **1998**, *39*, 8775; Lygo, B., Andrews, B. I., Crosby, J. and Peterson, J. A. *Tetrahedron Lett.* **2002**, *43*, 8015.
4. A number of alternative asymmetric phase-transfer catalysts have also been reported give high enantioselectivities in this type of reaction; for leading references see: Shibuguchi, T., Fukuta, Y., Akachi, Y., Sekine, A., Ohshima, T. and Shibasaki, M. *Tetrahedron Lett.* **2002**, *43*, 9539; Ooi, T., Kameda, M. and Maruoka, K. *J. Am. Chem. Soc.* **2003**, *125*, 5139; Thierry, B., Plaquevent, J. C. and Cahard, D. *Tetrahedron: Asymm*etry **2003**, *14*, 1671; Park, H. G., Jeong, B. S., Yoo, M. S., Lee, J. H., Park, B. S., Kim, M. G. and Jew, S. S. *Tetrahedron Lett.* **2003**, *44*, 3497; Lygo, B., Allbutt, B. and James, S. R. *Tetrahedron Lett.* **2003**, *44*, 5629.

3 Asymmetric Alkylation or Amination of Allylic Esters

CONTENTS

Catalysts for Fine Chemical Synthesis, Vol. 3, Metal Catalysed Carbon-Carbon Bond-Forming Reactions
Edited by S. M. Roberts, J. Xiao, J. Whittall, and T. Pickett
© 2004 John Wiley & Sons, Ltd   ISBN: 0-470-86199-1

3.1 SYNTHESIS AND APPLICATION IN PALLADIUM-CATALYSED ASYMMETRIC ALLYLIC SUBSTITUTION OF ENANTIOPURE CYCLIC β-IMINOPHOSPHINE LIGANDS

MARIA ZABLOCKA[a*], MAREK KOPROWSKI[a], JEAN-PIERRE MAJORAL[b*], MATHIEU ACHARD[c] and GÉRARD BUONO[c*]

[a]Centre of Molecular and Macromolecular Studies, Polish Academy of Sciences, Sienkiewicza 112, 90-363 Lodz, Poland
[b]Laboratoire de Chimie de Coordination, CNRS, 205 route de Narbonne, 31077 Toulouse Cedex, France
[c]ENSSPICAM, CNRS, UMR 6516, Faculté St Jérôme, av. Escadrille Normandie-Niemen, 13397 Marseille cedex 20, France

Asymmetric palladium(0)-catalysed substitution of racemic allylic substrates with malonate and benzylamine as nucleophiles were performed using new enantiopure cyclic β-iminophosphine ligands namely (2,6-dimethyl-phenyl)-(1-phenyl-2,3,3a,8a-tetrahydro-1H-1-phospha-cyclopenta[-α]inden-8-ylidene)-amines 1(Rp) or 1(Sp)[1,2]

3.1.1 SYNTHESIS OF (2,6-DIMETHYL-PHENYL)-(1-PHENYL-2,3,3a,8a-TETRAHYDRO-1H-1-PHOSPHA-CYCLOPENTA[α]INDEN-8-YLIDENE)-AMINES 1R$_p$

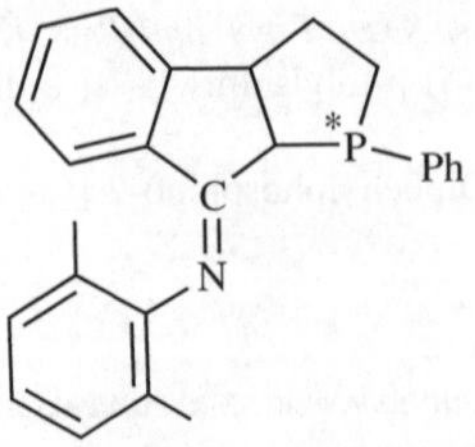

1(Rp) or 1(Sp)

Materials and equipment

- Phospholene 1(**R$_p$**) or (**S$_p$**)^2, 0.162 g, 1 mmol
- Diphenylzirconocene, 0.375 g, 1 mmol

- 2,6-Dimethylphenyl isocyanide **3**, 0.132 g, 1 mmol
- Dry toluene, 11 mL
- Dry pentane, 150 mL

- Celite® 5 g
- Schlenk tube, 100 mL
- Two double-tipped needles
- Magnetic stirrer plate
- Pressure filter funnel, diameter 5 cm

Procedure

1. Diphenylzirconocene Cp_2ZrPh_2 (0.375 g, 1 mmol) was placed in the tube with a magnetic stirrer bar, under argon. Toluene (5 mL) was added followed by addition *via* a double-tipped needle of a solution of phospholene **1(Rp)** or **1(Sp)** (0.162 g, 1 mmol) in 3 mL of toluene at room temperature. The resulting mixture was heated at 80 °C for 6 hours, then cooled to −20 °C. This solution contains the 8,8 dicyclopentadienyl-1-phenyl-1,2,3,3a,8,8a-hexahydro-1-phospha-8-zircona-cyclopenta[α]inden **2(Rp)** or **2(Sp)** which were used as prepared.

2. 2,6-Dimethylphenyl isocyanide (0.132 g, 1 mmol) in solution in 3 mL of toluene was added *via* a double tipped needle to the solution prepared above and stirred overnight at room temperature. Solvents were evaporated to dryness and the solid residue was extracted with pentane (3 × 50 mL) and filtered on Celite® 5 g. Enantiopure ligands **1(Rp)** or **1(Sp)** were obtained as yellow solids in 72% yield (0.276 g) after solvent removal.

$^{31}P\{^1H\}$ NMR (C_6D_6): δ 9.1 (s) ppm.

^1H NMR (C_6D_6): δ 1.18–2.31 (m, 4H, CH_2), 2.17 (s, 3H, CH_3), 2.80 (s, 3H, CH_3), 3.46 (dd, 1H, $J_{HH} = 3.4$ Hz, $J_{HP} = 6.8$ Hz, PCH), 3.63 (m, 1H, CH), 6.39–7.30 (m, 11H, CH_{arom}), 8.36 (d, 1H, $J_{HH} = 7.5$ Hz, CH_{arom}) ppm.

$^{13}C\{^1H\}$ NMR (C_6D_6): δ 20.0 (s, CH_3), 20.2 (s, CH_2), 23.1 (d, $J_{CP} = 17.1$ Hz, CH_2P), 33.4 (d, $J_{CP} = 4.4$ Hz, CH_2), 49.6 (s,CH), 55.1 (d, $J_{CP} = 26.8$ Hz, CHP), 123.5, 123.8, 124.0, 127.7, 128.5, 129.1 and 133.0 (s, CH_{arom}), 129.2 (d, $^2J_{CP} = 3.5$ Hz, o-PPh), 129.8 (s, p-PPh), 130.2 (d, $^3J_{CP} = 12.8$ Hz, m-PPh), 139.2 (d, $^3J_{CP} = 28.2$ Hz, i-PPh_2), 140.4 (s, CC = N), 151.8 (d, $J_{CP} = 1.6$ Hz, $CCC = N$), 152.0 (s, CMe), 175.0 (d, $J_{CP} = 14.4$ Hz, C = N) ppm.

MS (DCI/NH_3): m/z 370 [$M^+ + 1$].

Anal. Calcd for $C_{25}H_{24}PN$: C, 81.27; H, 6.54; N, 3.79. Found: C, 81.19; 6.50; N, 3.82.

3.1.2 SYNTHESIS OF (E)-METHYL 2-CARBOMETHOXY-3,5-DIPHENYLPENT-4-ENOATE

Materials and equipment

- Palladium allyl chloride dimer, 3 mg, 0.008 mmol
- Iminophosphine (**1R_P**), 12 mg, 0.033 mmol
- 1,3-Diphenylprop-3-en-1-yl acetate, 100 mg, 0.396 mmol
- Dimethyl malonate, 157 mg, 1.18 mmol
- Dry potassium acetate, 4 mg, 0.04 mmol
- Dry dichloromethane, 5 mL
- N,O-bis(trimethylsilyl)acetamide (BSA), 241 mg, 1.18 mmol
- Diethyl ether, 30 mL
- Anhydrous magnesium sulfate
- Ammonium chloride, saturated, 15 mL
- Ethyl acetate
- Pentane

- 25-mL two-necked round-bottomed flask
- Magnetic stirrer plate
- TLC plates, SIL G-60 UV$_{254}$
- Silica gel (Matrex 60A, 37–70 μm), 5 g

Procedure

In a 25-mL two-necked round-bottomed flask under argon atmosphere, a mixture of palladium allyl chloride dimer (3 mg, 0.008 mmol) and ligand **1(R_p)** (12 mg, 0.033 mmol), in dry dichloromethane (5 mL) was stirred at room temperature for 15 minutes. 1,3-Diphenylprop-3-en-1-yl acetate (100 mg, 0.396 mmol) was added and stirring was maintained for 5 minutes. Dimethyl malonate (157 mg, 1.18 mmol), N,O-bis(trimethylsilyl)acetamide (BSA) (241 mg, 1.18 mmol) and a catalytic amount of potassium acetate (4 mg, 0.04 mmol) were subsequently added. The resulting solution was stirred at 20 °C for 16 hours. Saturated ammonium chloride solution (15 mL) was added and the organic solution was extracted with diethyl ether (3 × 10 ml). The combined organic phases were dried over anhydrous magnesium sulfate and the solvent was removed *in vacuo* to afford a pale yellow oil. The crude product was purified by column chromatography (silica gel, pentane/ethyl acetate: 85/15) to afford 121 mg ((95%, yield) (57%, ee (*S*)) of pure product.

^{1}H NMR (CDCl$_3$, 500 MHz) δ 7.4–7.1 (m, 10H), 6.48 (d, J = 15.9 Hz, 1H), 6.33 (dd, J = 15.9, 8.5 Hz, 1H), 4.27 (dd, J = 10.9, 8.5 Hz, 1H), 3.95 (d, J = 10.9 Hz, 1H), 3.70 (s, 3H), 3.52 (s, 3H).

The enantiomeric excess was determined by HPLC analysis on a Chiralcel AD-H column at λ = 254 nm, 25 °C: eluent: hexane/isopropanol: 95/5; flow rate: 1 mL/min; t_S = 14.34 min; t_R = 19.39 min.

3.1.3 SYNTHESIS OF BENZYL(1,3-DIPHENYLPROP-2-ENYL)AMINE

$$\text{Ph} \diagdown \diagup \overset{\overset{\displaystyle NHCH_2Ph}{|}}{\underset{Ph}{}}$$

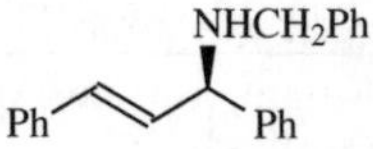

Materials and equipment

- Palladium allyl chloride dimer, 3 mg, 0.008 mmol
- Iminophosphine (R_P), 12 mg, 0.033 mmol
- 1,3-diphenylprop-3-en-1-yl acetate, 100 mg, 0.396 mmol
- Benzyl(1,3-diphenylprop-2-enyl)amine, 126 mg, 1.18 mmol
- Dry dichloromethane, 5 mL
- Diethyl ether, 30 mL
- Anhydrous magnesium sulfate
- Ethyl acetate
- Ammonium chloride, saturated, 15 mL
- Pentane

- 25-mL two-necked round-bottomed flask
- Magnetic stirrer plate
- TLC plates, SIL G-60 UV$_{254}$
- Silica gel (Matrex 60A, 37–70 μm), 5 g

Procedure

In a 25-mL two-necked round-bottomed flask under argon atmosphere, a mixture of palladium allyl chloride dimer (3 mg, 0.008 mmol) and ligand **1**(R_p) (12 mg, 0.033 mmol) in dry dichloromethane (5 mL) was stirred at room temperature for 15 minutes. 1,3-Diphenylprop-3-en-1-yl acetate (100 mg, 0.396 mmol) was added and stirring was maintained at $-10\,°C$ for 5 minutes. Benzyl(1,3-diphenylprop-2-enyl)amine (126 mg, 1.18 mmol) was added and the resulting solution was stirred at $-10\,°C$ for 24 hours. A solution of saturated ammonium chloride (15 mL) was added and the organic solution was extracted with diethyl ether $(3 \times 10\,mL)$. The combined organic phases were dried over anhydrous magnesium sulfate and the solvent was removed *in vacuo* to afford a crude oil. The crude product was purified by column chromatography (silica gel, pentane/ ethyl acetate: 85/15) to afford 85.5 mg ((72%, yield), (64%, ee(R)) of pure white solid.

^{1}H NMR (CDCl$_3$, 500 MHz) δ 7.4–7.1 (m, 10H), 6.58 (d, $J = 15.8$ Hz, 1H), 6.33 (dd, $J = 15.8, 7.4$ Hz, 1H), 4.41 (d, $J = 7.4$ Hz, 1H), 3.80 (d, $J = 13.4$ Hz, 1H), 3.77 (d, $J = 13.5$ Hz, 1H).

The enantiomeric excess was determined by HPLC analysis on a Chiralcel OD-H column at $\lambda = 254$ nm; eluent: hexane/isopropanol: 99/1; flow rate: 0.5 mL/min; $t_S = 20.98$ min; $t_R = 22.53$ min.

Table 3.1 Palladium-catalysed allylic substitutions.

$$\underset{(+/-)}{\overset{\overset{\displaystyle OAc}{|}}{Ph\diagup\diagdown Ph}} \xrightarrow[\text{[Pd(C}_3\text{H}_5)\text{Cl]}_2/\text{Ligand}]{\text{Nucleophile: NuH}} \overset{\overset{\displaystyle Nu}{|}}{Ph\diagup\diagdown Ph}$$

Entry	Nucleophile	Solvent	Temp.(°C)	Time (h)	Yield (%)	ee (%)
1	CH₂(COOMe)₂	CH₂Cl₂	20	16	95	57 (S)
2	CH₂(COOMe)₂	CH₂Cl₂	20	16	90	48 (S)
3	PhCH₂NH₂	Toluene	−10	24	96	57 (R)
4	PhCH₂NH₂	CH₂Cl₂	−10	24	72	64 (R)

3.1.4 CONCLUSION

The substitution of an allylic acetate group by malonate or benzylamine can be accomplished in a stereocontrolled fashion using ligand $1(\mathbf{R_p})$ (Table 3.1).

REFERENCES

1. Zablocka, M., Koprowski, M., Donnadieu, B., Majoral, J. P., Achard, M. and Buono, G. *Tetrahedron Letters* **2003**, *44*, 2413.
2. Cadierno, V., Zablocka, M., Donnadieu, B., Igau, A., Majoral, J. P. and Skowronska, A. *J. Am. Chem. Soc.*, **1999**, *121*, 11086.

3.2 (9H,9′H,10H,10′H,11H,11′H,13H,13′H,14H,14′H,15H,15′H-PERFLUOROTRICOSANE-12,12′-DIYL)BIS[(4S)-4-PHENYL-2-OXAZOLINE] AS A LIGAND FOR ASYMMETRIC PALLADIUM-CATALYSED ALKYLATION OF ALLYLIC ACETATES IN FLUOROUS MEDIA

Jérôme Bayardon and Denis Sinou*

Laboratoire de Synthèse Asymétrique, associé au CNRS, CPE Lyon, Université Claude Bernard Lyon 1, 43, boulevard du 11 novembre 1918, 69622 Villeurbanne cédex, France

Chiral bisoxazoline associated with palladium is a very efficient organometallic catalyst for the asymmetric allylic alkylation of allylic acetates and carbonates, allowing the formation of carbon-carbon as well as carbon-heteroatom bond in enantiomeric excesses higher than 95%.[1,2] However one of the problems in organometallic homogeneous catalysis is the separation of the catalyst, often toxic and very expensive, from the product(s) of the reaction. Very recently, a chiral fluorous bisoxazoline (Figure 3.1) has been shown to be an efficient ligand in the

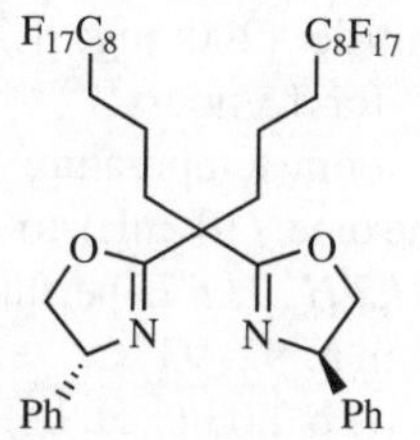

Figure 3.1 Chiral fluorous bixoxazoline.

asymmetric palladium-catalysed alkylation of allylic acetates, enantiomeric excesses as high as 94% being obtained.[3] Moreover, a simple elution on a FluoroFlash[TM] cartridge[4] allows the easy separation of the fluorous ligand from the product.

3.2.1 SYNTHESIS OF 2-IODO-1-(1H,1$'H$,2H,2$'H$,3H,3$'H$-PERFLUOROOCTYL)-3-PROPANOL

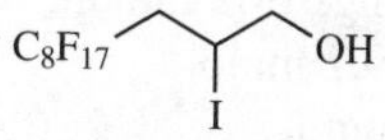

Materials and equipment

- Perfluorooctyl iodide, 14 g, 25.6 mmol
- Allyl alcohol, 2 mL, 30 mmol
- Azobisisobutyronitrile, 600 mg, 3.66 mmol
- Hexane, 10 mL

- 100-mL Schlenk tube with magnetic stirrer bar
- Magnetic heating stirrer plate
- One glass sintered funnel, diameter 5 cm
- One Buchner flask, 200 mL
- One 50-mL round-bottomed flask, equipped with a refrigerator
- Rotory evaporator

Procedure

1. A mixture of perfluorooctyl iodide (14.0 g, 35.6 mmol), allyl alcohol (2 mL, 30 mmol), and azobisisobutyronitrile (200 mg, 1.22 mmol) was stirred at 75 °C for 1 hour under an inert atmosphere of argon.
2. The solution was cooled to room temperature, a new portion of azobisisobutyronitrile (200 mg, 1.22 mmol) was added and the mixture was stirred again at 75 °C for 1 hour. The solution was again cooled at room temperature, a new

portion of azobisisobutyronitrile (200 mg, 1.22 mmol) was added and the solution was stirred at 75 °C for 15 hours.

3. The solution was cooled at room temperature to give a solid. This solid was recrystallised in refluxing hexane (10 mL) to give after filtration on a glass sintered funnel 3-(*1H,1'H,2H,2'H,3H,3'H*-perfluorooctyl)-1-propanol as a white solid (11.2 g, 72% yield);[5] m.p. 92–93 °C.

^{1}H NMR (300 MHz, CDCl$_3$) δ 2.03 (1 H, bs), 2.69–3.12 (2 H, m), 3.71–3.82 (2 H, m), 4.40–4.48 (1 H, m).

^{19}F (282.4, CDCl$_3$) δ −126.5 (2 F, m), −123.9 (2 F, m), −123.1 (2 F, m), −122.1 (6 F, m), −114.1 (2 F, m), −81.2 (3 F, t, *J* 10.3 Hz).

3.2.2   SYNTHESIS OF 3-(1*H*,1'*H*,2*H*,2'*H*,3*H*,3'*H*-PERFLUOROOCTYL)-1-PROPANOL

$$C_8F_{17}\diagdown\diagup\diagdown\diagup OH$$

Materials and equipment

- 2-Iodo-1-(1*H*,1'*H*,2*H*,2'*H*,3*H*,3'*H*-perfluorooctyl)-3-propanol, 14 g, 25.6 mmol
- Azobisisobutyronitrile, 300 mg, 1.62 mmol
- Tributyltin hydride, 8 mL, 29.7 mmol
- Butyronitrile 300 mg, 1.62 mmol
- Trifluorotoluene, 40mL
- Hexane, 10 mL

- 100-mL Schlenk tube with magnetic stirrer bar
- Magnetic heating stirrer plate
- One glass sintered funnel, diameter 5 cm
- One Buchner flask, 100 mL
- One 50-mL round-bottomed flask, equipped with a refrigerator
- Rotory evaporator

Procedure

1. To a mixture of 2-iodo-1-(1*H*,1'*H*,2*H*,2'*H*,3*H*,3'*H*-perfluorooctyl)-3-propanol (12.08 g, 20 mmol) and azobisisobutyronitrile (300 mg, 1.62 mmol) in trifluorotoluene (80 mL), HSnBu$_3$ (8 mL, 29.7 mmol) was slowly added under argon.

2. The solution was stirred at 80 °C for 4 hours and the solvent was removed using a rotary evaporator. The residue was recrystallised in refluxing hexane (10 mL) and the solid was filtered on a glass sintered funnel to afford the fluorinated alcohol as a solid (6.92 g, 72% yield);[5] m.p. 44–46 °C.

^{1}H NMR (300 MHz, CDCl$_3$) δ 1.40 (1 H, t, *J* 4.9 Hz), 1.83–1.92 (2 H, m), 2.13–2.29 (2 H, m), 3.72–3.78 (2 H, m).

^{19}F (282.4, CDCl$_3$) δ −126.6 (2 F, m), −123.9 (2 F, m), −123.2 (2 F, m), −122.3 (6 F, m), −114.7 (2 F, m), −81.2 (3 F, t, *J* 10.2 Hz).

3.2.3   SYNTHESIS OF 3-(1H,1′H,2H,2′H,3H,3′H-PERFLUOROOCTYL)-1-IODOPROPANE

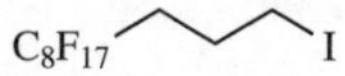

$$C_8F_{17}\diagdown\diagup\diagdown\diagup I$$

Materials and equipment

- 3-(1H,1′H,2H,2′H,3H,3′H-perfluorooctyl)-1-propanol, (fluorous alcohol) 6.1 g, 12.8 mmol
- Potassium iodide, 5.6 g, 34.0 mol
- 85% Phosphoric acid in water, 29.2 mL
- Phosphorus pentoxide, 14.1 g, 99.3 mmol
- Diethyl ether, 90 mL
- Aqueous solution 0.1 M of sodium thiosulfate, 30 mL
- Saturated aqueous solution of sodium chloride, 20 mL
- Anhydrous sodium sulfate
- Petroleum ether
- Silica gel (Merck 60, 40–63 μm mesh)

- TLC plates (Merck 60 F_{254}, 0.2 mm)
- One 100-mL round-bottom flask, equipped with a refrigerator
- Magnetic heating stirrer plate
- One glass sintered funnel, diameter 5 cm
- One Buchner flask, 100 mL
- Rotary evaporator
- One 250-mL separatory funnel
- One glass column, diameter 2.5 cm
- Two 250-mL Erlenmeyer flasks
- One filter paper

Procedure

1. To a mixture of 85% phosphoric acid in water (29.2 mL) and phosphorus pentoxide (14.1 g, 99.3 mmol) cooled to 0 °C potassium iodide (5.6 g, 34.0 mol) and fluorous alcohol (6.1 g, 12.8 mmol) were added rapidly. The mixture was stirred at 120 °C for 5 hours.
2. The solution was cooled to room temperature and water (30 mL) was added. The aqueous phase was extracted with diethyl ether (3 × 30 mL). The organic phase was washed with a 0.1 M aqueous solution of sodium thiosulfate (2 × 15 mL), then a saturated aqueous solution of sodium chloride (20 mL), and finally dried over anhydrous sodium sulfate.
3. The solution was filtered and the solvent evaporated in a rotary evaporator to afford a residue that was purified by column chromatography (silica gel, petroleum ether, $R_f = 0.86$) to give the iodo derivative as a solid (5.98 g, 79% yield);[5] m.p. 36–38 °C.

^{1}H NMR (300 MHz, CDCl$_3$) δ 2.11–2.29 (4 H, m), 3.25 (2 H, t, $J = 6.2$ Hz). ^{19}F (282.4, CDCl$_3$) δ −126.5 (2 F, m), −123.8 (2 F, m), −123.1 (2 F, m), −122.1 (6 F, m), −114.1 (2 F, m), −81.2 (3 F, t, J 10.3 Hz).

3.2.4   SYNTHESIS OF (9H,9′H,10H,10′H,11H,11′H,13H,13′H,-14H,14′H,15H,15′H-PERFLUOROTRICOSANE-12,12′-DIYL)-BIS-[(4S)-4-PHENYL-2-OXAZOLINE]

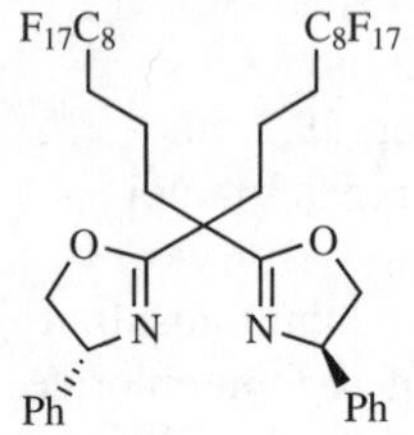

Materials and equipment

- 2,2′-Methylenebis[(4S)-4-phenyloxazoline], 0.3 g, 0.98 mmol
- Dry dimethylformamide (12 mL)
- 3-(1H,1′H,2H,2′H,3H,3′H-perfluorooctyl)-1-iodopropane, 1.33 g, 2.25 mmol
- Sodium hydride 60%, 120 mg
- Diethyl ether, 80 mL
- Petroleum ether
- Ethyl acetate
- Dried pentane, 10 mL
- Saturated aqueous solution of sodium chloride
- Anhydrous sodium sulfate

- Silica gel (Merck 60, 40–63 μm mesh)
- TLC plates (Merck 60 F$_{254}$, 0.2 mm)
- 25-mL Schlenk tube with magnetic stirrer bars
- Magnetic heating stirrer plate
- Two 250-mL Erlenmeyer flask
- One 250-mL separatory funnel
- Rotary evaporator
- One glass column, diameter 2.5 cm
- One filter paper

Procedure

1. A solution of 2,2′-methylenebis[(4S)-4-phenyloxazoline] (0.3 g, 0.98 mmol) in dry dimethylformamide (4 mL) was added slowly at 0 °C to a suspension of sodium hydride 60% washed with pentane (70.6 mg, 2.94 mmol) in dry dimethyl-formamide (8 mL), and the solution was stirred at room temperature for 1 hour.

2. A solution of $C_8F_{17}(CH_2)_3I$ (1.33 g, 2.25 mmol) in dry dimethylformamide (4 mL) was then added slowly and the mixture was heated at 80 °C for 16 hours.

3. After evaporation of half the solvent, water (20 mL) and diethyl ether (20 mL) were added. The organic phase was separated, and the aqueous phase was extracted with diethyl ether (3 × 20 mL). The combined ethereal solutions were washed with a saturated aqueous solution of sodium chloride (2 × 20 mL), and dried over sodium sulfate. Filtration of the solution followed by solvent evaporation in a rotary evaporator afforded a solid which was purified by column chromatography (silica gel, petroleum ether/AcOEt: 7/1, $R_f = 0.58$) to give the fluorous bisoxazoline as a white solid (0.56 g, 47% yield); m.p. 68–70 °C; $[\alpha]_D^{25} = -34.7$ (c 0.3, $C_6H_5CF_3$).

^{1}H NMR (300 MHz, CDCl$_3$) δ 1.61–1.64 (4 H, m), 2.03–2.13 (8 H, m), 4.17 (2 H, dd, J 8.3, 8.3 Hz), 4.68 (2 H, dd, J 9.6, 8.3 Hz), 5.27 (2 H, dd, J 9.6, 8.3 Hz), 7.16–7.21 (10 H, m).

^{19}F (282.4, CDCl$_3$) δ −126.9 (4 F, m), −124.0 (4 F, m), −123.4 (4 F, m), −122.4 (12 F, m), −114.7 (4 F, m), −81.7 (6 F, t, J 9.2 Hz).

3.2.5 SYNTHESIS OF (E)-METHYL 2-CARBOMETHOXY-3,5-DIPHENYLPENT-4-ENOATE

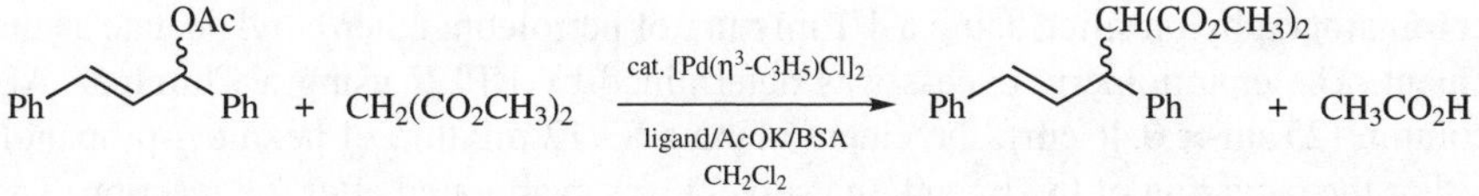

A typical allylic alkylation

Materials and equipment

- [Pd(η^3-C$_3$H$_5$)Cl]$_2$, 4.1 mg, 11.3 μmol
- (9H,9′H,10H,10′H,11H,11′H,13H,13′H,14H,14′H,15H,15′H-perfluorotricosane-12,12′-diyl)-bis-[(*4S*)-4-phenyl-2-oxazoline], 34.5 mg, 28.1 μmol
- 1,3-Diphenyl propen-2-enyl acetate, 113.4 mg, 0.45 mmol
- Dimethyl malonate, 0.15 mL, 1.35 mmol
- KOAc, 4.3 mg, 0.05 mmol
- BSA, 0.32 mL, 1.35 mmol
- Dichloromethane, 5 mL
- Diethyl ether 5 mL
- Saturated aqueous solution of ammonium chloride, 10 mL
- Anhydrous sodium sulfate
- Fluorocarbon-72, 6 mL

- Silica gel (Merck 60, 40–63 μm mesh)
- TLC plates (Merck 60 F$_{254}$, 0.2 mm)
- Two 25-mL Schlenk tube with magnetic stirrer bar
- Magnetic stirrer plate

- Two 50-mL Erlenmeyer flask
- One 100-mL separatory funnel
- Rotory evaporator
- One glass column, diameter 2.5 cm
- One filter paper

Procedure

The palladium precursor [Pd(η^3-C$_3$H$_5$)Cl]$_2$ (4.1 mg, 11.3 µmol) and the fluorous ligand (*9H,9'H,10H,10'H,11H,11'H,13H,13'H,14H,14'H,15H,15'H*-perfluorotricosane-12,12'-diyl)-bis-[(*4S*)-4-phenyl-2-oxazoline] (34.5 mg, 28.1 µmol) were dissolved in 1.5 mL dichloromethane in a Schlenk tube under argon. After being stirred for 1 hour at 50 °C, a solution of 1,3-diphenyl propen-2-enyl acetate (113.4 mg, 0.45 mmol) in dichloromethane (1.5 mL) was added. After 20 minutes, the solution was transfered in an Schlenk tube containing a solution of dimethyl malonate (0.15 mL, 1.35 mmol), KOAc (4.3 mg, 0.05 mmol), and BSA (0.32 mL, 1.35 mmol) in dichloromethane (2 mL). The solution was stirred at room temperature for the indicated time. The solution was then diluted with diethyl ether (5 mL) and the organic phase was washed with a saturated aqueous solution of ammonium chloride (2×5 mL). The solution was dried over sodium sulfate and the solvent was evaporated under reduced pressure to give a residue that was purified by column chromatography on silica using a 4/1 mixture of petroleum ether/ethyl acetate as the eluent. The enantiomeric excess was determined by HPLC using a Chiralpak AD column (25 cm $\times 0.46$ cm), the eluent being a 83/17 mixture of hexane/i-propanol.

For the recycling of the ligand, the solvent was evaporated after the reaction. The residue was extracted three times with fluorocarbon-72 (2 mL). The fluorous phase was washed with acetonitrile (2×2 mL) and the fluorous solvent was evaporated under reduced pressure to give practically quantitatively the fluorous bisoxazoline.

3.2.6   CONCLUSION

The synthesis of chiral fluorous bisoxazolines is very easy to reproduce. This ligand in association with palladium is an efficient catalyst for the asymmetric catalytic alkylation of 1,3-diphenylprop-2-enyl acetate with various carbonucleophiles. Table 3.2 gives some examples of such allylic alkylation.

Table 3.2 Asymmetric allylic alkylation of 1,3-diphenylprop-2-enyl acetate with various carbon nucleophiles using [Pd(η^3-C$_3$H$_5$)Cl]$_2$ in association with the fluorous bisoxazoline.

Entry	Nucleophile	Base	Solvent	t (h)	Yield (%)	ee (%)
1	CH$_2$(CO$_2$Me)$_2$	NaH	CH$_2$Cl$_2$	8	98	94 (S)
2	CH$_2$(CO$_2$Me)$_2$	NaH	BTF	20	100	90 (S)
3	CH$_2$(CO$_2$Me)$_2$	BSA/KOAc	CH$_2$Cl$_2$	24	89	94 (S)
4	CH$_2$(CO$_2$Me)$_2$	BSA/KOAc	BTF	40	93	92 (S)
5	CH$_3$CH(CO$_2$Me)$_2$	BSA/KOAc	CH$_2$Cl$_2$	24	93	90 (R)
6	AcNHCH(CO$_2$Me)$_2$	BSA/KOAc	CH$_2$Cl$_2$	90	83	93 (R)

REFERENCES

1. Ghosh, A. K., Mathivanan, P., and Cappiello, J. *Tetrahedron; Asymmetry* **1998**, *9*, 1.
2. Pfaltz, A. and Lautens, M. In *Comprehensive Asymmetric Catalysis*, Jacobsen, E. N. and Pfaltz, A. (Eds.) Springer, Berlin, **1999**, Vol. 2, 833.
3. Bayardon, J. and Sinou, D. *Tetrahedron Lett.* **2003**, *44*, 1449.
4. Fluoroflash™ fluorous silica gel cartridges are available from Fluorous Technologies, Inc: www.fluorous.com
5. Vincent, J. M., Rabion, A., Yachandra, V. K., and Fish, R. H. *Can. J. Chem.* **2002**, *79*, 888.

3.3 FACILE SYNTHESIS OF NEW AXIALLY CHIRAL DIPHOSPHINE COMPLEXES FOR ASYMMETRIC CATALYSIS

MATTHIAS LOTZ, GERNOT KRAMER, KATJA TAPPE and PAUL KNOCHEL*

Department Chemie, Ludwig-Maximilians-Universität München, Butenandtstr. 5–13, Haus F, D-81377 München, Germany

Molecules having axial chirality have found important applications as ligands in asymmetric catalysis.[1] Recently reported was the synthesis of a new class of chiral diphosphines such as (**1**)[2] using the chiral building block (**2**), which can be readily prepared in optically pure form according to Kagan.[3] The diphosphine (**1**) forms, upon complexation to a metallic salt, axially chiral complexes such as (**3**) (Figure 3.2), which have been shown to be efficient catalysts for allylic substitution reactions for 1,3-diphenylallyl acetate affording the expected allylic substitution product in high yield and 92–98% ee.[4]

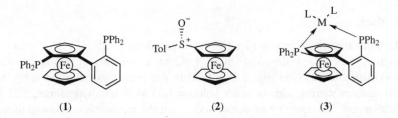

(**1**)   (**2**)   (**3**)

Figure 3.2 Ferrocenyl ligand (**1**), Sulfoxide (**2**) and mode of complexation (**3**).

3.3.1 SYNTHESIS OF (S_{FC})-1-[(S)-P-TOLYLSULFINYL]- 2-[(o-DIPHENYLPHOSPHINO)PHENYL]FERROCENE

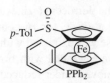

Materials and equipment

- (S)-Ferrocenyl-p-tolylsulfoxide, 793 mg, 2.45 mmol
- Lithiumdiisopropylamide solution (2 M in tetrahydrofuran), 1.35 mL, 2.70 mmol
- Zinc bromide solution (1.3 M in tetrahydrofuran), 2.51 mL, 3.26 mmol
- Pd(dba)$_2$, 61.2 mg, 5 mol%
- Tris-o-furylphosphine, 49.2 mg, 10 mol%
- (2-Iodophenyl)diphenylphosphine, 633 mg, 1.63 mmol
- Dry ice
- Dry tetrahydrofuran, 30 mL
- Diethyl ether, pentane
- Brine
- Anhydrous magnesium sulfate

- Silica gel (Matrex 60A, 40–63 μm, 230–400 mesh ASTM)
- TLC plates, SIL G-60 UV$_{254}$
- Two 50-mL round-bottomed flasks with nitrogen inlets and magnetic stirrer bars
- Two magnetic stirrer plates
- One Dewar flask
- One glass sintered funnel, diameter 7 cm
- One 250-mL Erlenmeyer flask
- One 250-mL round-bottomed flask
- One 250-mL separating funnel
- One glass column, diameter 3 cm
- Rotary evaporator

Procedure

1. Lithium diisopropylamide (2 M in tetrahydrofuran, 1.35 mL, 2.70 mmol, 1.1 equiv.) was added dropwise at −78 °C to a solution of (S)-ferrocenyl-p-tolylsulfoxide (**2**) (793 mg, 2.45 mmol) in dry tetrahydrofuran (15 mL). After 30 minutes stirring, zinc bromide solution (1.3 M in tetrahydrofuran, 2.51 mL, 3.26 mmol, 1.3 equiv.) was slowly added and the mixture was allowed to warm to room temperature. After 1 hour the solvent was evaporated *in vacuo* and the resulting residue was dissolved in dry tetrahydrofuran (10 mL). Pd(dba)$_2$ (61.2 mg, 5 mol%) and tris-o-furylphosphine (49.2 mg, 10 mol%) were dissolved in dry tetrahydrofuran (3 mL) and stirred for 10 minutes. A solution of (2-iodophenyl)diphenylphosphine (633 mg, 1.63 mmol, 0.67 equiv.) in dry tetra-hydrofuran (2 mL) was added and the mixture was stirred for an additional 10 minutes. The solution of the ferrocenyl zinc compound was added and the reaction mixture was stirred at 65 °C for 16 hours. The solution was quenched with water, the aqueous phase was extracted with diethyl ether (3 × 50 mL), the combined organic phases were washed with brine and dried over magnesium sulfate. The solvent was removed under reduced pressure.

2. The crude product was purified by column chromatography (silica gel, pentane/diethyl ether 1:2) (707 mg, 1.21 mmol, 74%).

^{1}H NMR (300 MHz, CDCl$_3$) δ 8.29–8.24 (1 H, m), 7.37–7.06 (14 H, m), 6.82–6.71 (3 H, m), 4.40–4.38 (1 H, m), 4.25–4.23 (1 H, m), 4.20 (5 H, s), 4.08–4.05 (1 H, m), 2.29 (3 H, s).

^{13}C NMR (75 MHz, CDCl$_3$) δ 140.9, 140.2, 140.1, 139.8, 138.3–137.5 (m), 134.4 (d, J 4.5 Hz), 133.9–133.3 (m), 129.0–127.5 (m), 124.7, 94.9, 90.5 (d, J 10.0 Hz), 74.0 (d, J 11.0 Hz), 70.8, 69.8, 67.7, 21.3.

^{31}P NMR (81 MHz, CDCl$_3$) δ −13.12.

3.3.2   SYNTHESIS OF (S_{FC})-1-DIPHENYLPHOSPHINO-2-[(o-DIPHENYLPHOSPHINO)PHENYL]FERROCENE

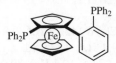

Materials and equipment

- (S_{Fc})-1-[(S)-p-Tolylsulfinyl]-2-[(o-diphenylphosphino)phenyl]ferrocene, 300 mg, 0.51 mmol
- t-BuLi (1.6 M in pentane), 0.64 mL, 1.03 mmol
- Chlorodiphenylphosphine, 0.32 mL, 1.80 mmol
- Dry ice
- Dry tetrahydrofuran, 8 mL
- Saturated aqueous ammonium chloride solution
- Diethyl ether, pentane
- Brine
- Anhydrous magnesium sulfate

- Silica gel (Matrex 60A, 40–63 μm, 230–400 mesh ASTM)
- TLC plates, SIL G-60 UV$_{254}$
- One 25-mL round-bottomed flask with nitrogen inlets and magnetic stirrer bar
- One magnetic stirrer plate
- One Dewar flask
- One glass sintered funnel, diameter 7 cm
- One 250-mL Erlenmeyer flask
- One 250-mL round-bottomed flask
- One 250-mL separating funnel
- One glass column, diameter 1.5 cm
- Rotary evaporator

Procedure

1. To a solution of (S_{Fc})-1-[(S)-p-Tolylsulfinyl]-2-[(o-diphenylphosphino)phenyl]-ferrocene (300 mg, 0.51 mmol) in dry tetrahydrofuran (8 mL) was slowly added

t-BuLi (1.6 M in pentane, 0.64 mL, 1.03 mmol, 2.0 equiv.) at −78 °C. After stirring for 5 minutes, chlorodiphenylphosphine (0.32 mL, 1.80 mmol, 3.5 equiv.) was added dropwise, the mixture was stirred for 5 minutes at −78 °C and for 30 minutes at room temperature. The reaction was quenched with saturated aqueous ammonium chloride solution, the aqueous phase was extracted with diethyl ether (3 × 50 mL), the combined organic phases were washed with brine and dried over magnesium sulfate. The solvent was removed under reduced pressure.

2. The crude product was purified by column chromatography (silica gel, pentane/diethyl ether 50:1). (260 mg, 0.41 mmol, 81%).

^{1}H NMR (300 MHz, CDCl$_3$) δ 8.36–8.32 (1 H, m), 7.56–7.50 (2 H, m), 7.37–7.27 (9 H, m), 7.18–7.12 (2 H, m), 7.08–6.96 (7 H, m), 6.81–6.77 (1 H, m), 6.67–6.61 (2 H, m), 4.28–4.26 (1 H, m), 4.19–4.16 (1 H, m), 3.95 (5 H, s), 3.76–3.75 (1 H, m).

^{13}C NMR (75 MHz, CDCl$_3$) δ 142.7 (d, *J* 30.8 Hz), 139.4 (d, *J* 4.1 Hz), 139.2 (d, *J* 6.7 Hz), 138.4 (d, *J* 9.4 Hz), 137.9 (d, *J* 12.9 Hz), 137.4 (d, *J* 14.3 Hz), 135.4–127.0 (m), 95.7 (dd, *J* 24.5, 10.1 Hz), 80.0 (d, *J* 8.7 Hz), 74.3 (dd, *J* 12.3, 2.9 Hz), 71.3 (d, *J* 4.1 Hz), 70.3, 68.8.

^{31}P NMR (81 MHz, CDCl$_3$) δ −14.35, −21.69.

3.3.3 CONCLUSION

In summary, a new simple method for the preparation of axially chiral diphosphine complexes has been developed. The presence of the planar chiral ferrocenyl unit avoids the need of resolution since only one diastereomer is formed in a metal complexation. The new ligand is useful for asymmetric palladium-catalysed allylic substitutions with malonates and amino derivatives (Schemes 1 and 2).

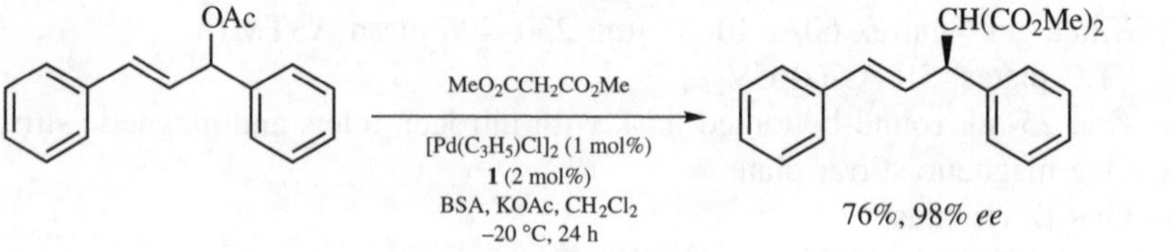

Scheme 1

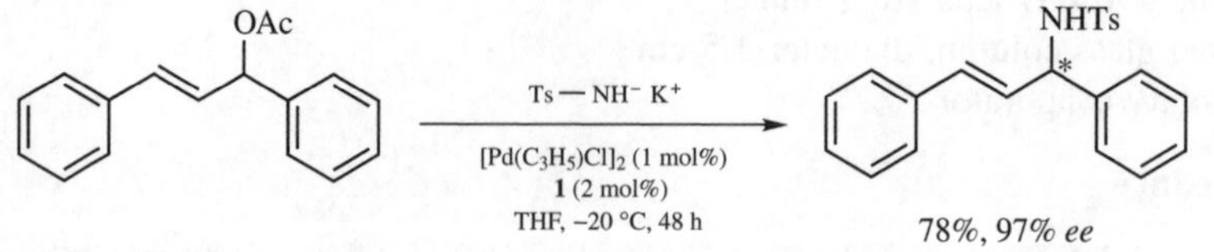

Scheme 2

REFERENCES

1. Noyori, R. *Asymmetric Catalysis in Organic Synthesis*, Wiley, New York, **1994**.
2. Lotz, M., Kramer, G. and Knochel, P. *Chem. Commun.* **2002**, 2546.
3. Guillaneux, D. and Kagan, H. B. *J. Org. Chem.* **1995**, 60, 2502.
4. Lotz, M., Knochel, P., Monsees, A., Riermeier, T., Kadyrov, R. and Almena, J. *Ger. Pat. No.* DE 10211250.

3.4 CHIRAL FERROCENYL-IMINO PHOSPHINES AS LIGANDS FOR PALLADIUM-CATALYSED ENANTIOSELECTIVE ALLYLIC ALKYLATIONS

PIERLUIGI BARBARO,[*,a] CLAUDIO BIANCHINI,[a] GIULIANO GIAMBASTIANI[*,a] and ANTONIO TOGNI[b]

[a]*Istituto di Chimica dei Composti Organo Metallici (ICCOM - CNR), Area di Ricerca di Firenze, Via Madonna del Piano, 50019, Sesto Fiorentino, Firenze, Italy*
[b]*Department of Chemistry and Applied Biosciences, Swiss Federal Institute of Technology, ETH Hönggerberg, CH-8093 Zürich, Switzerland*

Palladium-catalysed asymmetric allylic alkylation reactions employing soft carbon nucleophiles represent one of the most useful synthetic protocols to enantioselectively generate carbon-carbon bonds.[1] A variety of bidentate chiral ligands with mixed-donor atom sets, including chiral-planar ferrocenyl ligands, have been developed during the last decades for this purpose.[2] The synthesis has recently been reported of new class of ferrocenyl imino-phosphine ligands (**1a–e**) bearing a phosphorus donor atom on a side-chain stereocenter (Figure 3.3).[3] Palladium(II)-allyl complexes of the new ligands have been isolated and tested as catalyst precursors in the asymmetric allylic alkylation of 1,3-diphenylprop-2-en-1-yl acetate with dimethyl malonate (Figure 3.4). Quantitative yields and enantiomeric excesses as high as 80% have been obtained.

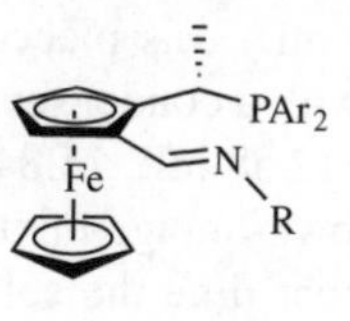

Ar = Ph

R = **1a**) 2,6-(Me)$_2$C$_6$H$_3$, **1b**) C$_6$H$_5$, **1c**) C$_6$H$_{11}$, **1d**) C$_2$H$_5$, **1e**) CH$_3$

Figure 3.3

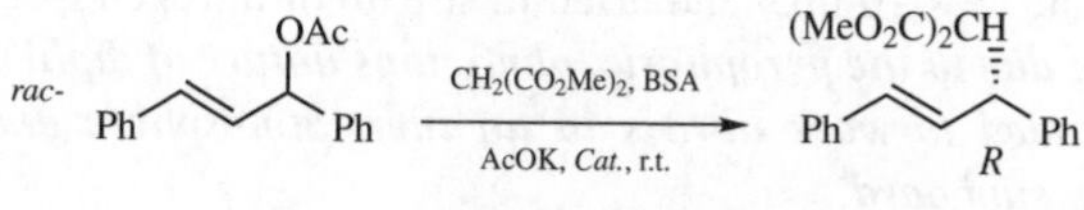

Figure 3.4

3.4.1   SYNTHESIS OF THE PRECURSOR(*R*)-1-[(*S*)-2-BROMOFERROCENYL]ETHYLDIPHENYLPHOSPHINE

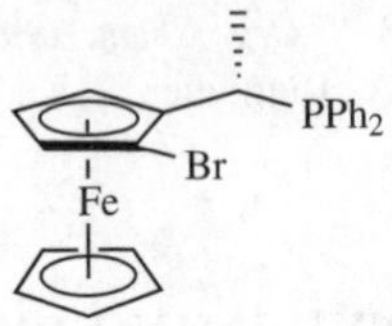

Materials and equipment

- *N,N*-Dimethyl-(*R*)-1-[(*S*)-2-bromoferrocenyl]ethylamine,[4] 4.84 g, 14.40 mmol
- Diphenylphosphine, 2.58 mL, 15.84 mmol
- Acetic acid (>99%), 10 mL
- *n*-Pentane, 50 mL
- Dichloromethane, 5 mL

- One 100-mL two-neck, round-bottomed flask with a magnetic stirring bar
- Two 100-mL round-bottomed flasks
- One 10-mL gas-tight syringe
- One 25-mL glass pipette
- One condenser equipped with a gas inlet
- One magnetic stirring hotplate equipped with a temperature controller
- One oil bath
- One rotary evaporator
- One sintered glass frit equipped with a gas inlet

Procedure

1. A solution of *N,N*-dimethyl-(*R*)-1-[(*S*)-2-bromoferrocenyl]ethylamine (4.84 g, 14.40 mmol) in acetic acid (10 mL) was placed under nitrogen in a 100-mL round-bottomed flask equipped with a condenser and a magnetic stirring bar. To this solution, diphenylphosphine (2.58 mL, 15.84 mmol) was added *via* syringe under nitrogen. The resulting brown-orange mixture was stirred at 110 °C for 4 hours under nitrogen and after that time the solvent was removed in *vacuo* at 100 °C.
2. After cooling to room temperature, the residue was crystallised by dissolving in dichloromethane (5 mL), followed by precipitation with *n*-pentane (30 mL) under nitrogen. The solid compound was collected on a sintered glass frit, washed with *n*-pentane (3 × 10 mL) and dried in stream of nitrogen (5.29 g, 77% yield).

    Attention: due to the pyrophoric, obnoxious nature of diphenylphosphine, it is recommended to work always in an inert atmosphere and under a well vented fume cupboard.

    ^{31}P{^{1}H} NMR (161.98 MHz, 294 K, CDCl$_3$): δ = 7.68 (s, Ph$_2$*P*).

^{1}H NMR (400.13 MHz, 294 K, CDCl$_3$): δ = 7.61 (m, 2H, PhH), 7.42 (m, 3H, PhH), 7.24 (m, 1H, PhH), 7.35 (m, 2H, PhH), 7.04 (m, 2H, PhH), 4.37 (m, 1H, HCp), 4.21 (s, 5H, HCp$'$), 4.14 (t, 1H, HCp, J_{HP} = 2.7 Hz), 4.04 (m, 1H, HCp), 3.48 (qnt, 1H, CHMe, J_{HH} = J_{HP} = 6.7 Hz), 1.54 (dd, 3H, CH_3, J_{HH} = 6.8, J_{HP} = 13.2 Hz).

^{13}C$\{^1$H$\}$ NMR (100.61 MHz, 294 K, CDCl$_3$): δ = 91.06 (d, Cp, J = 16.3 Hz), 80.32 (s, Cp), 70.84 (s, C$_5$H$_5$), 68.95 (s, HCp), 65.65 (s, HCp), 64.91 (d, HCp, J = 5.8 Hz), 29.62 (d, CHMe, J = 14.8 Hz), 18.81 (d, CH_3, J = 17.5 Hz).

3.4.2   SYNTHESIS OF KEY PRECURSOR (R)-1-[(S)-2-FORMYLFERROCENYL]ETHYLDIPHENYLPHOSPHINE

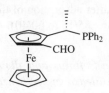

Materials and equipment

- (R)-1-[(S)-2-Bromoferrocenyl]ethyldiphenylphosphine, 1.77 g, 3.71 mmol
- n-BuLi (2.3 M in hexanes), 1.90 mL, 4.45 mmol
- Dry tetrahydrofuran, 16 mL
- Dry N,N-dimethylformamide, 1.44 mL, 18.55 mmol
- Acetone, dry ice
- Brine, 15 mL
- Diethyl ether, 140 mL
- Petroleum ether (b.p. 40–60 °C), ethyl acetate

- Two 100-mL round-bottomed flask
- One 100-mL round-bottomed flask equipped with a side stopcock, a magnetic stirring bar and a natural rubber septum
- Two 5-mL gas-tight syringes
- One 25-mL glass pipette
- One bench-top hemispherical 150 mL Dewar flask
- One 250-mL separatory funnel with a polypropylene stopper
- One 100-mL Erlenmeyer flask
- One magnetic stirring plate
- One rotary evaporator
- One chromatography glass column (i.d. 25 mm, length 50 cm) equipped with a gas inlet
- Silica gel (Merck 6 0, 230–400 mesh ASTM)
- Precoated plastic TLC plates, SIL G/UV254

Procedure

1. In a round-bottomed flask equipped with a side stopcock and rubber septum, (R)-1-[(S)-2-Bromoferrocenyl]ethyldiphenylphosphine (1.77 g, 3.71 mmol) was dissolved in dry tetrahydrofuran (16 mL) under nitrogen. To this solution n-BuLi (2.3M in hexanes) (1.90 mL, 4.45 mmol) was added dropwise *via* syringe at −60 °C with stirring. Dry dimethylformamide (1.44 mL, 18.55 mmol) was then added dropwise under nitrogen atmosphere. The reaction mixture was allowed to stand at −60 °C for 30 minutes, then was warmed at −40 °C and treated with brine (15 mL). After warming to room temperature, the organic layers were extracted with degassed diethyl ether (4 × 35 mL). The solvent was removed *in vacuo*.

2. The crude product obtained was purified by flash-chromatography on silica gel under nitrogen atmosphere after adsorption of the compound on a silica gel pad (petroleum ether/ethyl acetate: 85/15 to 60/40). A red-orange solid was obtained (1.26 g, 80% yield).

Note: Freshly dried tetrahydrofuran and dimethylformamide must be used. Tetrahydrofuran can be distilled over Na/benzophenone. N,N-Dimethylformamide can be distilled on K_2CO_3 under high vacuum and stored on activated 4Å molecular sieves.

$^{31}P\{^1H\}$ NMR (161.98 MHz, 294 K, $CDCl_3$): $\delta = 9.09$ (s, Ph_2P).

^1H NMR (400.13 MHz, 294 K, $CDCl_3$): $\delta = 9.46$ (s, 1H, HC = O), 7.62 (m, 2H, PhH), 7.48 (m, 3H, PhH), 7.21 (m, 1H, PhH), 7.12 (m, 2H, PhH), 7.01 (m, 2H, PhH), 4.60 (m, 1H, *H*Cp), 4.55 (m, 1H, *H*Cp), 4.53 (m, 1H, *H*Cp), 4.25 (s, 5H, *H*Cp′), 3.86 (qnt, 1H, C*H*Me, $J_{HH} = J_{HP} = 6.4$ Hz), 1.60 (dd, 3H, C*H₃*, $J_{HH} = 7.0$, $J_{HP} = 14.9$ Hz).

$^{13}C\{^1H\}$ NMR (100.61 MHz, 294 K, $CDCl_3$): $\delta = 168.23$ (s, C = O), 95.26 (d, Cp, $J = 15.4$ Hz), 71.67 (d, Cp, $J = 4.6$ Hz), 71.27 (s, HCp), 69.90 (s, C_5H_5), 68.74 (s, HCp), 68.17 (s, Cp), 29.57 (d, *C*HMe, $J = 14.8$ Hz), 18.59 (d, *C*H₃, $J = 20.0$ Hz).

3.4.3   SYNTHESIS OF (R)-1-[(S)-2-FERROCENYLIDENE-ETHYL-IMINE]ETHYLDIPHENYLPHOSPHINE

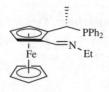

Materials and equipment

- (R)-1-[(S)-2-Formylferrocenyl]ethyldiphenylphosphine, 0.27 g, 0.61 mmol
- Ethylamine, 5 mL, 75.4 mmol

- Petroleum ether (b.p. 40–60 °C), ethyl acetate
- One 25-mL round-bottomed flask equipped with a side stopcock and a magnetic stirring bar
- One 100-mL round-bottomed flask
- One 10-mL glass pipette
- One oil bath
- One magnetic stirring hotplate equipped with a temperature controller
- One chromatography glass column (i.d. 10 mm, length 50 cm) equipped with a gas inlet
- Neutral aluminium oxide type 507 C
- Merck aluminium oxide 60 F_{254} neutral (type E) TLC plates
- One rotary evaporator

Procedure

1. (R)-1-[(S)-2-Formylferrocenyl]ethyldiphenylphosphine (0.270 g, 0.613 mmol), and ethylamine (5 mL) were placed in a 25-mL round-bottomed flask under nitrogen. The mixture was stirred at 40 °C for 6 hours. The excess of the amine was then evaporated under a stream of nitrogen to give a brown-orange compound.

2. After cooling to room temperature, the crude product was rapidly purified under a nitrogen atmosphere by flash-chromatography over neutral aluminium oxide (petroleum ether/ethyl acetate: 90/10) to afford the pure compound as an orange solid (0.27 g, 98% yield).

   $^{31}P\{^1H\}$ NMR (161.98 MHz, 294 K, CDCl$_3$): $\delta = 7.43$ (s, Ph$_2P$).

   ^1H NMR (400.13 MHz, 294 K, CDCl$_3$): $\delta = 7.51$ (m, 2H, PhH), 7.48 (s, 1H, CH=N), 7.44 (m, 3H, PhH), 7.16 (m, 1H, PhH), 7.12 (m, 2H, PhH), 6.94 (m, 2H, PhH), 4.66 (m, 1H, HCp), 4.45 (t, 1H, HCp, $J_{HP} = 2.5$ Hz), 4.27 (m, 1H, HCp), 4.19 (s, 5H, HCp'), 3.60 (qnt, 1H, CHMe, $J_{HH} = J_{HP} = 6.6$ Hz), 3.13 (m, 2H, CH$_2$CH$_3$), 1.60 (dd, 3H, CH$_3$, $J_{HH} = 7.0$, $J_{HP} = 14.3$ Hz), 1.10 (t, 3H, CH$_2$CH$_3$, $J_{HH} = 7.3$).

   $^{13}C\{^1H\}$ NMR (100.61 MHz, 294 K, CDCl$_3$): $\delta = 157.89$ (s, C=N), 92.91 (d, Cp, $J = 16.2$ Hz), 78.78 (s, Cp), 69.43 (s, C$_5$H$_5$), 68.83 (s, 2C, HCp), 66.32 (s, HCp), 56.04 (s, CH$_2$CH$_3$), 29.76 (d, CHCH$_3$, $J = 13.9$ Hz), 18.65 (d, CHCH$_3$, $J = 19.9$ Hz), 16.46 (s, CH$_2$CH$_3$).

   $[\alpha]_D^{23}$: $-387°$ (c 1.06, CHCl$_3$).

3.4.4 CONCLUSION

The cationic Pd(II)-π-allyl complexes of the formula [(Ligand)Pd(η^3-C$_3$H$_5$)]PF$_6$ (Ligand = ferrocenyl imino-phosphine **1a-e**) were isolated in the solid state by the literature procedure[5,6] using [(η^3-C$_3$H$_5$)PdCl]$_2$ as starting material.[7] The complexes obtained were used as catalyst precursors for the asymmetric allylic

Table 3.3 Asymmetric allylic alkylation of (rac)-(E)-3-acetoxy-1,3-diphenyl-1-propene catalysed by [(**1d**)Pd $(\eta^3$-$C_3H_5)$]PF$_6$.

Entry	Solvent	ee (%)
1	CH$_2$Cl$_2$	70.1
2	Toluene	79.6
3	THF	67.4
4	DMF	75.3

Experimental conditions: molar ratio substrate/catalyst/dimethyl malonate/ N,O-bis(trimethylsilyl)acetamide/potassium acetate = 100/1/200/200/10. Catalyst 0.006 mmol, solvent 4 mL, room temperature.

alkylation of (rac)-(E)-3-acetoxy-1,3-diphenyl-1-propene[8] with dimethyl malonate, under standard reaction conditions.[5,9] Complete conversion to the allylated product with (R)-configuration was observed for all catalysts in about 8–12 hours. The best results in term of enantioselectivity were obtained using the ligand **1d** (Table 3.3). ee's as high as 79.6% were observed in toluene.

REFERENCES

1. (a) Consiglio, G. and Waymouth, R. M. *Chem. Rev.* **1989**, *89*, 257. (b) Trost, B. M. and D. L. Van Vranken *Chem. Rev.* **1996**, *96*, 395. (c) *Catalytic Asymmetric Synthesis*; 2nd ed., Ojima I., Ed.; VCH Publishers: New York, **2000**. (d) Noyori R. *Asymmetric Catalysis in Organic Synthesis*, John Wiley & Sons, New York, **1994**. (e) Blaser H.U., Pugin B., and Spindler F. in *Applied Homogeneous Catalysis by Organometallic Complexes*, Cornils B. and Herrmann W.A., Eds., VCH, Weinheim, **1996**. (f) Tsuji J. *Palladium Reagents and Catalysts, Innovation in Organic Synthesis*, John Wiley & Sons, Chichester, **1995**. (g) Heck R. F. *Palladium Reagents in Organic Synthesis*, Academic Press, New York, **1985**. (h) Trost B. M. and Verhoeven T. R. in *Comprehensive Organometallic Chemistry*, Wilkinson G., Stone F. G. A. and Abel E. W., Eds.; Pergamon Press, Oxford, **1982**; *8*, pp 799–938.

2. (a) Hayashi, T. in *Ferrocenes: Homogeneous Catalysis, Organic Synthesis, Materials Science*, Togni A. and Hayashi T., Eds., VCH, Weinheim, **1995**, 105–142. (b) Colacot, T. J. *Chem. Rev.* **2003**, *103*, 3101. (c) Togni, A. and Venanzi, L. M. *Angew. Chem. Int. Ed. Engl.* **1994**, *33*, 497. (d) Togni, A. Bieler, N., Burckhardt, U., Köllner, C., Pioda, G., Schneider, R. and Schnyder, A. *Pure Appl. Chem.* **1999**, *71*, 1531. (e) Richards and C. J., Locke, A. *Tetrahedron: Asymmetry* **1998**, *9*, 2377. (f) Dai, L. X., Tu, T., You, S. L., Deng, W. P., and Hou, X. L. *Acc. Chem. Res.* **2003**, *36*, 659.

3. Barbaro, P., Bianchini, C., Giambastiani, G. and Togni A. *Tetrahedron Lett.* **2003**, *44*, 8279.

4. (a) Köllner, C., Pugin, B., and Togni, A. *J. Am. Chem. Soc.*, **1998**, *120*, 10274. (b) Schneider, R., Köllner, C., Weber, I., and Togni, A. *Chem. Commun.*, **1999**, 2415. (c) CIBA-GEIGY AG, *patent WO96/32400*, **1996**.

5. Barbaro, P., Pregosin, P. S., Salzmann, R., Albinati, A. and Kunz, R. W. *Organometallics*, **1995**, *14*, 5160.

6. Bianchini, C., Barbaro, P., and Scapacci G. *J. Organomet. Chem.* 2001, *621*, 26.
7. Tatsuno, Y. Yoshida, T. and Otsuka, S.; *Inorg. Synth.* **1990**, *28*, 342.
8. Barbaro, P., Currao, A., Herrmann, J., Nesper, R., Pregosin, P. S. and Salzmann, R. *Organometallics* **1996**, *15*, 1879.
9. (a) von Matt, P., Lloyd-Jones, G. C., Minidis, A. B. E., Pfaltz, A., Macko, L., Neuburger, M., Zehnder, M., Rüegger, H., and Pregosin, P. S. *Helv. Chim. Acta* **1995**, *78*, 265. (b) Togni, A., Breutel, C., Schnyder, A., Spindler, F., Landert, H. and Tijani, A. *J. Am. Chem. Soc.* **1994**, *116*, 4062.

4 Suzuki Coupling Reactions

CONTENTS

Catalysts for Fine Chemical Synthesis, Vol. 3, Metal Catalysed Carbon-Carbon Bond-Forming Reactions
Edited by S. M. Roberts, J. Xiao, J. Whittall, and T. Pickett
© 2004 John Wiley & Sons, Ltd   ISBN: 0-470-86199-1

4.1   PALLADIUM-CATALYSED BORYLATION AND SUZUKI COUPLING (BSC) TO OBTAIN β-BENZO[B]THIENYL-DEHYDROAMINO ACID DERIVATIVES

ANA S. ABREU, PAULA M. T. FERREIRA and MARIA-JOÃO R. P. QUEIROZ*

Departamento de Química, Universidade do Minho, Campus de Gualtar 4710-057 Braga, Portugal

The one-pot borylation and Suzuki coupling (BSC) was successfully applied to the synthesis of *ortho*-disubstituted biaryls, using pinacolborane and 2-(dicyclohexyl-phosphane)biphenyl as ligand.[1] The component to be borylated requires an electron donating group while the Suzuki coupling component needs an electron withdrawing group. The BSC reaction has been applied to the synthesis of 2-methyl-2′-nitro diaryl compounds in the benzo[b]thiophene series, precursors of methylated thienocarbazoles.[2] In the last few years there has been interest in the preparation of benzo[b]thienyldehydroamino acids for biological and/or photochemical applications.[3, 4] Recently the BSC reaction has been applied to the synthesis of this type of compounds using *ortho*-methoxylated or methylated bromobenzo[b]thiophenes as the components to be borylated and pure stereoisomers of β-bromodehydroamino acid derivatives.[5] Thus the scope of the BSC reaction has been extended to non-aromatic Suzuki coupling components.

4.1.1 SYNTHESIS OF THE *E* AND *Z* ISOMERS OF THE METHYL ESTER OF *N-tert*-BUTOXYCARBONYL-β-BROMODEHYDROAMINOBUTYRIC ACID

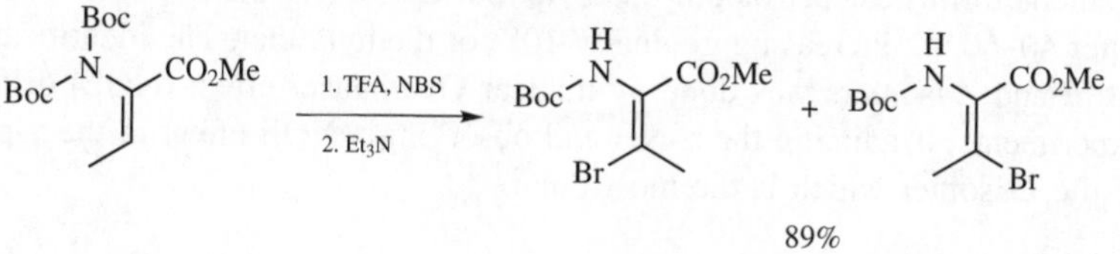

Z-isomer   E-isomer

9   :   1

Materials and equipment

- Methyl ester of *N,N*-bis(*tert*-butoxycarbonyl)-(Z)-dehydroaminobutyric acid,[6] 1.58 g, 5 mmol
- Dichloromethane, CH_2Cl_2, 50 mL
- Trifluoroacetic acid, TFA (99.5%), 1 mL
- *N*-bromosuccinimide, NBS (99%), 1.07 g, 6 mmol
- Triethylamine, Et_3N (99%), 2.10 mL, 15 mmol
- Water, 60 mL
- Brine, 30 mL
- Anhydrous magnesium sulfate
- Silica gel 60 (0.040–0.063 mm), 45 g
- TLC plates, Silica G-60 UV_{254}
- Petroleum ether, 40–60 °C
- Diethyl ether
- 250-mL round-bottomed flask with magnetic stirrer bar
- Magnetic stirrer plate
- One 250-mL separating funnel
- One glass column, diameter 3 cm.
- Rotatory evaporator

Procedure

1. The methyl ester of *N,N*-bis(*tert*-butoxycarbonyl)-(Z)-dehydroaminobutyric acid[6] (1.58 g, 5 mmol) was placed in a 250-mL round-bottomed flask equipped with a magnetic stirrer bar and was dissolved in dichloromethane (50 mL). To this, TFA (1 mL) was added slowly with vigorous stirring and the reaction was monitored by TLC. When no starting material was detected (**ca** 1 hour), *N*-bromosuccinimide (1.07 g, 6 mmol) was added and the mixture was left stirring at room temperature overnight, protected from light. Triethylamine (2.10 mL, 15 mmol) was then added and stirring continued for an additional hour. Water was added to the reaction mixture (30 mL) and the organic phase was separated and washed with water (30 mL) and brine (30 mL). The organic phase was dried (magnesium sulfate), filtered and the removal of solvent under reduced pressure

gave an oil which showed on ¹H-NMR to be a mixture in 89% yield of E and Z isomers (1:9).

2. The diastereomers were separated by column chromatography using solvent gradient from neat petroleum ether 40–60 °C to 30% diethyl ether/petroleum ether 40–60 °C, increasing gradually 10% of diethyl ether. The identification of the E and Z isomers was done by nuclear Overhauser effect (NOE) difference experiments, irradiating the α-NH and observing a NOE effect in the γ-protons of the E-isomer which is the most polar.[3]

4.1.2 SYNTHESIS OF THE METHYL ESTER OF N-tert-BUTOXYCARBONYL-(Z)-[β-(2,3,7-TRIMETHYLBENZO[b]THIEN-6-YL]DEHYDROAMINOBUTYRIC ACID

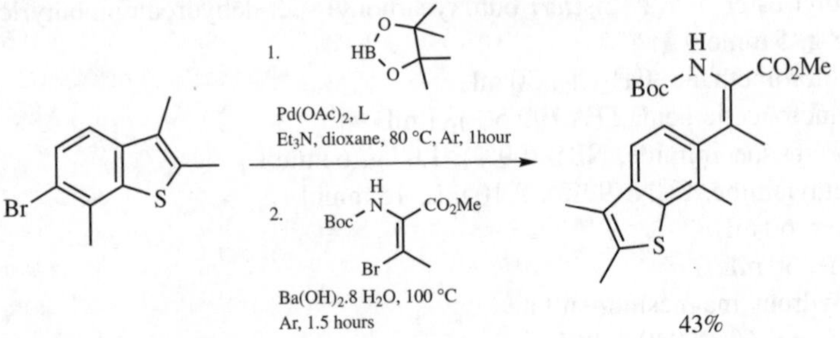

L = 2-(dicyclohexylphosphane)biphenyl

Materials and equipment

- 6-Bromo-2,3,7-trimethylbenzo[b]thiophene,[7] 0.128 g, 0.5 mmol
- Triethylamine, Et₃N (99%), 0.280 mL, 2 mmol
- Palladium acetate, Pd(OAc)₂ (98%), 6 mg, 0.025 mmol
- 2-(Dicyclohexylphosphane)biphenyl, (98%), 35 mg, 0.1 mmol
- 4,4,5,5-tetramethyl-1,3,2-dioxaborolane, (97%), 0.22 mL, 1.5 mmol
- 1,4-Dioxane anhydrous, (99.8%) 2 mL
- Methyl ester of N-tert-butoxycarbonyl-(Z)-β-bromodehydroaminobutyric acid, 0.14 g, 0.5 mmol
- Barium hydroxide octohydrate, Ba(OH)₂.8 H₂O (98%), 0.473 g, 1.5 mmol
- Water, 0.2 mL
- Ethyl acetate
- Water for extraction
- Anhydrous magnesium sulphate
- Silica gel 60 (0.040–0.063 mm), 30 g
- TLC plates, Silica G-60 UV₂₅₄
- Petroleum ether 40–60 °C
- Diethyl ether
- Double vacuum and argon glass line

- Vacuum oil pump
- Argon
- Heat gun
- Schlenk tube
- Oil bath
- Magnetic stirrer plate
- One 100-mL separating funnel
- One glass column, diameter 2 cm.
- Rotary evaporator

Procedure

1. A Schlenk tube with a magnetic stirrer bar was dried in a double vacuum and Argon glass line, under vacuum using the heat gun externally and was charged under argon with 6-bromo-2,3,7-trimethylbenzo[b]thiophene (0.128 g, 0.5 mmol), dioxane (2 mL), Et$_3$N (0.280 mL, 2 mmol), Pd(OAc)$_2$ (6 mg, 0.025 mmol), 2-(dicyclohexylphosphane)biphenyl (35 mg, 0.1 mmol) and 4,4,5,5-tetra-methyl-1,3,2-dioxaborolane (0.22 mL, 1.5 mmol). The tube was closed and the mixture was heated under argon, for 1 hour at 80 °C.

2. After cooling, the tube was opened under argon and the methyl ester of N-tert-butoxycarbonyl-(Z)-β-bromodehydroaminobutyric acid (0.14 g, 0.5 mmol), Ba(OH)$_2$.8 H$_2$O (0.473 g, 1.5 mmol) and water (0.2 mL) were added. The tube was closed, saturated with argon and the mixture was then heated for 1.5 hours at 100 °C.

3. After cooling, water (2 mL) was added and the mixture was poured into a separating funnel washing the Schlenk tube with ethyl acetate. More water (20 mL) and ethyl acetate (25 mL) were added and the phases were separated. The organic phase was dried (magnesium sulphate), filtered and removal of solvent under reduced pressure gave a brown solid which was submitted to column chromatography using solvent gradient from neat petroleum ether 40–60 °C to 30% diethyl ether/petroleum ether 40–60 °C, increasing gradually 10% of diethyl ether, to give the product as a white solid (84 mg, 43%). Recrystallisation from n-hexane gave colourless crystals; m.p. 87–88 °C.

   Found C, 64.69; H, 7.03; N, 3.53; S, 8.23%, calcd for C$_{21}$H$_{27}$NO$_4$S C, 64.76; H, 6.99; N, 3.60; S, 8.23%.

   δ_H (300MHz, CDCl$_3$) 1.38 (s, 9H, CH$_3$ Boc), 2.20 (s, 3H, CH$_3$), 2.31 (s, 3H, ArCH$_3$), 2.38 (s, 3H, ArCH$_3$), 2.52 (s, 3H, ArCH$_3$), 3.88 (s, 3H, OCH$_3$), 5.53 (s, 1H, NH), 7.07 (broad d, 1H, ArH, J 8 Hz), 7.49 (d, 1H, ArH, J 8 Hz); δ_C (75.4MHz, CDCl$_3$) 11.40 (CH$_3$), 13.78 (CH$_3$), 17.05 (CH$_3$), 20.80 (CH$_3$), 28.00 (CH$_3$ Boc), 51.94 (OCH$_3$), 80.66 (OC(CH$_3$)$_3$) 119.51 (CH), 123.77 (CH), 125.15 (C), 127.17 (C), 128.09 (C), 132.22 (C), 133.53 (C), 134.22 (C), 139.28 (C), 140.27 (C), 152.91 (C=O), 165.57 (C=O).

   The stereochemistry of the product was determined by NOE difference experiments irradiating the α-NH and observing a NOE effect in the ArCH$_3$ group at 2.38 ppm.

4.1.3   CONCLUSION

With this methodology it is possible to synthesise in a one-pot procedure new benzo[b]thienyldehydroamino acids avoiding the lithiation step and the transmetalation to boron. The dehydroamino acid derivative acts as the coupling component having an electron withdrawing group despite the slightly donating effect of the carbamate group, which affects to some extent the product yield.

REFERENCES

1. Baudoin, O., Guénard, D. and Guéritte, F. *J. Org. Chem.*, **2000**, *65*, 9268–9271.
2. Ferreira, I. C. F. R., Queiroz, M.-J. R. P. and Kirsch, G. *Tetrahedron Lett.*, **2003**, *44*, 4327–4329.
3. Silva, N. O., Abreu, A. S., Ferreira, P. M. T., Monteiro L. S. and Queiroz, M.-J. R. P. *Eur. J. Org. Chem.*, **2002**, 2524–2528.
4. Abreu, A. S., Silva, N. O., Ferreira, P. M. T. and Queiroz, M.-J. R. P. *Eur. J. Org. Chem.*, **2003**, 1537–1544.
5. Abreu, A. S., Silva, N. O., Ferreira, P. M. T. and Queiroz, M.-J. R. P. *Tetrahedron Lett.*, **2003**, *44*, 6007–6009.
6. Ferreira, P. M. T., Maia, H., Monteiro, L. S. and Sacramento, J. *J. Chem. Soc. Perkin Trans.*, **1999**, *1*, 3697–3703.
7. Ferreira, I. C. F. R., Queiroz, M.-J. R. P. and Kirsch, G. *J. Heterocycl. Chem.*, **2001**, *38*, 749–754.

4.2   PALLADIUM-CATALYSED CROSS-COUPLING REACTIONS OF 4-TOSYLCOUMARINS AND ARYLBORONIC ACIDS: SYNTHESIS OF 4-ARYLCOUMARIN COMPOUNDS

JIE WU, LISHA WANG, REZA FATHI, and ZHEN YANG*

Harvard Institute of Chemistry and Cell Biology (ICCB), Harvard University, 250 Longwood Avenue, SGM 604, Boston, Massachusetts 02115, USA

The prominence of coumarins in natural products and biologically active molecules has promoted considerable efforts toward their synthesis. As a 'privileged' scaffold, 4-substituted coumarin shows interesting biological properties, especially for their anti-HIV and antibiotic activities.[1] Very recently, a new palladium-based catalytic system for successful coupling of the 4-tosylcoumarin and aryl boronic acids was developed. 4-Tosylcoumarin has been identified as an ideal synthon for synthesizing 4-substituted coumarins in terms of its stability, cost-effectiveness and easy preparation compared to its corresponding triflate.[2] The new catalytic system is quite efficient for generation of diversified 4-arylcoumarins in excellent yields.[3]

4.2.1 SYNTHESIS OF 4-TOSYLCOUMARINS

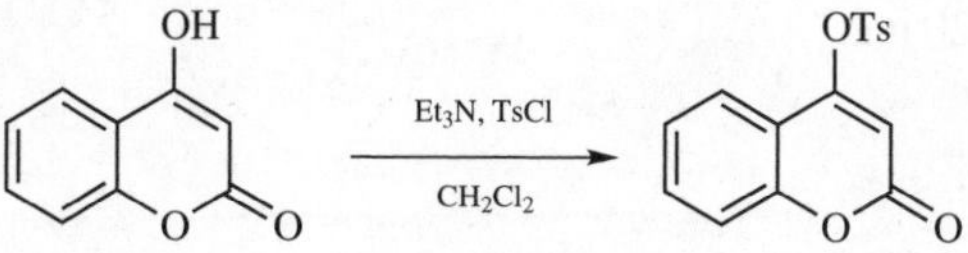

Materials and equipment

- 4-Hydroxycoumarin, 1.62 g, 10 mmol
- Triethylamine, 1.67 mL, 12 mmol
- p-Toluenesulfonyl chloride, 1.90 g, 10 mmol
- Methylene chloride, 20 mL

- Silica gel (Matrex 60A, 37–70 um), 30 g
- TLC plates, SIL G-60 UV254
- Column
- 50-mL round-bottomed flask with magnetic stirrer bar
- Tubes
- Filter paper
- Rotary evaporator

Procedure

4-Hydroxycoumarin (1.62 g, 10 mmol) and p-toluenesulfonyl chloride (1.90 g, 10 mmol) were placed in a 50-mL round-bottom flask equipped with a magnetic stirrer bar. Methylene chloride (20 mL) and triethylamine (1.67 mL, 12 mmol) were added. The reaction mixture was stirred at room temperature overnight. Following completion of the reaction as monitored by TLC, the reaction mixture was filtered. The filtrate was concentrated under vacuum and the residue was purified by flash chromatography to give the 4-tosylcoumarin as white solid (3.10 g, 98%).

4.2.2 SYNTHESIS OF 4-ARYLCOUMARIN

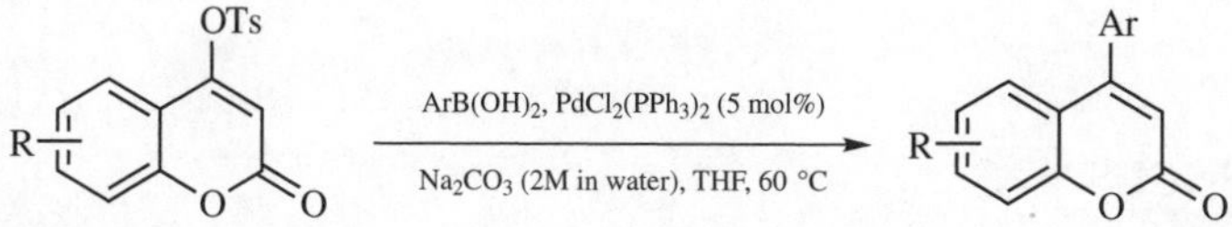

Materials and equipment

- 4-Tosylatecoumarin, 0.25 mmol
- Arylboronic acid, 0.50 mmol
- Bis(triphenylphosphine)dichloropalladium (II), 9 mg, 5 mol%

- Tetrahydrofuran, 2.0 mL
- Sodium carbonate (aqueous, 2 M), 2.0 mL
- Ethyl acetate, 20 mL
- Water, 5.0 mL
- Brine, 5.0 mL

- Silica gel (Matrex 60A, 37–70 um), 20 g
- Column
- TLC plates, SIL G-60 UV254
- 10-mL round-bottomed flask with magnetic stirrer bar
- Tubes
- Separating funnel
- Rotary evaporator

Procedure

A mixture of 4-tosylcoumarin (0.25 mmol), arylboronic acid (0.50 mmol) and bis(triphenylphosphline) dichloropalladium (II) $PdCl_2(PPh_3)_2$ (9 mg, 5 mol%), was added into a reaction tube under argon atmosphere. Then tetrahydrofuran (2.0 mL) and aqueous sodium carbonate (2.0 mL, 2M solution) were added. The reaction mixture was heated with stirring at 60 °C overnight. Following completion of the reaction, as monitored by TLC, the reaction mixture was cooled, diluted with ethyl acetate (20 mL), washed with water (5.0 mL), brine (5.0 mL) and dried over sodium sulfate. The solvent was removed under vacuum and the residue was purified by flash chromatography to give the corresponding products.

4.2.3   CONCLUSION

The procedure is very easy to reproduce and the palladium-catalysed cross-couplings may be applied to a wide range of substituted 4-tosylcoumarin and arylboronic acids. The method successfully broadens existing approaches for the cross-couplings of tosylate with boronic acids by using palladium as the catalyst. Table 4.1 demonstrates diverse tosylates that can undergo palladium-catalyzed cross-couplings with various arylboronic acids successfully.

REFERENCES

1. For example, see: (a) Murray, R. D. H., Méndez, J. and Brown, S. A. *The natural coumarines: Occurrence, Chemistry, and Biochemistry*; John Wiley & Sons, New York, 1982. (b) Xie, L., Takeuchi, Y., Cosentino, L. M., McPhail, A. T. and Lee, K.-H. *J. Med. Chem.* **2001**, *44*, 664.
2. Wu, J., Liao, Y. and Yang, Z. *J. Org. Chem.* **2001**, *66*, 3642.
3. Wu, J., Wang, L., Fathi, R. and Yang, Z. *Tetrahedron Lett.* **2002**, *43*, 4395.

Table 4.1 Palladium-catalysed cross-couplings of substituted 4-tosylcoumarin and arylboronic acids.

Entry	4-Tosylcoumarin	Arylboronic acids	Yield (%)
1	R = H	Phenyl	63
2	R = H	2-OMe-C_6H_4	80
3	R = H	4-OMe-C_6H_4	82
4	R = H	4-CF_3-C_6H_4	69
5	R = H	2-F-C_6H_4	62
6	R = H	3-CHO-C_6H_4	51
7	R = 7-OMe	Phenyl	91
8	R = 7-OMe	2-OMe-C_6H_4	81
9	R = 6-Me	Phenyl	68
10	R = 6-Me	2-OMe-C_6H_4	76
11	R = 6-OMe	2-OMe-C_6H_4	72

4.3 CYCLOPROPYL ARENES, ALKYNES AND ALKENES FROM THE IN SITU GENERATION OF B-CYCLOPROPYL-9-BBN AND THE SUZUKI-MIYAURA COUPLING OF ARYL, ALKYNYL AND ALKENYL BROMIDES

RAMON E. HUERTAS and JOHN A. SODERQUIST*

Department of Chemistry, University of Puerto Rico, Box 23346, San Juan, Puerto Rico, USA 00931-3346

The double hydroboration of propargyl bromide with 9-borabicyclo[3.3.1]nonane (9-BBN-H) proceeds smoothly to afford the corresponding 1,1-diboryl-3-bromo-propane adduct which is quantitatively converted to the B-hydroxy-B-cyclopropyl-9-BBN and bis-hydroxy-9BBN complexes in equal molar amounts. Addition of this mixture to aryl, alkynyl and alkenyl bromides containing a catalytic (3%) amount of Pd[PPh_3]$_4$, after 8–16 hours at reflux temperature, provides the corresponding cyclopropylated arenes, alkynes and alkenes in good to excellent yields.[1]

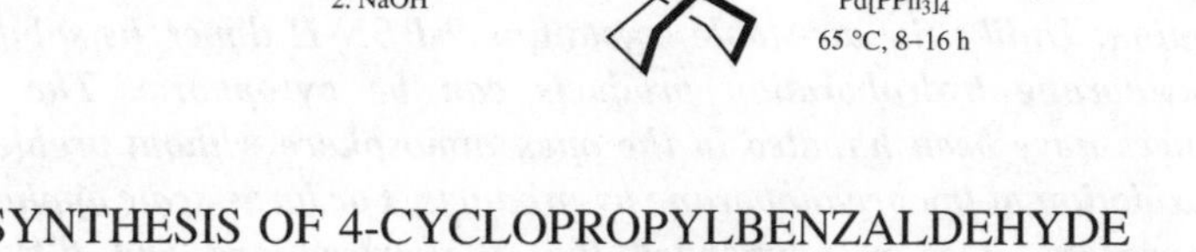

4.3.1 SYNTHESIS OF 4-CYCLOPROPYLBENZALDEHYDE

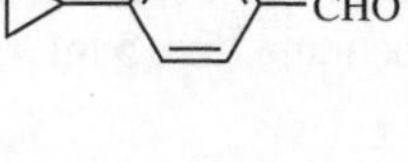

Materials and equipment

- Propargyl bromide (from fractional distillation of an 80 wt% mixture in toluene, >98%), 0.66 g, 5.6 mmole
- 9-Borabicyclo[3.3.1]nonane (>98%), 1.35 g, 11.1 mmol
- Dry tetrahydrofuran (THF), 15 mL
- Sodium hydroxide solution (deoxygenated under a nitrogen purge), 55.5 mL of 3.0 M, 166.5 mmol
- 4-Bromobenzaldehyde (99%), 0.92 g, 5.0 mmol
- Pd(PPh$_3$)$_4$ (99%), 0.182 g, 0.16 mmol
- Pentane, 25 mL
- Pentane/ether (98:2)

- One each, 10- and 25-mL round-bottomed flasks fitted with septum-fitted side arms, magnetic stir bars, heating mantles, spiral reflux condensers and vented to a fume hood through a Schlenk-type connection and oil bubbler
- Transfer needles and syringes
- Nitrogen gas source
- Silica gel (60 Å, 70–230 mesh)
- Chromatography column, 100 mL
- Separating funnel, septum-fitted, 50 mL
- One round-bottomed flask, 250 mL
- Rotary evaporator

Procedure

1. The 9-BBN-H dimer (1.35 g, 5.65 mmol) was placed into a dry 10-mL round-bottomed flask containing a magnetic stirring bar; the reflux condenser was added, septum-fitted and the assembly was thoroughly purged with nitrogen gas vented through an oil bubbler to the fumehood. Dry tetrahydrofuran (5 mL) was added *via* syringe followed by propargyl bromide (0.66 g, 5.6 mmol). The mixture was heated at reflux (65 °C) for 2 hours. The mixture was allowed to cool to room temperature. Deoxygenated aqueous sodium hydroxide (5.5 ml of 3.0 M, 16.5 mmol) was added and the mixture was stirred for 1 hour. Using a slight positive pressure of nitrogen, this mixture was transferred to the second 25-mL flask assembly (as above) containing 4-bromobenzaldehyde (0.92 g, 5.0 mmol) and Pd(PPh$_3$)$_4$ (0.182 g, 0.16 mmol) in THF (10 mL). The stirred mixture was heated at reflux temperature for 18 hours, allowed to reach room temperature. *Attention: Unlike the air-stable crystalline 9-BBN-H dimer, its solutions and trialkylborane hydroboration products can be pyrophoric. The coupling mixtures have been handled in the open atmosphere without problems from the oxidation of the organoborane by-products. For large-scale applications of this process, it is recommended that the extractions and filtrations be conducted under a nitrogen atmosphere using septum-sealed equipment.[2]*
2. The open mixture was transferred to a separating funnel containing water (10 mL) and extracted with pentane (5 × 5 mL). The combined organic material

was washed with aqueous sodium hydroxide (5×25 mL of 3.0 M) followed by water (20×100 mL). The organic material was filtered through silica (*ca.* 70 mL dry volume) using pentane/ether (98:2, *ca.* 200 mL) as the eluent. Concentration at reduced pressure furnishes 0.57 g (79%) of 4-cyclopropylbenzaldehyde as a colorless liquid.

^{1}H NMR (300 MHz, CDCl$_3$). 9.94 (s, 1H), 7.76 (d, $J = 8.2$ Hz, 2H), 7.19 (d, $J = 8.2$ Hz, 2H), 1.92–2.01 (m, 1H), 1.00–1.12 (m, 2H), 0.78–0.83 (m, 2H); ^{13}C NMR (75 MHz, CDCl$_3$). 191.8, 152.1, 134.1, 129.9, 125.9, 16.0, 10.6; IR (neat) 3090, 3010, 2930, 2830, 2730, 2060, 1920, 1700, 1600, 1570, 1512, 1460, 1420, 1400, 1360, 1310, 1220, 1170, 1110, 1040, 1020, 900, 840, 820, 720; MS *m/z* (relative abundance) 146 (M$^+$, 98), 145 (M-1)(92), 115 (100), 91 (68).

4.3.2 CONCLUSION

The present procedure is an extremely simple method to cyclopropylate a wide variety of aryl, alkynyl and vinyl bromides. The essential hydroxyborate complex for the Suzuki-Miyaura coupling is easily generated *in situ* thereby avoiding the need to isolate the pyrophoric trialkylborane intermediate. This methodology which is based upon related observations with alkynylations,[3,4] greatly simplifies the coupling process. Some additional examples of this process, including a dicyclo-propylation of a *gem*-dibromide (entry 8) are given in Table 4.2.

Table 4.2 The cyclopropylation of aryl, alkynyl and alkenyl bromides with *B*-cyclopropyl-9-BBN generated *in situ*.

Entry	R	Yield (%)
1	Ph	61
2	*m*-C$_6$H$_4$(Me)	63
3	*o*-C$_6$H$_4$(OMe)	92
4	C(Ph)=CH$_2$	84
5	α-naphthyl	85
6	*cis*-CH=CHBu	68 (>99% Z)
7	*trans*-CH=CHBu	60 (>99% E)
8	PhCH=C($-$)$_2$	58

REFERENCES

1. Soderquist, J. A., Huertas, R. and Leon-Colon, G. *Tetrahedron Lett.*, **2000**, *41*, 4251.
2. Brown, H. C., Midland, M. M., Levy, A. B. and Kramer, G. W. *Organic Synthesis via Boranes*; Wiley Interscience: New York, 1975.
3. Soderquist, J. A., Matos, K. and Ramos, J. *Tetrahedron Lett.*, **1995**, *36*, 2401.
4. Soderquist, J. A., Rane, A. M., Matos, K. and Ramos, J. *Tetrahedron Lett.*, **1995**, *36*, 6847.

4.4   ONE-POT SYNTHESIS OF UNSYMMETRICAL 1,3-DIENES THROUGH PALLADIUM-CATALYSED SEQUENTIAL BORYLATION OF A VINYL ELECTROPHILE BY A DIBORON AND CROSS-COUPLING WITH A DISTINCT VINYL ELECTROPHILE

TATSUO ISHIYAMA* and NORIO MIYAURA*

Division of Molecular Chemistry, Graduate School of Engineering, Hokkaido University, Sapporo 060-8628, Japan

Regio- and stereospecific synthesis of 1,3-dienes is of great importance in synthetic organic chemistry due to frequent occurrence of these fragments in biologically active natural products, as well as their use in numerous transformations such as the Diels-Alder reaction. Among available methods for preparing 1,3-dienes, palladium-catalysed cross-coupling of vinylboron compounds with vinyl electrophiles has attracted considerable attention because of its practical usefulness.[1] Although vinylboron compounds have been generally obtained by hydroboration of alkynes or by transmetalation between trialkylborates with vinyllithium or magnesium reagents,[2] these methods have inherent limitations. To overcome such limitations, palladium-catalysed borylation of vinyl electrophiles by bis(pinacolato)diboron has recently been developed; this provides an efficient and convenient route to vinylboron compounds.[3] This new borylation protocol allows a one-pot, two-step procedure for synthesis of various types of unsymmetrical 1,3-dienes through a borylation-coupling sequence.[3c-e]

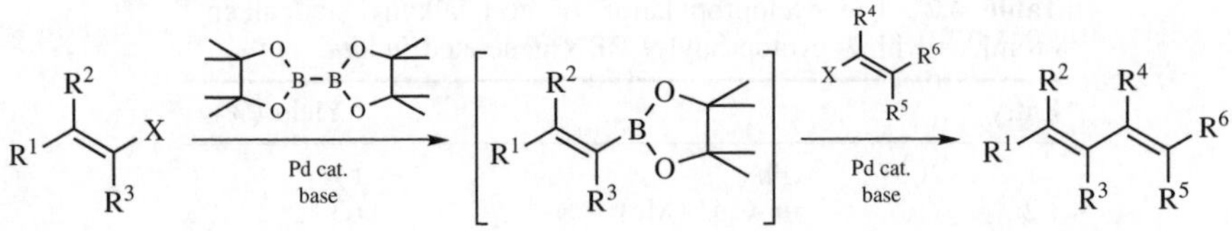

X = halogen, OTf

4.4.1   SYNTHESIS OF 2-(1-CYCLOPENTENYL)-1-DECENE

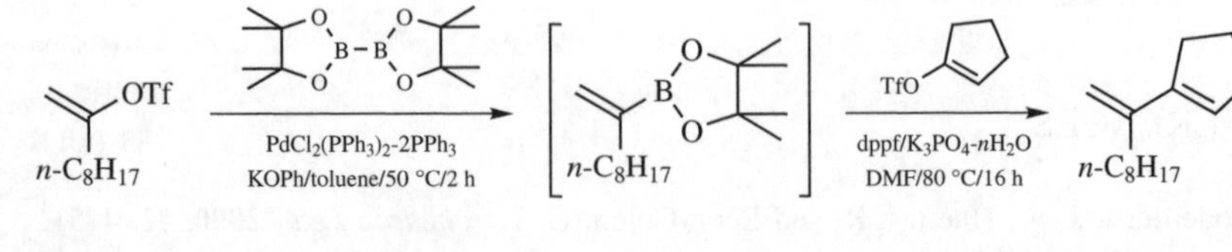

Materials and equipment

- 1-Decen-2-yl triflate[4] (99%), 317 mg, 1.1 mmol

- Bis(pinacolato)diboron[5] (99%), 279 mg, 1.1 mmol
- Dichlorobis(triphenylphosphine)palladium(II) (99%), 21 mg, 0.03 mmol
- Triphenylphosphine (99%), 16 mg, 0.06 mmol
- Potassium phenoxide[6] (99%), 198 mg, 1.5 mmol
- Dry toluene, 4 mL
- 1-Cyclopentenyl triflate[4] (99%), 216 mg, 1.0 mmol
- 1,1'-Bis(diphenylphosphino)ferrocene (97%), 17 mg, 0.03 mmol
- Potassium phosphate n-hydrate, 637 mg, ca. 3.0 mmol
- Dry N,N-dimethylformamide, 4 mL
- Hexane, 400 mL
- Water, 50 mL
- Anhydrous magnesium sulfate
- Silica gel (Kanto Chemical Co. Inc., 60 A, 63–210 μm, spherical, neutral), 60 g

- One 25-mL, two-necked, round-bottomed flask equipped with a magnetic stirring bar and a rubber septum
- One reflux condenser
- Nitrogen inlet equipped with an oil bubbler
- Magnetic stirrer plate
- One oil bath
- One 100-mL separating funnel
- One 50-mL Erlenmeyer flask
- One Büchner funnel, diameter 4 cm
- Filter paper
- Two 100-mL recovery flasks
- Diaphragm pump
- Rotary evaporator
- One glass column, diameter 2 cm, length 30 cm
- Twenty 20-mL culture tubes
- One 25-mL pear-shaped flask
- Vacuum pump
- One vacuum trap

Procedure

1. A 25-mL, two-necked, round-bottomed flask was fitted with a magnetic stirring bar, a rubber septum, and a reflux condenser to which a nitrogen inlet and an oil bubbler were attached. The flask and condenser were pre-dried in an oven at 120 °C for 1 hour, assembled while hot, and allowed to cool under a stream of nitrogen. The septum was removed and the flask was charged with bis(pinacolato)diboron (279 mg, 1.1 mmol), dichlorobis(triphenylphosphine)palladium(II) ($PdCl_2(PPh_3)_2$) (21 mg, 0.03 mmol), triphenylphosphine (PPh_3) (16 mg, 0.06 mmol), and potassium phenoxide (KOPh) (198 mg, 1.5 mmol). The septum was again placed on the flask, and the flask was purged with nitrogen for 1 minute. Dry toluene (4 mL) and 1-decen-2-yl triflate (317 mg, 1.1 mmol) were

added, and the flask was immersed in an oil bath that was maintained at 50 °C. The mixture was stirred at that temperature for 2 hours to give a solution of 2-(4,4,5,5-tetramethyl-1,3,2-dioxaborolan-2-yl)-1-decene.

Notes: Bis(pinacolato)diboron is air-stable and commercially available.

In the borylation of vinyl triflates β-substituted by a carbonyl group, use of potassium carbonate (K_2CO_3) (207 mg, 1.5 mmol) in dry 1,4-dioxane (4 mL) at 80 °C gives better results.

The borylation is usually completed within 3 hours.

2. To the solution of 2-(4,4,5,5-tetramethyl-1,3,2-dioxaborolan-2-yl)-1-decene were added 1,1'-bis(diphenylphosphino)ferrocene (dppf) (17 mg, 0.03 mmol) and potassium phosphate *n*-hydrate (K_3PO_4-nH_2O) (637 mg, ca. 3.0 mmol) under a gentle stream of nitrogen by removing the septum. The septum was again placed on the flask. Dry *N,N*-dimethylformamide (DMF) (4 mL) and 1-cyclopentenyl triflate (216 mg, 1.0 mmol) were added and the flask was purged with nitrogen for 1 minute. The mixture was stirred at 80 °C for 16 hours.

Notes: When the cross-coupling results in lower yields, addition of dichloro-[1,1'-bis(diphenylphosphino)ferrocene]palladium(II) ($PdCl_2(dppf)$) (22 mg, 0.03 mmol) in place of dppf is effective.

In the cross-coupling with vinyl triflates β-substituted by a carbonyl group, use of dry 1,4-dioxane (4 mL) in place of dry DMF gives better yields.

The cross-coupling usually completed within 16 hours.

3. The resulting mixture was removed from the oil bath, allowed to cool to room temperature, and poured into a 100-mL separating funnel. The reaction flask was rinsed with hexane (3 × 10 mL). The rinses and water (50 mL) were added to the separating funnel, the funnel was shaken, the layers were separated, and the organic extracts were dried over anhydrous magnesium sulfate in a 50-mL Erlenmeyer flask. The drying agent was filtered off under reduced pressure by using a Büchner funnel (diameter 4 cm), a filter paper, a 100-mL recovery flask, and a diaphragm pump, and was washed with hexane (3 × 10 mL). The filtrate was concentrated on a rotary evaporator to give a crude product as an oil. The oil was placed on the top of a glass column (diameter 2 cm, length 30 cm) of silica gel (60 g) and eluted with hexane (**ca.** 300 mL). A total of twenty 15 mL fractions were collected in 20 mL tubes. Fractions containing desired product were poured into a 100-mL recovery flask and each of the tubes was rinsed with hexane (1 mL). After concentration by a rotary evaporator, the product was transferred to a 25-mL pear-shaped flask by using hexane (5 × 1 mL). Removal of hexane by a rotary evaporator and further concentration by a vacuum pump (0.1 mm) equipped with a vacuum trap at room temperature for 1 hour yielded 2-(1-cyclopentenyl)-1-decene as a colorless oil (167 mg, 81% yield based on 1-cyclopentenyl triflate).

^{1}H NMR (400 MHz, CDCl$_3$) δ 0.88 (3 H, t, *J* 6.6 Hz), 1.20–1.40 (10 H, m), 1.40–1.55 (2 H, m), 1.90 (2 H, tt, *J* 7.6, 7.6 Hz), 2.26 (2 H, t, *J* 7.7 Hz), 2.40–2.50 (4 H, m), 4.89 (2 H, br s) and 5.78 (1 H, br s).

^{13}C NMR (100 MHz, CDCl$_3$) δ 14.11, 22.69, 23.10, 29.04, 29.32, 29.52, 29.72, 31.91, 32.70, 33.30, 34.19, 110.99, 126.76, 143.57 and 144.77.
exact mass calcd for C$_{15}$H$_{26}$ 206.2035, found 206.2032.

4.4.2 CONCLUSION

The method provides an efficient and convenient route to access a wide variety of unsymmetrical 1,3-dienes, because the borylation proceeds in high yields under mild conditions and is applied to the preparation of 1-alken-2-ylboron, cyclic vinylboron, and functionalized vinylboron compounds which cannot be directly obtained by the conventional methods based on hydroboration and transmetalation. Table 4.3 shows generality of the process.

Table 4.3 Synthesis of 1,3-dienes through borylation-coupling sequence.[a]

Entry	Product[b]	Yield (%)[c]	Entry	Product[b]	Yield (%)[c]
1	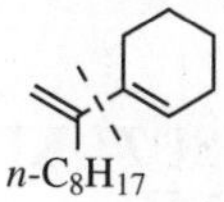 n-C$_8$H$_{17}$	(96)	6	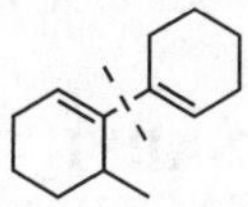	79
2	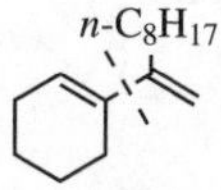 n-C$_8$H$_{17}$	(62)[d]	7	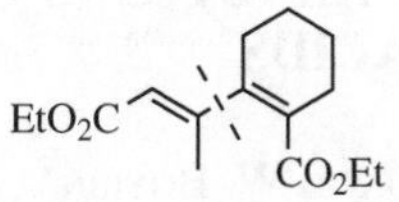 EtO$_2$C CO$_2$Et	(76)[d,e,f]
3	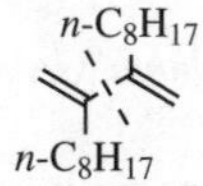 n-C$_8$H$_{17}$ n-C$_8$H$_{17}$	(99)[d]	8	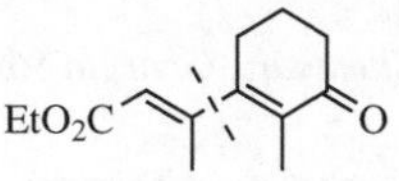 EtO$_2$C O	76[d,e]
4	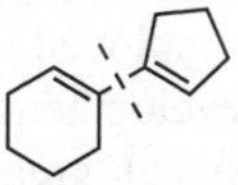	74	9	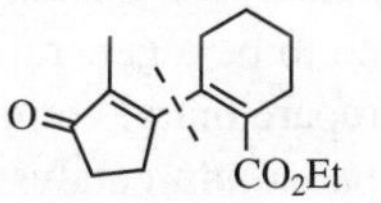 O= CO$_2$Et	77[d,e,g]
5	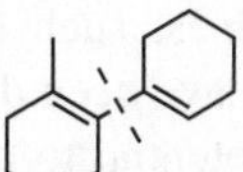	81			

[a]Borylation of a vinyl triflate (1.1 mmol) with bis(pinacolato)diboron (1.1 mmol) in toluene (4 mL) at 50 °C for 1–3 hours in the presence of PdCl$_2$(PPh$_3$)$_2$-2PPh$_3$ (0.03 mmol) and KOPh (1.5 mmol) was followed by cross-coupling with a vinyl triflate (1.0 mmol) at 80 °C for 16 hours by using dppf (0.03 mmol), K$_3$PO$_4$-nH$_2$O (ca. 3.0 mmol), and DMF (4 mL).
[b]Left part of dotted line comes from a vinyl triflate used for the borylation and right part from a vinyl triflate used for the cross-coupling.
[c]Isolated yields based on vinyl triflates used for the cross-coupling and GC yields are in parentheses.
[d]PdCl$_2$(dppf) (0.03 mmol) was used in place of dppf.
[e]1,4-Dioxane (4 mL) was used in place of DMF for the cross-coupling.
[f]The cross-coupling was carried out for 5 hours.
[g]The borylation was conducted at 80 °C by using K$_2$CO$_3$ (1.5 mmol) in dioxane (4 mL).

REFERENCES

1. Reviews: (a) Miyaura, N. and Suzuki, A. *Chem. Rev.* **1995**, *95*, 2457. (b) Miyaura, N. *Top. Curr. Chem.* **2002**, *219*, 11. (c) Suzuki, A. and Brown, H. C. *Organic Syntheses Via Boranes*, Aldrich Chemical Company, Inc., Milwaukee, **2003**, Vol. 3.
2. Reviews: (a) Nesmeyanov, A. N. Sokolik, R. A. *Methods of Elemento-Organic Chemistry*, North-Holland, Amsterdam, **1967**, Vol. 1. (b) Pelter, A., Smith, K. and Brown, H. C. *Borane Reagents*, Academic Press, London, **1988**. (c) Miyaura, N. and Maruoka, K. In *Synthesis of Organometallic Compounds*, John Wiley & Sons, Chichester, **1997**, 345.
3. (a) Takahashi, K., Takagi, J., Ishiyama, T. and Miyaura, N. *Chem. Lett.* **2000**, 126. (b) A review: Ishiyama, T. and Miyaura, N. *J. Organomet. Chem.* **2000**, *611*, 392. (c) Takagi, J., Takahashi, K., Ishiyama, T. and Miyaura, N. *J. Am. Chem. Soc.* **2002**, *124*, 8001. (d) Takagi, J., Kamon, A., Ishiyama, T. and Miyaura, N. *Synlett* **2002**, 1880. (e) Ishiyama, T., Takagi, J., Kamon, A. and Miyaura, N. *J. Organomet. Chem.* in press.
4. (a) Stang, P. J., Hanack, M. and Subramanian, L. R. *Synthesis* **1982**, 85. (b) Ritter, K. *Synthesis* **1993**, 735.
5. (a) Nöth, H. *Z. Naturforsch.* **1984**, *39b*, 1463. (b) Ishiyama, T., Murata, M., Ahiko, T.-a. and Miyaura, N. *Org. Synth.* **2000**, *77*, 176.
6. Kornblum, N. and Lurie, A. P. *J. Am. Chem. Soc.* **1959**, *81*, 2705.

4.5 Pd(OAc)₂/2-ARYL OXAZOLINE CATALYSED SUZUKI COUPLING REACTIONS OF ARYL BROMIDES AND BORONIC ACIDS

BIN TAO and DAVID W. BOYKIN*

Department of Chemistry, Georgia State University, University Plaza, Atlanta, GA 30303, USA

The palladium-catalysed Suzuki coupling reaction of aryl halides with arylboronic acids has proved to be a general and convenient synthetic tool employed in organic chemistry to prepare biaryl compounds.[1] The discovery and development of active and efficient palladium-catalyst systems have been the focus of great interest recently. New catalytic systems based on palladium-oxazolines, such as 2-aryl oxazolines and 2,2′-(1,3-phenylene)bisoxazoline (Figure 4.1) have been developed for the coupling reaction. These catalytic systems have the potential to overcome

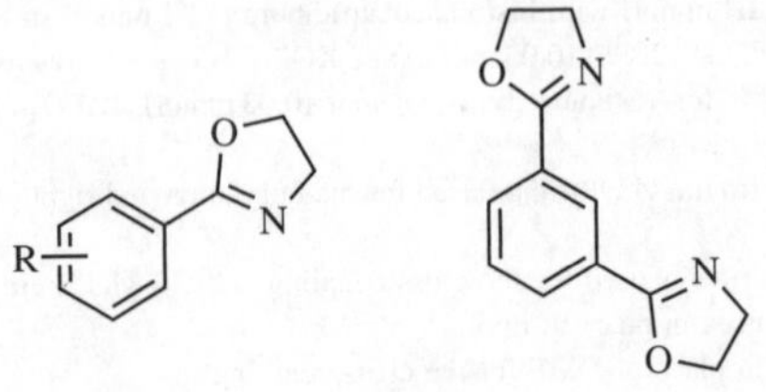

Figure 4.1   2-Aryl oxazolines and 2,2′-(1,3-phenylene)bisoxazoline.

the air-sensitive problem associated with the traditional phosphine-containing catalysts and possess the similar high reactivity at the same time.

4.5.1 SYNTHESIS OF 4-METHOXYBIPHENYL

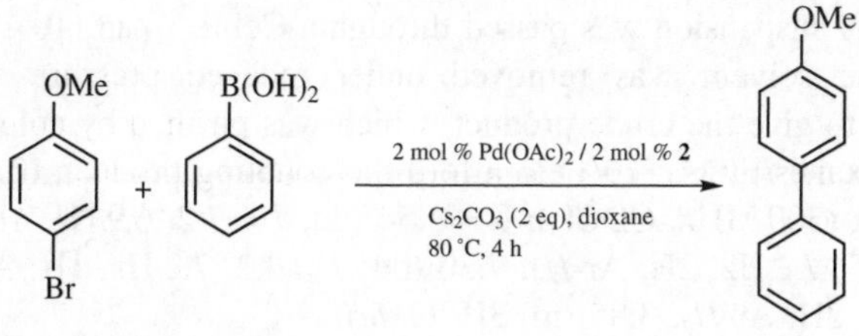

$$\text{2 mol \% Pd(OAc)}_2 \text{ / 2 mol \% } \mathbf{2}$$
$$\text{Cs}_2\text{CO}_3 \text{ (2 eq), dioxane}$$
$$80\,^{\circ}\text{C, 4 h}$$

Materials and equipment

- 4-Bromoanisole (99%), 0.187 g, 1.0 mmol
- Phenylboronic acid (97%), 0.183 g, 1.5 mmol
- Palladium acetate (98%), 4.5 mg, 0.02 mmol
- 2,2′-(1,3-Phenylene) bisoxazoline, 4.3 mg, 0.02 mmol
- Cesium carbonate (99%), 0.65 g, 2.0 mmol
- 1,4-Dioxane (anhydrous), 4 mL
- Diethyl ether, 150 mL
- Hexanes, 200 mL
- Ethyl acetate, 20 mL
- Dry nitrogen gas

- Magnetic stirrer plate
- One oil bath
- Magnetic stirrer bar
- 50-mL two-necked, round-bottomed flask
- Celite®, 10 g
- Silical gel (60–230A), 20 g
- TLC plates, G-60 UV$_{254}$
- One glass sintered funnel, diameter 4 cm
- Three 125-mL Erlenmeyer flasks
- One Buchner funnel, diameter 4 cm
- One Buchner flask, 250 mL
- One recovery flask, 250 mL
- One glass column, diameter 2 cm
- Rotary evaporator

Procedure

1. Under a nitrogen atmosphere, palladium acetate (4.5 mg, 0.02 mmol), 2,2′-(1,3-phenylene)bisoxazoline (4.3 mg, 0.02 mmol) and 1,4-dioxane (4 mL) were

added to a 50-mL two-necked reaction flask equipped with a magnetic stirring bar at room temperature, followed by addition of 4-bromoanisole (0.187 g, 1.0 mmol), phenylboronic acid (0.183 g, 1.5 mmol) and cesium carbonate (0.65 g, 2.0 mmol). After stirring at room temperature for 5 minutes, the mixture was heated to 80 °C for 4 hours. It was cooled to room temperature and diethyl ether (100 mL) was added to the mixture.

2. The ethereal suspension was passed through a Celite® pad (10 g) in a Buchner funnel. The solvent was removed under reduced pressure with a rotary evaporator to give the crude product, which was purified by column chromatography (hexanes/EtOAc 10/1) to afford the coupling product. (0.177 g, 96%).

^{1}H NMR (300 MHz, CDCl$_3$): δ = 7.54 (dd, J = 7.2, 6.9 Hz, 4H, Ar-H), 7.38 (dd, J = 7.5, 7.2 Hz, 2H, Ar-H), 7.30 (dd, J = 7.2, 7.2 Hz, 1H, Ar-H), 6.97 (d, J = 6.9 Hz, 2H, Ar-H), 3.85 (m, 3H, OMe).

^{13}C NMR (75 MHz, CDCl$_3$): δ = 159.1, 140.8, 133.7, 128.7, 128.1, 126.7, 126.6, 114.1(Ar-Cs), 55.3 (OMe).

Note: *This procedure has been scaled up to produce 1.70 g of the coupling product using the same amount of the catalyst (0.2 mol%).*

4.5.2   CONCLUSION

2,2′-(1,3-Phenylene)bisoxazoline, easily prepared from 1,3-dicyanobenzene and aminoethanol in the presence of catalytic amount of potassium carbonate,[2] is stable in air, unlike phosphines. This catalytic system can be applied to the coupling reactions of a range of substrates as illustrated in the Table 4.4.[3]

Table 4.4   Suzuki coupling reactions of aryl bromides with aryl boronic acids catalysed by (**2**).^{a}

$$R^1 \!-\!\!\langle\ \rangle\!-\!Br \ + \ (HO)_2B\!-\!\langle\ \rangle\!-\!R^2 \xrightarrow[\substack{Cs_2CO_3 \ (2\ eq),\ dioxane \\ 80\,°C,\ 4\ h}]{2\ mol\ \%\ Pd(OAc)_2\ /\ 2\ mol\ \%\ \mathbf{2}} R^1\!-\!\langle\ \rangle\!-\!\langle\ \rangle\!-\!R^2$$

Entry	R^1	R^2	Yield (%)
1	4-Cl	H	81
2	3-NO$_2$	H	76
3	4-NO$_2$	H	96
4	4-Me	H	77
5	3-Me	H	89
6	ArBr = 6-Br-2-naphthol	H	75
7	3-NO$_2$	4-NO$_2$	93
8	4-Me	4-OMe	82
9	4-Me	2-Me	77
10	2-CN	3,4-Methylenedioxo	95

^{a}**2** = 2,2′-(1,3-phenylene)bisoxazoline.

REFERENCES

1. Miyaura, N. and Suzuki, A. *Chem. Rev.* **1995,** *95,* 2457.
2. Schumacher, D. P., Clark, J. E., Murphy, B. L. and Fischer, P. A. *J. Org. Chem.* **1990,** *55,* 5291.
3. Tao, B. and Boykin, D. W. *Tetrahedron Lett.* **2002,** *43,* 4955.

4.6 PALLADIUM-CATALYSED REACTIONS OF HALOARYL PHOSPHINE OXIDES: MODULAR ROUTES TO FUNCTIONALISED LIGANDS

COLIN BAILLIE, LIJIN XU and JIANLIANG XIAO*

Liverpool Centre for Materials and Catalysis, Department of Chemistry, University of Liverpool, L69 7ZD, UK

Phosphines play an extremely important role in homogeneous catalysis, with the choice of ligand often being the crucial factor determining the success of a reaction. Traditionally phosphine compounds are prepared by standard reactions using organometallic reagents.[1] These are often air and moisture sensitive and can even be pyrophoric. Furthermore, due to the low functional group tolerance of these synthetic procedures, the resulting phosphines are limited in their structural diversity. Catalytic methods for phosphine synthesis are an increasingly attractive alternative.[2] Indeed, it has recently been shown that the Heck reaction of $OPPh_{3-n}$-$(4-C_6H_4Br)_n$ (n = 1–3) can be readily used to prepare phosphines with designer electronic or solubility properties, e.g. those that are soluble in supercritical carbon dioxide or water.[3] Other palladium-catalysed reactions, e.g. Suzuki coupling, Buchwald-Hartwig amination and carbonylation are equally effective for the preparation of novel *para*-substituted arylphosphines starting with the phosphine oxides.[3d] Herein it is shown that biaryl and heterobiaryl monodentate phosphine ligands can be easily accessed *via* the Suzuki coupling of $OPPh_2(2-C_6H_4Br)$ with arylboronic acids (Figure 4.2). Ligands of this type are usually prepared *via* traditional organometallic routes and have been shown to be useful in homogeneous catalysis, co-ordination chemistry and material science.[4,5]

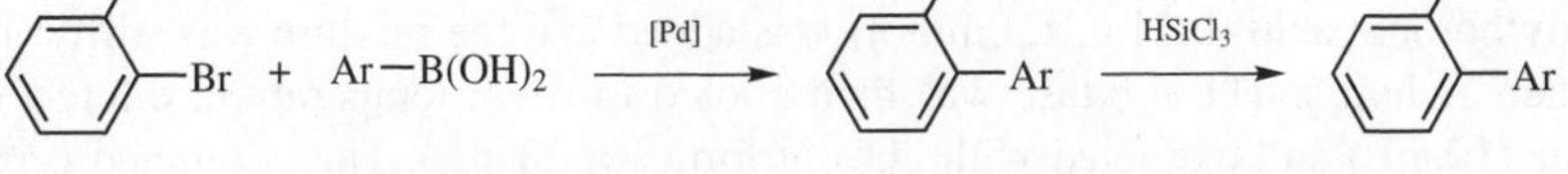

Figure 4.2 Suzuki coupling of a phosphine oxide.

4.6.1   SYNTHESIS OF 2-DIPHENYLPHOSPHINYL-2′-METHOXYBIPHENYL VIA SUZUKI COUPLING

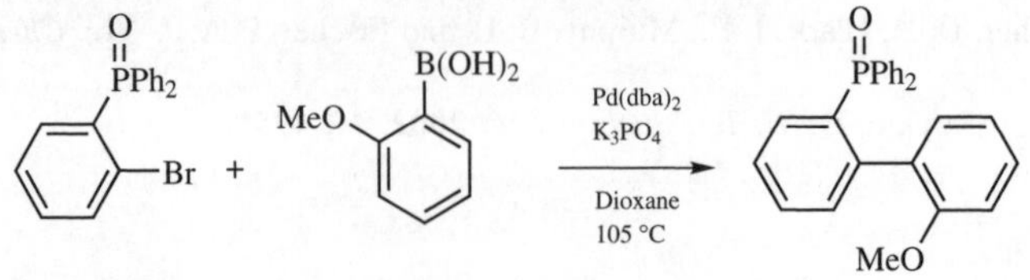

Materials and equipment

- $OPPh_2(o\text{-}C_6H_4Br)$,[6] 0.50 g, 1.4 mmol
- 2-Methoxyphenylboronic acid, 2×0.21 g, 2×1.40 mmol
- Bis(dibenzylideneacetone)palladium(0) $(Pd(dba)_2)$, 24 mg, 0.042 mmol
- Triphenylphosphine (PPh_3), 44 mg, 0.17 mmol
- Potassium phosphate, 0.59 g, 2.8 mmol
- Dry dioxane
- Ethyl acetate
- Hexane
- Chloroform
- Anhydrous magnesium sulfate

- Silica gel (Matrex 60A, 37–70 μm)
- TLC plates, 60 F_{254}
- 250-mL Conical flasks
- 100-mL Separating funnel
- 100-mL Round-bottomed flask
- 50-mL Schlenk tube
- Magnetic stirrer bar
- Magnetic stirrer plate and temperature control
- Glass sintered funnel
- Glass column, diameter 5 cm
- Rotary evaporator

Procedure

To a Schlenk tube were charged $OPPh_2(o\text{-}C_6H_4Br)$ (0.50 g, 1.4 mmol) and 2-methoxyphenylboronic acid (0.21 g, 1.4 mmol) together with $Pd(dba)_2$ (24 mg, 0.042 mmol), PPh_3 (44 mg, 0.17 mmol) and potassium phosphate (0.59 g, 2.8 mmol) in 5 mL of dioxane under an atmosphere of argon. The Schlenk tube was stirred at 105 °C for 24 hours, at which time a second batch of 2-methoxyphenylboronic acid (0.21 g, 1.4 mmol) was added and the mixture was stirred for a further 24 hours. The mixture was then cooled to room temperature, diluted with water (10 mL) and extracted with chloroform (3 × 20 mL). The combined organic extracts were washed with brine, dried over anhydrous magnesium sulfate and evaporated *in vacuo*. The crude product was purified by flash chromatography (EtOAc/hexane: 2/1). Crystallization from EtOAc/hexane yielded 0.51 g (95%) of the title compound as white crystals.

^{1}H NMR (400 MHz, CDCl$_3$) δ7.70–7.15(m, 15H), 7.05 (ddd, 1H, $J = 8.0$ Hz, 8.0 Hz, 1.8 Hz), 6.80 (ddd, 1H, $J = 7.5$ Hz, 7.5 Hz, 0.8 Hz), 6.35 (d, 1H, $J = 8.1$ Hz), 3.40 (s, 3H).

^{13}C NMR (100 MHz, CDCl$_3$) δ 144.0 (d, $J_{CP} = 8.0$ Hz), 134.4, 133.7 (d, $J_{CP} = 111.1$ Hz), 133.5 (d, $J_{CP} = 104.7$ Hz), 133.1, 132.5 (d, $J_{CP} = 8.8$ Hz), 131.6 (d, $J_{CP} = 2.4$ Hz), 131.3 (d, $J_{CP} = 9.6$ Hz), 131.0 (d, $J_{CP} = 2.4$ Hz), 129.5, 129.0 (d, $J_{CP} = 4.0$ Hz), 128.5 (d, $J_{CP} = 12.0$ Hz), 127.9 (d, $J_{CP} = 12.0$ Hz), 126.9 (d, $J_{CP} = 12.0$ Hz), 119.7, 109.6, 54.6.

^{31}P NMR (162 MHz, CDCl$_3$) δ 28.2.

Anal. Calcd for C$_{25}$H$_{21}$PO$_2$: C, 78.10; H, 5.52. Found: C, 77.50; H, 5.60.

4.6.2 SYNTHESIS OF 2-DIPHENYLPHOSPHINO-2′-METHOXYBIPHENYL

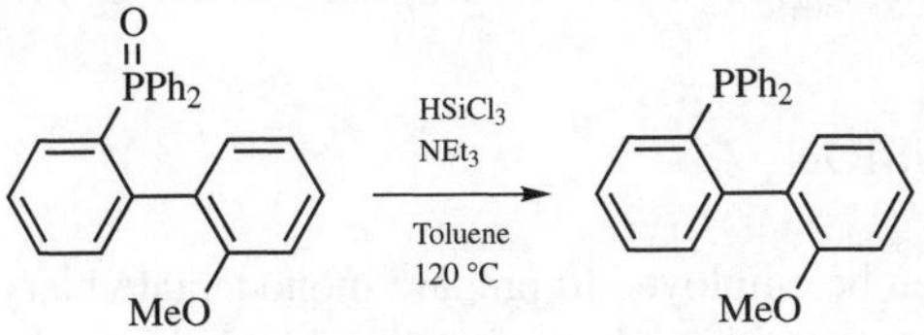

Materials and equipment

- 2-Diphenylphosphinyl-2′-methoxybiphenyl, 0.42 g, 1.1 mmol
- Triethylamine, 0.84 mL, 6.1 mmol
- Trichlorosilane, 0.56 mL, 5.5 mmol
- Dry toluene
- Ethyl acetate
- Hexane
- Sodium hydrogencarbonate

- Celite$^{®}$
- Silica gel (Matrex 60A, 37–70 μm)
- TLC plates, 60 F$_{254}$
- 250-mL Conical flasks
- 100-mL Round-bottomed flask
- 50-mL Schlenk tube
- Magnetic stirrer bar
- Magnetic stirrer plate and temperature control
- Glass sintered funnel
- Glass column, diameter 5 cm
- Rotary evaporator

Procedure

A 10 mL toluene solution of 2-diphenylphosphinyl-2′-methoxybiphenyl (0.42 g, 1.1 mmol) was frozen in liquid nitrogen, to which trichlorosilane (0.56 mL, 5.5 mmol) and triethylamine (0.84 mL, 6.1 mmol) were added. The mixture was stirred at 120 °C under argon overnight. After cooling to room temperature, a saturated sodium hydrogen carbonate aqueous solution (1 mL) was added, and the mixture was further stirred for 5 minutes. This was then filtered through a pad of alumina and evaporated *in vacuo* to give a crude oily product. Purification by flash chromatography (hexane/EtOAc: 9/1), and crystallization in hexane gave the title compound as white crystals. Yield: 0.36 g, 74%.

^{1}H NMR (400 MHz, CDCl$_3$) δ 7.43–7.39 (m, 1H), 7.34–7.16 (m, 14H), 7.13–7.10 (m, 1H), 6.98–6.94 (m, 1H), 6.81–6.79 (m, 1H), 2.36 (s, 3H).

^{13}C NMR (100 MHz, CDCl$_3$) δ 146.4, 138.0, 135.2, 134.4, 134.4, 134.2, 133.9 (d, $J_{CP} = 19.2$ Hz), 130.8 (d, $J_{CP} = 3.2$ Hz), 130.5 (d, $J_{CP} = 5.6$ Hz), 129.2, 128.9, 128.7 (d, $J_{CP} = 7.2$ Hz), 128.5 (d, $J_{CP} = 7.2$ Hz), 128.4, 124.8, 124.1, 55.1.

^{31}P NMR (162 MHz, CDCl$_3$) δ −12.3.

Anal. Calcd for C$_{25}$H$_{21}$PS: C, 78.09; H, 5.52. Found: C, 77.89; H, 5.50.

4.6.3 CONCLUSION

Suzuki coupling can be employed to prepare monodentate biarylphosphine oxides. Reduction by treatment with trichlorosilane yields the free phosphines with good yields. The coupling reaction applies not only to normal arylboronic acids but also to those based on heterocycles, thus providing an easy entry to stereoelectronically variable biarylphosphine and hemilabile P-X (X = N, O, S) ligands. Table 4.5 highlights the *ortho*-substituted biarylphosphines accessed through this methodology.

REFERENCES

1. Quin, L. D. *A Guide to Organophosphorus Chemistry*, John Wiley & Sons, New York, **2000**.
2. Baillie, C. and Xiao, J. *Curr. Org. Chem.* **2003**, *7*, 477.
3. (a) Chen, W., Xu, L. and Xiao, J. *Org. Lett.* **2000**, *2*, 2675. (b) Baillie, C., Chen. W. and Xiao, J. *Tetrahedron Lett.* **2000**, *41*, 9085. (c) Hu, Y., Chen, W., Xu, L. and Xiao, J. *Organometallics* **2001**, *14*, 3206. (c) Chen, W., Xu, L., Hu, Y., Banet, A. M. and Xiao, J. *Tetrahedron* **2002**, *58*, 3889. (d) Xu, L., Mo, J., Baillie, C. and Xiao, J. *J. Organomet. Chem.* **2003**, *687*, 301.
4. Littke, A. F. and Fu, G. C. *Angew. Chem. Int. Ed.* **2002**, *41*, 4176.
5. Slone, C. S., Weinberger, D. A. and Mirkin, C, A. *Prog. Inorg. Chem.* **1999**, *48*, 233.
6. This compound could be obtained in >90% yield by coupling diphenylphosphine with 1 equiv of 1,2-bromoiodobenzene in the presence of 1.1 equiv of NaOAc and 0.5% Pd(OAc)$_2$ in DMAc at 130 °C,[7] followed by oxidation with H$_2$O$_2$.
7. Machnitzki, P., Nickel, T., Stelzer, O. and Landgraf, C. *Eur. J. Inorg. Chem.* **1998**, *7*, 1029.
8. The Boc protection group was cleaved during the coupling reaction. Boc = 1,1-dimethylethoxycarbonyl.

Table 4.5 Suzuki coupling of $OPPh_2(o\text{-}C_6H_4Br)$ with various arylboronic acids and reduction to corresponding free phosphines.

Arylboronic acid	Yield of phosphine oxides (%)	Yield of phosphines (%)
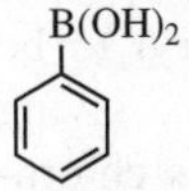	83	75
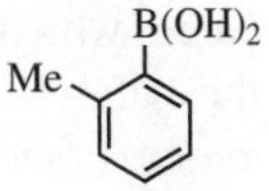	74	71
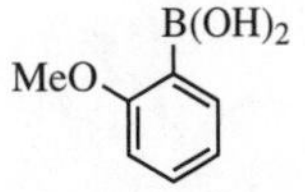	95	74
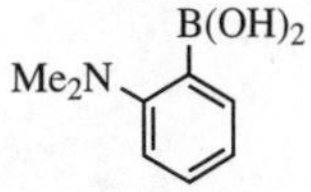	66	72
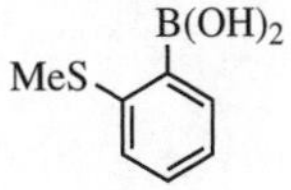	75	78
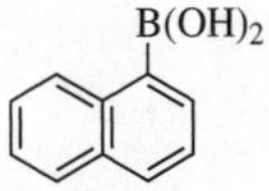	71	70
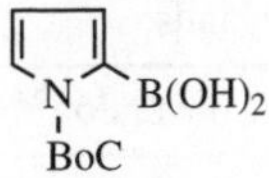	83[8]	75
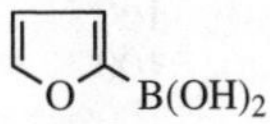	90	80
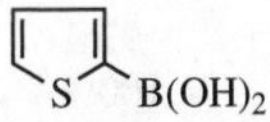	81	74

4.7 BULKY ELECTRON RICH PHOSPHINO-AMINES AS LIGANDS FOR THE SUZUKI COUPLING REACTION OF ARYL CHLORIDES

MATTHEW L. CLARKE* and J. DEREK WOOLLINS

School of Chemistry, University of St. Andrews, St. Andrews, Fife, KY16 9ST, UK

The Suzuki coupling reaction is a very powerful method to construct functionalised biaryls. Up until the late 1990s, these reactions were generally carried out with less readily available, more expensive aryl bromides and iodides. Since that time, several ligand/catalyst systems have emerged that allow these couplings to take place with a range of aryl chloride substrates.[1–9] A combination of Pd$_2$dba$_3$. CHCl$_3$ and *N*-diisopropylphosphino-*N*-methylpiperazine, (1) has been found to act as a catalyst for the Suzuki cross coupling of a range of aryl chlorides with phenyl boronic acid, allowing moderate to excellent yields of the desired biaryl to be formed. The cyclohexyl substituted ligand (2) is even more effective, giving essentially complete conversion even with unactivated substrates such as chloro-toluene (Figure 4.3 and Table 4.6). A procedure has also been used in which a toluene solution of this ligand is prepared and used directly in the catalytic reactions without isolation to further improve the convenience of the process (see Section 4.7.3).***

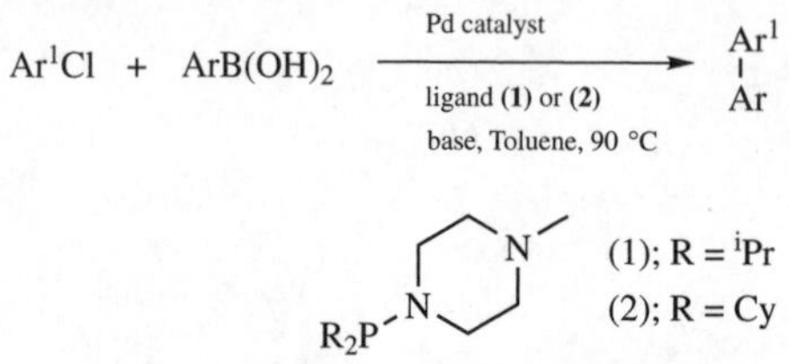

$$Ar^l Cl \ + \ ArB(OH)_2 \ \xrightarrow[\text{base, Toluene, 90 °C}]{\substack{\text{Pd catalyst} \\ \text{ligand (1) or (2)}}} \ \begin{array}{c} Ar^l \\ | \\ Ar \end{array}$$

(1); R = ^{i}Pr
(2); R = Cy

Figure 4.3 Suzuki coupling reaction and ligands used in this study.

Table 4.6 Substrate scope of Suzuki coupling reaction using St. Andrews ligands.

Entry	Ligand^a	Ar$'$	Base	Conversion^{b,c}
1	(1)	p-F$_3$C-C$_6$H$_4$	K$_3$PO$_4$	100%
2	(1)	m-NO$_2$-C$_6$H$_4$^{d}	K$_3$PO$_4$	100%
3	(1)	p-NC-C$_6$H$_4$	K$_3$PO$_4$	100%
4	(2)	p-Me-C$_6$H$_4$	CsF	93%
5	(2)	p-CH$_3$CO-C$_6$H$_4$^{e}	CsF	100%
6	(2)	m-CHO-C$_6$H$_4$^{e}	CsF	100%
7	(2)	p-F$_3$C-C$_6$H$_4$^{f}	CsF	100%

a: reactions were carried out using 1 mol% Pd$_2$dba$_3$.CHCl$_3$, with L:Pd ratio of 2:1, toluene, 90 °C, 16 hours unless stated. b: Conversions calculated by GCMS using naphthalene as an internal standard.
c: some of the reactions contained ~2% of biphenyl isomers. d: reaction time = 6 hours. e: 0.5 mol% catalyst.
f: 0.2 mol% catalyst.

4.7.1 SYNTHESIS OF N-DI-ISOPROPYLPHOSPHINO-*N*-METHYL PIPERAZINE (1)

Materials and equipment

All reagents were used as received except solvents and triethylamine which were distilled over the appropriate drying agent.

- *N*-Methyl-piperazine, 0.140 mL, 0.127 g, 1.263 mmols, 1 equiv.
- Di-isopropylchlorophosphine or dicyclohexylchlorophosphine (1 equiv.)
- Dry triethylamine, 0.194 mL, 0.141g, 1.389 mmols, 1.1 equiv.
- Dry diethylether, 40 mL

- Two Schlenk-type flasks, 250 mL
- Filter stick, sinter porosity 3 (or cannula, filter paper and PTFE tape).
- Disposable syringes, 50 mL, 3 × 1 mL
- Steel needles

Procedure

1. A dry Schlenk flask equipped with a rubber septum is evacuated and purged with nitrogen two or three times prior to the addition of dry diethylether, *N*-methyl-piperazine and then triethylamine.

   Note: It is essential that dry solvents and triethylamine are used in the ligand syntheses. Failure to do this results in the formation of secondary phosphine-oxide by-products.

2. Neat di-isopropylchlorophosphine was added dropwise to this solution *via* a syringe. This immediately gives a white precipitate of Et_3NHCl and a solution of the desired ligand.
3. After about an hour, the suspension was filtered (using a cannula or filter stick) and solvent removed *in vacuo*. This gave the desired ligand (colourless oil) in essentially pure form, and in near quantitative yield. This ligand is stable indefinitely if stored under an atmosphere of dry nitrogen.

   Selected Data: ^{31}P NMR (121.4 MHz; $CDCl_3$): δ 82.8. M. S. (E.I.) 216.1763; MH+ req's; 216.1755.

   N-(Dicyclohexylphosphino)-*N*-methylpiperazine (**2**) was prepared similarly although reaction times were generally increased (about 6 hours or overnight for convenience).

   Selected Data: ^{31}P NMR (121.4 MHz; $CDCl_3$): δ 75.8. M. S. (E.I.) 296.2381; MH+ req's; 296.2376.

4.7.2 SUZUKI COUPLING REACTIONS USING ISOLATED LIGAND AND $Pd_2dba_3.CHCl_3$ AS CATALYST

Materials and equipment

- Ligand (**2**), 36 mg, 0.121 mmols, 4 mol%
- $Pd_2dba_3.CHCl_3$, 31 mg, 0.030 mmols, 1 mol%
- Naphthalene, 0.389 g, 3.03 mmols, 1 equiv.
- Cesium fluoride, 1.39 g, 9.12 mmols, 3 equiv.
- Phenyl boronic acid, 0.556 g, 4.55 mmols, 1.5 equiv
- Dry toluene, 10 mL

- One Schlenk flask equipped with magnetic stirrer bar, and rubber septum
- Stirrer hotplate with oil bath
- Syringes and needles
- For product isolation: One chromatography column, Silica Gel, normal grade petroleum ether/diethyl ether (80/20 mixture), TLC plates.

Procedure

1. To a Schlenk flask containing ligand (**2**) (36 mg, 0.121 mmols, 4 mol%) under nitrogen atmosphere was added dry toluene (10 mL). This was then degassed by two freeze/thaw cycles.
2. This solution was then added to a second Schlenk flask containing $Pd_2dba_3.CHCl_3$ (31 mg, 0.030 mmols, 1 mol%), naphthalene internal standard and cesium fluoride (1.39 g, 9.12 mmols, 3 equiv.).
3. 4-Chlorotoluene was then added by syringe (0.385 g, 3.04 mmols). A sample for GC analysis was taken at this time before introduction of solid phenyl boronic acid (0.556 g, 4.55 mmols, 1.5 equiv.).
4. The flask was then heated to 90 °C for the time specified (Table 4.6). The reactions were sampled periodically by removing 0.05 mL by syringe, and analysing by GC (conversion against internal standard). The product can be isolated with yields only slightly lower than the conversions using column chromatography (Eluent petroleum ether/diethyl ether 80/20).

4.7.3   IN SITU LIGAND PREPARATION AND APPLICATION IN SUZUKI COUPLING OF 3-FLUOROBENZENE WITH PHENYLBORONIC ACID

Materials and equipment

- N-Methyl-piperazine, 0.210 mL, 0.189 g, 1.895 mmols
- Di-cyclohexylchlorophosphine, 0.294 g, 1.263 mmols
- Dry triethylamine, 0.194 mL, 0.141 g, 1.389 mmols
- Dry toluene, 1 × 60 mL, 1 × 15 mL
- Allyl-palladium chloride dimer, 8.0 mg, 0.0219 mmols, 0.2 mol%
- 1-Chloro-3-fluorobenzene, 1.71 mL, 1.43 g, 10.93 mmols
- Phenyl boronic acid, 1.87 g, 15.31 mmols, 1.4 equiv.
- Potassium phosphate (tribasic), 7.0 g, 30.60 mmols, 2.8 equiv.

- Two Schlenk-type flasks, ca. 250 mL
- Disposable syringes, 50 mL, 3 × 1 mL, 5 mL
- Steel needles
- Stirring bars (one medium, one large)

Procedure

1. A dry Schlenk flask equipped with a rubber septum is evacuated and purged with nitrogen two or three times prior to the addition of dry toluene, N-methyl-

piperazine and then triethylamine. An excess of N-methyl-piperazine is used here as it is proposed that it also functions as reductant for the Pd(II) precursor (*via* allylic amination) used in the catalysis.

2. Neat di-cyclohexyl-chlorophosphine (0.294 g, 1.263 mmols) was added drop-wise to this solution *via* a syringe. This suspension was stirred overnight and then left to settle giving a solution of *ca.* 6 mg of ligand/mL concentration.

3. A second Schlenk tube containing allyl-palladium chloride dimer (8.0 mg, 0.0219 mmols, 0.2 mol%), 1-chloro-3-fluorobenzene (1.71 mL, 1.43 g, 10.93 mmols) and a magnetic stirring bar was then placed under an inert atmosphere before the addition of dry toluene (15 mL).

4. A portion of the ligand solution (3.6 mL of 6 mgml^{-1} solution, *ca.* 0.6 mol%) was then added to this second flask using a syringe.

5. The flask was then briefly opened for the addition of solid phenyl boronic acid (1.87 g, 15.31 mmols, 1.4 equiv.) and potassium phosphate (7.0 g, 30.60 mmols, 2.8 equiv.). The reaction solution was then carefully evacuated, placed under an atmosphere of nitrogen, and then heated at 80 °C for about 30 hours. Monitoring by ^{19}F NMR and g.c.m.s. shows that after this time, conversion to the desired biphenyl ($\delta_{product}$ (C_6D_6, internal standard) $= -113.0$ ppm; ($\delta_{reactant} = -110$ ppm) is >90%.

4.7.4 CONCLUSION

In summary, bulky, electron rich, $P^\wedge N$ ligands have been prepared and shown to belong to the rare class of ligands that catalyse the Suzuki reaction of aryl chlorides. Co-ordination chemistry studies[9] and experiments run using related ligands suggest that the presence of the second amine function is critical in forming the desired reactive mono-ligated palladium species. The ligands are very simply prepared (even *in situ*) and applied to the Suzuki coupling of a range of aryl chlorides with phenyl boronic acid. However, if the reactions were run at lower temperatures (50 °C) conversions were lower, which suggests that under the conditions studied this is a less reactive catalytic system than that developed by Buchwald and co-workers.[2, 3]

REFERENCES

1. A. F. Littke, and G. C. Fu, *Angew. Chem. Int. Ed. Engl.*, **1998**, *37*, 3387.
2. D. Old, J. P. Wolfe, and S. L. Buchwald, *J. Am. Chem. Soc.*, **1998**, *120*, 9722.
3. J. P. Wolfe, R. A. Singer, B. H. Yang and S. L. Buchald, *J. Am. Chem. Soc.*, **1999**, *121*, 9550.
4. X. Bei, H. Turner, W. H. Weinberg, and A. S. Gurran, *J. Org. Chem*, **1999**, *64*, 6797.
5. G. Y. Li, *Ang. Chem. Int. Ed. Engl.*, **2001**, *40*, 1513.
6. S. Gibson, G. R. Eastham, D. F. Foster, R. P. Tooze, D. J. Cole-Hamilton, *J. Chem. Soc. Chem. Commun.*, **2001**, 779.
7. T. E. Pickett and C. J. Richards, *Tetrahedron Lett.*, **2001**, *42*, 3767.
8. (a) A. Furstner and A. Leitner, *Synlett*, **2001**, 290. (b) See Section 4.8.

9. M. L. Clarke, D. J. Cole-Hamilton and J. D. Woollins, *J. Chem. Soc., Dalton Trans.*, **2001**, 2721. (b) M. L. Clarke, A. M. Z. Slawin, J. D. Woollins, *Polyhedron*, **2003**, 22, 19.

4.8 ARYLATION OF KETONES, ARYL AMINATION AND SUZUKI-MIYAURA CROSS COUPLING USING A WELL-DEFINED PALLADIUM CATALYST BEARING AN *N*-HETEROCYCLIC CARBENE LIGAND

NICOLAS MARION, OSCAR NAVARRO, ROY A. KELLY III and STEVEN P. NOLAN*

Department of Chemistry, University of New Orleans, 2000 Lakeshore Dr., New Orleans, LA, 70148, USA

Palladium-mediated cross couplings have become a very important method for the construction of carbon-carbon and carbon-nitrogen bonds.[1] Within this context, the use of N-heterocyclic carbenes (NHC) as supporting ligands in these transformations and in homogeneous catalysis has proven significantly beneficial due to increased stability (thermal and air stability) as well as increased reactivity, properties imparted by the ligand.[2] Recently developed is a well-defined, air-stable palladium catalyst that has displayed excellent catalytic activity in a number of reactions.[3] The use of an NHC-bearing palladium catalyst is reported for larger-scale synthesis of biaryl molecules using three different catalytic protocols: ketone arylation[4], aryl amination[5] and Suzuki-Miyaura[6] cross-coupling.

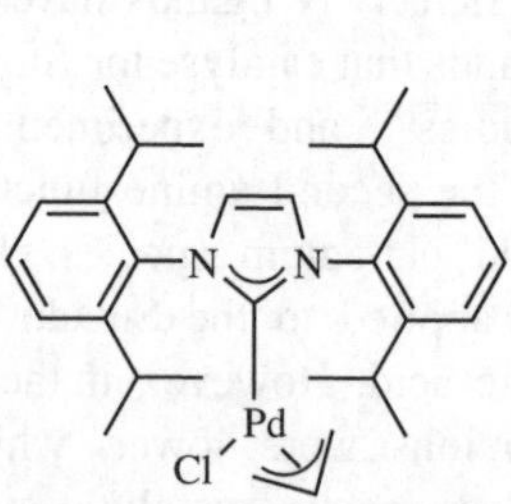

The (IPr)Pd(allyl)Cl catalyst[7]

4.8.1 SYNTHESIS OF 1,2-DIPHENYL-ETHANONE BY KETONE ARYLATION

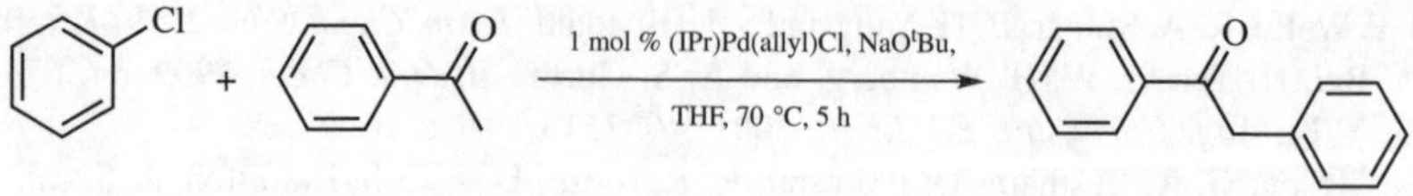

Materials and equipment

- Acetophenone (99%), 0.597 ml, 5.10 mmol
- Chlorobenzene (>99%), 0.519 ml, 5.10 mmol
- (IPr)Pd(allyl)Cl, 0.0291 g, 5.10×10^{-5} mmol
- Sodium *tert*-butoxide (97%), 0.539 g, 5.61 mmol
- Dry tetrahydrofuran (THF), 11 ml
- Diethyl ether
- Ethyl acetate
- Hexanes
- Anhydrous magnesium sulfate
- Silica gel
- TLC plates

- One 50-ml reaction vial with septum-fitted screw cap and magnetic stirring bar
- One 250-ml separating funnel
- Magnetic stirring plate and oil bath
- Filter paper and funnel
- One glass column, diameter 3 cm
- One 250-ml Erlenmeyer flask
- One 100-ml round-bottomed flask
- Rotary evaporator

Procedure

1. In a glove box, the palladium catalyst (29.1 mg, 5.1×10^{-5} mol), sodium *tert*-butoxide (539 mg, 5.61 mmol), and THF were loaded into a reaction vial. (Alternatively, the catalyst and alkoxide can be loaded in air into the reaction vial and the vial is then flushed with nitrogen or argon prior to addition of anhydrous THF.) The vial was taken out of the glove box and acetophenone (0.597 mL, 5.10 mmol) and chlorobenzene (0.519 ml, 5.10 mmol) were injected through a septum. The vial was heated to 70 °C with stirring for 5 hours. The reaction progress was monitored by GC.
2. Water was added to quench the reaction. The reaction mixture was poured into a separating funnel, extracted with diethyl ether and dried over magnesium sulfate. The magnesium sulfate was removed by filtration using a filter paper. The filtrate was then dried over silica. The product was purified by column chromatography using silica gel and a 10% ethyl acetate/hexanes solution (0.85 g, 85% yield).
   ^{1}H NMR, CDCl$_3$, δ: 8.05 (m, 2H), 7.6 (m, 3H), 7.3 (m, 5H), 4.3 (s, 2H).

4.8.2 SYNTHESIS OF DIBUTYL-p-TOLYL-AMINE BY ARYL AMINATION

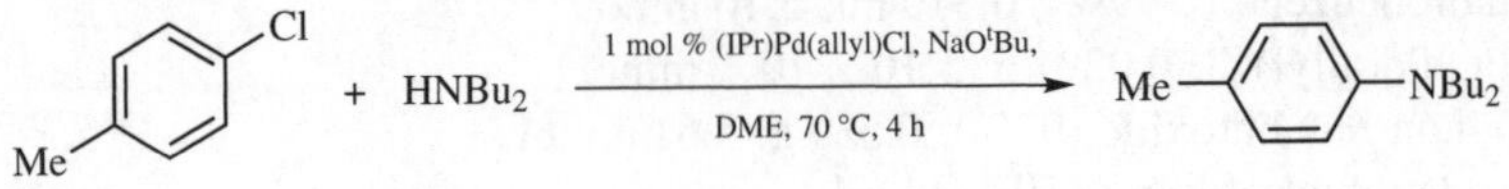

Materials and equipment

- 4-Chlorotoluene (98%), 0.539 ml, 4.56 mmol
- Dibutylamine (99.5+%), 0.768 ml, 4.56 mmol
- (IPr)Pd(allyl)Cl, 26.0 mg, 4.6×10^{-5} mol
- Sodium *tert*-butoxide (97%), 0.481 g, 5.01 mmol
- Dry DME, 11 ml
- Diethyl ether
- Ethyl acetate
- Hexanes
- Anhydrous magnesium sulfate
- Silica gel
- TLC plates

- One 50-ml reaction vial with septum-fitted screw cap and magnetic stirring bar
- One 250-ml separating funnel
- Magnetic stirring plate and oil bath
- Filter paper and funnel
- One glass column, diameter 3 cm
- One 250-ml Erlenmeyer flask
- One 100-ml round-bottom flask
- Rotary evaporator

Procedure

1. In a glove box, the palladium catalyst (26.0 mg, 4.6×10^{-5} mol), sodium *tert*-butoxide (0.481 g, 5.01 mmol), and DME were loaded into a reaction vial. (Alternatively, the catalyst and alkoxide can be loaded in air into the reaction vial and the vial then flushed with nitrogen or argon prior to addition of anhydrous DME.) The vial was taken out of the glove box and 4-chlorotoluene (0.539 ml, 4.56 mmol) and dibutylamine (0.768 ml, 4.56 mmol) were injected through the cap septum. The vial was heated to 70 °C with stirring for 4 hours. Reaction progress was monitored by GC.
2. Water was added to quench the reaction. The reaction mixture was poured into a separating funnel, extracted with diethyl ether and dried over magnesium sulfate. The magnesium sulfate was removed by filtration using a filter paper. The filtrate was then dried over silica. The product was purified by column

chromatography using silica gel and a 10% ethyl acetate/hexanes solution (0.97 g, 97% yield).

^{1}H NMR, CDCl$_3$. δ: 7.03 (d, 2H), 6.6 (d, 2H), 3.25 (t, 2H), 2.25 (s, 3H), 1.6 (m, 2H), 1.35 (m, 2H), 1.0 (t, 3H).

4.8.3 SYNTHESIS OF 4-METHOXYBIPHENYL

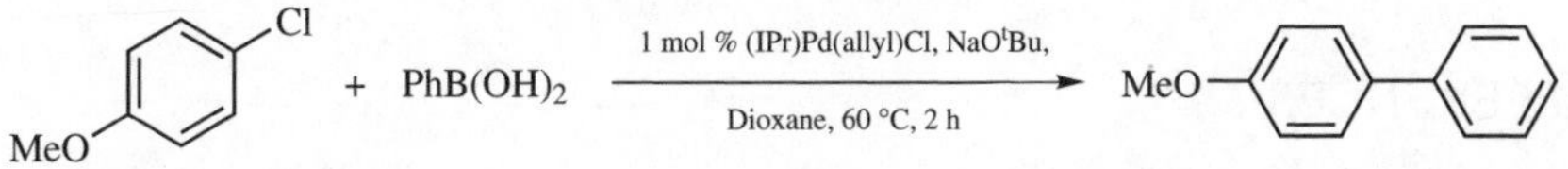

Materials and equipment

- 4-Chloroanisole (99%), 0.610 mL, 5 mmol
- Phenylboronic acid (97%), 732 mg, 6 mmol
- (IPr)Pd(allyl)Cl, 57 mg, 1.0×10^{-5} mol
- Sodium *tert*-butoxide (97%), 1.440 g, 15 mmol
- Dry 1,4-dioxane, 7.5 ml
- Ethyl acetate
- Hexanes
- Silica gel
- TLC plates

- One 50-ml Schlenk flask, magnetic stirring bar and rubber septum
- Magnetic stirring plate and oil bath
- One glass column, diameter 3 cm
- One 250-ml Erlenmeyer flask
- One 50-ml round-bottomed flask
- Rotary evaporator

Procedure

1. In a glove box, the palladium catalyst (57 mg, 1.0×10^{-5} mol), NaO^{t}Bu (1.440 g, 15 mmol) and phenylboronic acid (0.732 g, 6 mmol) were added in turn to a 50-mL Schlenk flask containing a magnetic stirring bar, and sealed with a rubber septum. The flask was removed from the glove box where 4-chloroanisole (0.610 ml, 5 mmol) and 1,4-dioxane (7.5 mL) were injected into the Schlenk flask in this order. The Schlenk flask was then placed in an oil bath over a magnetic stirring plate set at 60 °C for 2 hours with stirring. Reaction progress was monitored by GC.

2. The product was purified by flash chromatography using silica gel and a 10% ethyl acetate/hexanes solution (1.69 g, 92% yield).

   ^{1}H NMR, CDCl$_3$. δ: 3.8 (s, 3H); 6.92 (dd, 2H, $J = 8$ Hz, 2 Hz); 7.17–7.60 (m, 7H).

4.8.4   CONCLUSION

Simple synthetic protocols for carbon-carbon and carbon-nitrogen bond formation using an air-stable palladium catalyst bearing an NHC ligand have been presented. The (IPr)Pd(allyl)Cl catalyst can be prepared on a large scale in high yield, better yet it is commercially available. Initial catalyst testing was performed on a small scale but the ease with which the described chemistry can be scaled up has now been shown.

REFERENCES

1. Littke, A. F. and Fu, G. C. *Angew. Chem. Int. Ed.*, **2002**, *41*, 4176–4211.
2. For reviews see: (a) Hermann W. *Angew. Chem. Int. Ed.* **2002**, *41*, 1290–1309. (b) Hillier, A. C. and Nolan, S. P. *Platinum Metals Rev.* **2002**, *46*, 50–64. (c) Hillier, A. C., Grasa, G. A., Viciu, M. S., Lee, H. M., Yang, C. and Nolan, S. P. *J. Organomet. Chem.* **2002**, *653*, 69–82. (d) Jafarpour, L. and Nolan, S. P. *Adv. Organomet. Chem.* **2000**, *46*, 181–222. (e) Weskamp, T., Bohm, V. P. W. and Hermann, W. A. *J. Organomet. Chem.* **2000**, *600*, 12–22.
3. (a) Viciu, M. S., Germaneau, R. F., Navarro, O., Stevens, E. D. and Nolan, S. P. *Organometallics* **2002**, *21*, 5470–5472.(b) Viciu, M. S., Germaneau, R. F. and Nolan, S. P. *Org. Lett.* **2002**,*4*, 4053–4056.
4. (a) Hamann, B. C. and Hartwig, J. F. *J. Am. Chem. Soc.* **1997**, *119*, 12382–12383. (b) Palucki, M. and Buchwald, S. L. *J. Am. Chem. Soc.* **1997**, *119*, 11108–11109. (c) Satoh, T., Kawamura, Y., Mirua, M. and Nomura, M. *Angew. Chem. Int. Ed. Engl.* **1997**, *36*, 1740–1742. (d) Terao, Y., Fukuoka, Y., Satoh, T., Miura, M. and Nomura, M. *Tetrahedron Lett.* **2002**, *43*, 101–104. (e) Fox, J. M., Huang X., Chieffi, A. and Buchwald, S. L. *J. Am. Chem. Soc.* **2000**, *122*, 1360–1370.
5. (a) Wolfe, J. P., Wagaw, S., Marcoux, J.-F. and Buchwald, S. L. *Acc. Chem. Res.* **1998**, *31*, 805–818. (b) Hartwig, J. F. *Acc. Chem. Res.* **1998**, *31*, 852–860.
6. (a) For a review, see: Miyaura, N. and Suzuki, A. *Chem. Rev.* **1995**, *95*, 2457–2483. (b) Suzuki, A. *Metal-Catalyzed Cross-Coupling Reactions*; Diederich, F., Stang, P. J. (Eds.) Wiley-VCH, Weinheim, 1998, 49–97, and references therein. (c) Hamann, B. C. and Hartwig, J. F. *J. Am. Chem. Soc.* **1998**, *120*, 7369–7370. (d) Reetz, M. T., Lohmer, G. and Schwickardi, R. *Angew. Chem., Int. Ed. Engl.* **1998**, *37*, 481–483. (e) Littke, A. F. and Fu, G. C. *J. Org. Chem.* **1999**, *64*, 10–11.
7. The (IPr)Pd(allyl)Cl catalyst can be purchased from Strem Chemicals.

5 Heck Coupling Reactions

CONTENTS

Catalysts for Fine Chemical Synthesis, Vol. 3, Metal Catalysed Carbon-Carbon Bond-Forming Reactions
Edited by S. M. Roberts, J. Xiao, J. Whittall, and T. Pickett
© 2004 John Wiley & Sons, Ltd   ISBN: 0-470-86199-1

5.1   PALLADIUM-CATALYSED MULTIPLE AND ASYMMETRIC ARYLATIONS OF VINYL ETHERS CARRYING CO-ORDINATING NITROGEN AUXILIARIES: SYNTHESIS OF ARYLATED KETONES AND ALDEHYDES

PETER NILSSON* and MATS LARHED

Organic Pharmaceutical Chemistry, Department of Medicinal Chemistry, Uppsala University, P.O. Box 574, SE-751 23 Uppsala, Sweden

The intermolecular Heck arylation has found wide utility, but the reaction is essentially limited to monoarylations of olefins where the success is largely dependent on the choice of reaction conditions and on the sensitivity of the reaction to steric and electronic factors. In a series of papers, the concept of chelation to control the selectivity of intermolecular vinyl ether arylations has recently been reported.[1–5] The observed increase in reactivity using chelating olefins with a coordinating, metal-directing nitrogen auxiliary suggested that these reactions rely on transient formation of intermediate π-complexes of type (a) and (b). Accordingly, vinylic substitutions of sterically hindered olefins are facilitated, thus permitting multiarylations as well as asymmetric arylations to be carried out.

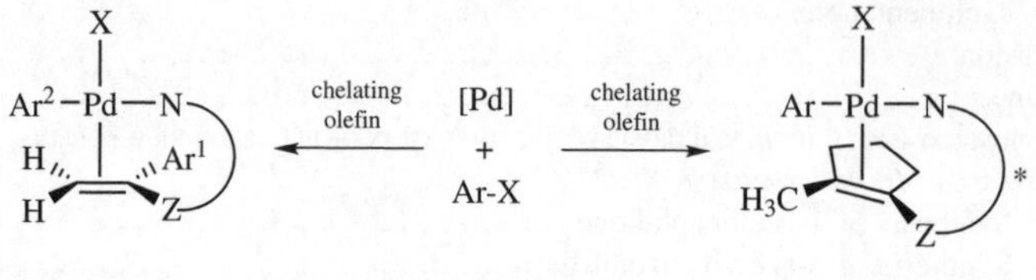

(a)   (b)

Employing a type (a) enol ether with a two-carbon oxygen-nitrogen tether allows for chelation-controlled substitution of both terminal vinylic hydrogens, furnishing a tetra-substituted olefin product.[4] Using a prochiral enol ether of type (b) with a *N*-methyl proline derived coordinating auxiliary, an all-carbon quaternary centre is smoothly produced in excellent optical purity after syn β-hydrogen elimination in the opposite direction.[5]

5.1.1   TRIARYLATION[4]: SYNTHESIS OF *N,N*-DIMETHYL-2-[1,2,2-(TRIARYL)ETHENYLOXY]ETHANAMINES (2) WITH SUBSEQUENT HYDROLYSIS FURNISHING 1,2,2-TRIARYL ETHANONES, (3) (TABLE 5.1)

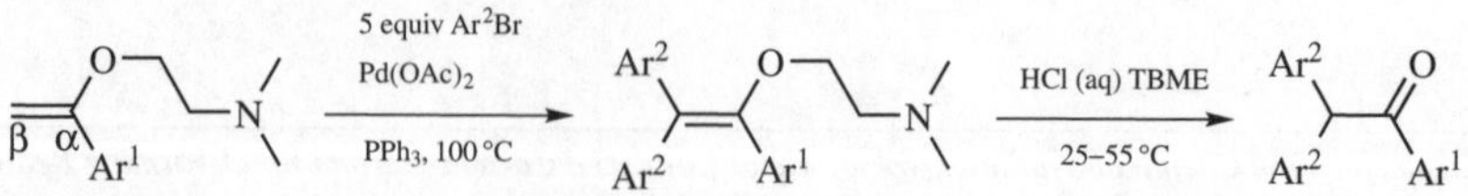

(1)   (2)   (3)

Table 5.1 Procedure 1: Heck Triarylation.

Entry	Ar^1Br	Ar^2Br		Yield of 3 (%)[a]
1	Ph	Ph	**3a**	65
2	p-MeO-C_6H_4	Ph	**3b**	66
3	p-Me-C_6H_4	Ph	**3c**	45
4	p-Cl-C_6H_4	Ph	**3d**	50
5	p-MeOC-C_6H_4	Ph	**3e**	23
6	p-Me-C_6H_4	p-Me-C_6H_4	**3f**	59
7	p-MeO-C_6H_4	p-MeO-C_6H_4	**3g**	21

[a] Based on the starting material Ar^1Br utilized in the α-arylation reaction. >95% purity by GC-MS.

Materials and equipment

Internal α-arylation, synthesis of (1):

- Palladium acetate, 0.150 mmol, 0.0334 g
- 1,3-bis(diphenylphosphino)propane (DPPP), 0.330 mmol, 0.136 g
- Aryl bromide, 5.0 mmol
- Dimethyl-(2-vinyloxy-ethyl)-amine (**4**), 10 mmol, 1.15 g
- Thallium acetate, 5.50 mmol, 1.45 g
- Water, 1.1 mL
- Potassium carbonate, 6.0 mmol, 0.83 g
- Dimethylformamide (DMF), 20 mL
- Sodium hydroxide, 0.1 M
- Diethyl ether
- Aqueous potassium carbonate, 10%
- Anhydrous potassium carbonate
- Thick-walled tube fitted with teflon-lined screw cap

Double terminal β-arylation, synthesis of (2):

- Palladium acetate, 2 × 0.300 mmol, 2 × 0.067 g
- Triphenylphosphine, 0.600 mmol, 0.157 g
- Aryl bromide, 25.0 mmol
- Anhydrous sodium acetate, 5.0 mmol, 0.41 g
- Anhydrous potassium carbonate, 25 mmol, 3.5 g
- Anhydrous DMF, 20 mL
- Sodium hydroxide, 0.1 M
- Diethyl ether
- Sodium chloride (aq, sat)
- Triethylamine, pentane, ethyl acetate
- Thick-walled tube fitted with teflon-lined screw cap
- Aluminium oxide, 150 mesh, 58 Å, Aldrich no 19,997-4, deactivated with 6% water
- TLC plates, aluminium oxide 60 F_{254} neutral (type E)

Hydrolysis, synthesis of (3):

- HCl, 6 M
- *t*-Butyl methyl ether (TBME)

Procedure

1. *Internal α-arylation[6,7]: Synthesis of* (**1a**) (*Ar*^{1} = *Ph*). A thick-walled tube was charged under a nitrogen atmosphere with dimethyl-(2-vinyloxy-ethyl)-amine (**4**) (10.0 mmol), Pd(OAc)$_2$ (0.150 mmol, 0.0334 g), DPPP (0.330 mmol, 0.136 g), bromobenzene (5.00 mmol), TlOAc (5.50 mmol, 1.45 g), water 1.1 mL, potassium carbonate (6.0 mmol, 0.83 g) and DMF (20 mL). The tube was closed, and the contents were magnetically stirred and heated at 80 °C for 5 hours. After cooling, the reaction mixture was diluted with diethyl ether and was washed with two portions of 0.1 M sodium hydroxide. The combined aqueous phases were additionally extracted with diethyl ether. The organic phases were combined, washed with 10% potassium carbonate (aqueous), dried with potassium carbonate (solid) and concentrated under reduced pressure until no excess (**4**) remained (GC-MS), producing (**1a**).

   Double terminal β,β-*arylation, synthesis of* (**2a**): A thick-walled tube was charged under a nitrogen atmosphere with the crude internally arylated olefin (**1a**) (5 mmol), Pd(OAc)$_2$ (0.300 mmol, 0.067 g), PPh$_3$ (0.600 mmol, 0.157 g), bromobenzene (25.0 mmol), NaOAc (5.0 mmol, 0.41 g), potassium carbonate (25 mmol, 3.5 g) and DMF (20 mL). An extra addition of Pd(OAc)$_2$ (0.300 mmol, 0.067 g) after 18 hours were performed. After heating with stirring at 100 °C for 48 hours the tube was cooled and a portion of diethyl ether was added. The organic mixture was transferred to a separating funnel and washed twice with 0.1 M sodium hydroxide. Additional extraction of the aqueous phases was performed with diethyl ether. The combined organic portions were thereafter washed with brine, dried with potassium carbonate (solid) and evaporated under reduced pressure. The triarylated enol ether (**2a**) (Ar^2 = Ph) was purified using an alumina column (pentane/ethyl acetate (39/1) with 2 vol% triethylamine) in 65% yield (1.12 g, >95% by GC-MS, white crystals).

   N,N-dimethyl-2-[1,2,2-(triphenyl)ethenyloxy]ethanamine (2a)

   ^{1}H NMR (270 MHz, CDCl$_3$) δ 7.34 – 7.14 (m, 10H), 7.08 – 7.03 (m, 3H), 6.98 – 6.94 (m, 2H), 3.70 (t, J = 6.1 Hz, 2H), 2.50 (t, J = 6.1 Hz, 2H), 2.15 (s, 6H);
   ^{13}C NMR δ 152.2, 141.3, 141.0, 135.5, 131.3, 130.2, 129.8, 129.7, 127.8, 127.7, 127.6, 126.3, 125.9, 125.8, 68.2, 58.5, 45.7.
   MS m/z (relative intensity 70 eV) 343 (M$^+$, 1), 165 (5), 72 (100).
   High Resolution MS calcd for C$_{24}$H$_{25}$NO: M$^+$ 343.1936, Found: 343.1938.

2. *Synthesis of ketone (3a)*. The ketone (**3a**) (Ar^1 = Ar^2 = Ph) was obtained after hydrolysis of **2a** using 6 M HCl/TBME (24 hours) and preparative straight-phase HPLC (eluent: hexane/ethyl acetate (99/1)) in 65% yield (>95% by GC-MS) calculated from the starting phenyl bromide in the first Heck arylation of the sequence.

5.1.2 TERMINAL DIARYLATION[4]: SYNTHESIS OF *N,N*-DIMETHYL-2-[2,2-DIARYLETHENYLOXY]ETHANAMINE (6) WITH SUBSEQUENT HYDROLYSIS FURNISHING DIARYL ETHANALS (7) (TABLE 5.2)

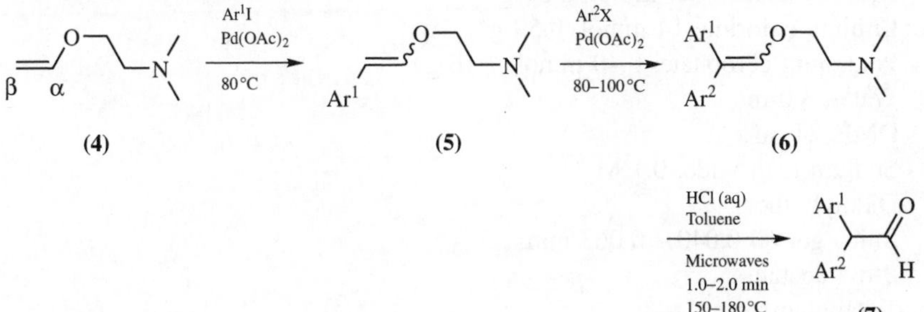

Table 5.2 Procedure 2: Heck β,β-Diarylation.

Entry	Ar^1I	Ar^2X		Yield of 6 (%)[a]		Yield of 7 (%)[b]
1	p-Me-C$_6$H$_4$	m-MeO-C$_6$H$_4$	**6a**	65	**7a**	93
2	Ph	p-Me-C$_6$H$_4$	**6b**	63	**7b**	98
3	Ph	o-Me-C$_6$H$_4$	**6c**	57	**7c**	91
4	p-Me-C$_6$H$_4$	p-Br-C$_6$H$_4$	**6d**	50	**7d**	83
5	p-MeOC-C$_6$H$_4$	Ph	**6e**	40	**7e**	80
6	p-Me-C$_6$H$_4$	p-MeOC-C$_6$H$_4$	**6f**	37	**7f**	80
7	Ph	p-C$_6$H$_5$OC-C$_6$H$_4$	**6g**	50	**7g**	82

[a] Based on the starting material Ar^1X utilized in the first β-arylation reaction as a mixture of Z and E isomers. >95% purity by GC-MS.
[b] Calculated from **6**. >95% purity by GC-MS.

Materials and equipment

First β-arylation:

- Palladium acetate, 0.210 mmol, 0.047 g
- Aryl iodide, 7.0 mmol
- Dimethyl-(2-vinyloxy-ethyl)-amine (**4**), 14.0 mmol, 1.61 g
- Sodium acetate, 8.4 mmol, 0.69 g
- Lithium chloride, 14 mmol, 0.59 g
- Potassium carbonate, 8.40 mmol, 1.16 g
- Water, 3.0 mL
- DMF, 28 mL
- Sodium hydroxide, 0.1 M
- Diethyl ether
- Thick-walled tube fitted with teflon-lined screw cap
- Anhydrous potassium carbonate

Second β-arylation:

- Palladium acetate, 0.210 mmol, 0.047 g
- Aryl iodide or aryl bromide, 8.4 mmol
- Crude mono β-arylated (**5**), 7 mmol
- Sodium acetate, 8.4 mmol, 0.69 g
- Lithium chloride, 14 mmol, 0.59 g
- Potassium carbonate, 8.40 mmol, 1.16 g
- Water, 3.0 mL
- DMF, 24 mL
- Sodium hydroxide, 0.1 M
- Diethyl ether
- Silica gel 60 0.040 – 0.063 mm
- Ethyl acetate
- Triethylamine
- Thick-walled tube fitted with teflon-lined screw cap
- Anhydrous potassium carbonate

Microwave assisted hydrolysis:

- Smith process vial, 5 mL
- Concentrated HCl, 0.25 mL
- Toluene, 1.0 mL
- Water, 0.25 mL
- Sodium hydroxide, 2 M
- Diethyl ether
- Smith microwave synthesiser (Biotage AB)

Procedure

1. *First β-arylation: synthesis of* (**5a**) *($Ar^1 = Ph$).* The reactants were dissolved or dispersed in DMF (28 mL) and added under nitrogen to a thick-walled tube in the following order: iodobenzene (7.0 mmol), (**4**) (14.0 mmol, 1.61 g), Pd (OAc)$_2$ (0.210 mmol, 0.047 g), NaOAc (8.4 mmol, 0.69 g), lithium chloride (14 mmol, 0.59 g), potassium carbonate (8.40 mmol, 1.16 g) and water (3.0 mL). The tube was closed and the contents were magnetically stirred and heated at 80 °C overnight. The reaction was interrupted when GC-MS analysis showed that the starting phenyl iodide was consumed. After cooling, the reaction mixture was diluted with diethyl ether and was washed twice with 0.1 M sodium hydroxide. The combined aqueous phases were additionally extracted twice with diethyl ether. The ethereal phases were combined and dried with potassium carbonate (solid). After evaporation of the solvent the remaining yellowish oil was concentrated under reduced pressure until no non-arylated vinyl ether (**4**) remained (GC-MS).

2. *Second β-arylation: synthesis of* (**6b**) *($Ar^1 = Ph$, $Ar^2 = p$-tolyl).* The reactants were dissolved or dispersed in DMF (24 mL) and added under nitrogen to a

thick-walled tube in the following order: 4-iodotoluene (8.4 mmol), crude β-monoarylated product (5a) ($\leq$7 mmol), Pd(OAc)$_2$ (0.210 mmol, 0.047 g), NaOAc (8.4 mmol, 0.69 g), lithium chloride (14 mmol, 0.59 g), potassium carbonate (8.40 mmol, 1.16 g) and water (3.0 mL). The tube was then closed and the contents were magnetically stirred and heated at 80 °C for 24 hours. The reaction was interrupted when GC-MS analysis showed no further improvements in conversion of (5a). After cooling, the reaction mixture was diluted with diethyl ether and was washed twice with 0.1 M sodium hydroxide. The combined aqueous phases were additionally extracted twice with diethyl ether. The ethereal phases were combined and dried with potassium carbonate (solid). After evaporation of the solvent and silica column chromatography (eluent EtOAc/Et$_3$N, 9/1) the product (6b) was obtained as an yellow/brown viscous oil in 63% yield as an E/Z mixture.

N,N-Dimethyl-2-[2-(4-methylphenyl)-2-phenylethenyloxy]ethanamine (6b)

^{1}H NMR (400 MHz, DMSO-D$_6$, E/Z = 3/7) δ 7.32 − 7.02 (m, 9H), 6.72 − 6.68 (m, 1H), 4.00 (m, 2H), 2.50 (m, 2H), 2.31 − 2.28 (m, 3H), 2.22 − 2.14 (m, 6H). MS (70 eV) m/z (relative intensity) 281 (M$^+$, 7), 72 (100), 58 (63).

Anal calcd for C$_{19}$H$_{23}$NO: C, 81.10; H, 8.24. Found: C, 80.1; H, 8.1. High Resolution MS calcd for C$_{19}$H$_{23}$NO: [M+H]$^+$ 282.1858, Found: 282.1859.

3. Microwave-assisted hydrolysis: synthesis of β,β-arylated acetaldehyde (7b). The enol ether (6b) was placed in a Smith process vial together with 0.25 mL water, 0.25 mL concentrated HCl and 1.0 mL toluene. The tube was sealed and positioned in a Smith synthesiser rack. After microwave irradiation at 180 °C for 2 minutes and subsequent cooling the reaction mixture was carefully neutralized with 2 M sodium hydroxide and extracted with diethyl ether. After drying with potassium carbonate (solid) and evaporation of the solvent the crude aldehyde (7b) (98%) was obtained as a yellow oil (> 95% pure by GC-MS). NMR and IR spectra were quickly recorded on this air and light sensitive compound.

5.1.3 ASYMMETRIC HECK ARYLATION[5]: SYNTHESIS OF 2-ARYL-2-METHYL CYCLOPENTANONE (10) (TABLE 5.3)

(8)　　　　　　　　　　(9)　　　　　　　　　　(10)

Materials and equipment

- Palladium acetate, 0.018 mmol, 0.004 g

Table 5.3 Procedure 3: Asymmetric Heck arylation.

Entry	ArX		Yield of 10 (%)[a]	ee (%)[b]
1	o-MeO-C$_6$H$_4$-I	10a	67	98
2	p-Me-C$_6$H$_4$-I	10b	54	93
3	o-Me-C$_6$H$_4$-I	10c	50	94
4	Ph-I	10d	61	94
5	m-MeO-C$_6$H$_4$-I	10e	68	93
6	1-Naphthyl-I	10f	45	90
7	1-Naphthyl-Br	10g	49	91
8	p-C$_6$H$_5$OC-C$_6$H$_4$-I	10h	47[c]	97
9	p-C$_6$H$_5$OC-C$_6$H$_4$-Br	10i	78	94

[a] Cumulative two-step yield from ArX after silica column chromatography (>95% purity by GC-MS). [b] ee of (+) isomer of **10** as determined by chiral HPLC or chiral GC. [c] Yield calculated after intermediate isolation of **9h** (53%) and subsequent hydrolysis (88%).

- 1-Methyl-2-(S)-(2-methyl-cyclopent-1-enyloxymethyl)-pyrrolidine (**8**), 0.78 mmol, 0.152 g
- Aryl halide, 0.60 mmol
- Sodium acetate, 0.72 mmol, 0.059 g
- Lithium chloride, 1.2 mmol, 0.051 g
- Potassium carbonate, 0.72 mmol, 0.100 g
- Water, 0.2 mL
- DMF, 2.0 mL
- 0.1 M sodium hydroxide
- Diethyl ether, pentane
- Silica gel 60 0.040 – 0.063 mm
- Concentrated HCl
- Chiracel OD-H column
- Degassed isohexane
- Degassed 2-propanol
- Anhydrous potassium carbonate

Procedure

One-pot synthesis of (**10a**). The reactants were added to a reaction vial in the following order: 2-iodoanisole (0.60 mmol), enol ether (**8**) (0.78 mmol, 0.152 g), Pd(OAc)$_2$ (0.018 mmol, 0.004 g), NaOAc (0.72 mmol, 0.059 g), lithium chloride (1.20 mmol, 0.051 g), potassium carbonate (0.72 mmol, 0.100 g), DMF (2 mL) and water (0.2 mL). The tube was closed and the contents were magnetically stirred and heated for 24 hours at 70°C. The reaction was interrupted when GC-MS analysis showed no further improvements in conversion of (**8**). After cooling, 0.5 mL concentrated HCl was added and the contents stirred for 30 minutes. After full hydrolysis of (**9a**) (GC-MS) the reaction mixture was diluted with 10 mL 0.5 M HCl and extracted five times with diethyl ether. The ethereal phases were washed twice with 0.1 M sodium hydroxide and dried with potassium carbonate (solid).

Evaporation of the solvent and silica column chromatography (hexane-diethyl ether, 70:30) afforded pure product (**10a**) (67% yield). The optical purity was calculated (98% ee) after separation of the enantiomers on HPLC using a Chiracel OD-H column (isohexane/2-propanol 99.5:0.5): $t_r = 52$ min (+), $t_r = 57$ min (−); flowrate $= 0.4$ mL/min; $[\alpha]_D^{23} = +39°$.

CONCLUSION

The described procedures are easy to carry out even at multi-gram scale and do not require complicated handling. The scope of the reactions are also quite broad since a wide range of aryl halides with different functional groups can be efficiently coupled. Furthermore, the Heck arylation conditions are mild, using only a weak base (potassium carbonate) and moderate reaction temperatures. Alternative methods for the preparation of triarylated ketones and diarylated aldehydes,[8] are mainly those which relies on direct palladium-catalysed coupling of carbon nucleophiles (enolates) with aryl halides in the presence of a strong base.[9–12]

A different method for enantioselective synthesis of chiral quaternary carbon utilizing palladium catalysis and an axially chiral *P,O*-ligand has been reported by Buchwald and co-workers.[13] This impressive application produces high enantiomeric purities, but requires a protection/deprotection procedure at the cyclopentanone ring to avoid arylation at the non-methylated α-position. In contrast to this procedure, reporting a significant drop in ee when employing sterically demanding ortho-substituted aryl halides, the chelation-controlled asymmetric approach affords excellent enantiomeric excess and good yields also with hindered aryl substrates.

REFERENCES

1. Andersson, C. M., Larsson, J. and Hallberg, A. *J. Org. Chem.* **1990**, *55*, 5757.
2. Larhed, M., Andersson, C. M. and Hallberg, A. *Acta Chem. Scand.* **1993**, *47*, 212.
3. Larhed, M., Andersson, C. M. and Hallberg, A. *Tetrahedron* **1994**, *50*, 285.
4. Nilsson, P., Larhed, M. and Hallberg, A. *J. Am. Chem. Soc.* **2001**, *123*, 8217.
5. Nilsson, P., Larhed, M. and Hallberg, A. *J. Am. Chem. Soc.* **2003**, *125*, 3430.
6. Grigg, R., Loganathan, V., Sukirthalingam, S. and Sridharan, V. *Tetrahedron Lett.* **1990**, *31*, 6573.
7. Cabri, W., Candiani, I., Bedeschi, A., Penco, S. and Santi, R. *J. Org. Chem.* **1992**, *57*, 1481.
8. Terao, Y., Fukuoka, Y., Satoh, T., Miura, M. and Nomura, M. *Tetrahedron Lett.* **2002**, *43*, 101.
9. Hamann, B. C. and Hartwig, J. F. *J. Am. Chem. Soc.* **1997**, *119*, 12382.
10. Palucki, M. and Buchwald, S. L. *J. Am. Chem. Soc.* **1997**, *119*, 11108.
11. Hamada, T. and Buchwald, S. L. *Org. Lett.* **2002**, *4*, 999.
12. Fox, J. M., Huang, X. H., Chieffi, A. and Buchwald, S. L. *J. Am. Chem. Soc.* **2000**, *122*, 1360.
13. Hamada, T., Chieffi, A., Ahman, J. and Buchwald, S. L. *J. Am. Chem. Soc.* **2002**, *124*, 1261.

5.2   PALLADIUM-CATALYSED HIGHLY REGIOSELECTIVE ARYLATION OF ELECTRON-RICH OLEFINS

LIJIN XU, JUN MO and JIANLIANG XIAO*

Liverpool Centre for Materials and Catalysis, Department of Chemistry, University of Liverpool, Liverpool L69 7ZD, UK

The palladium-catalysed Heck reaction of aryl or vinyl halides with olefins has been widely used in synthetic chemistry. The reaction works impressively well with a wide range of electron-deficient and neutral olefins, generally affording β-arylated products. In the case of electron-rich olefins, however, a mixture of regioisomers is usually obtained under standard Heck reaction conditions (Figure 5.1).[1] This

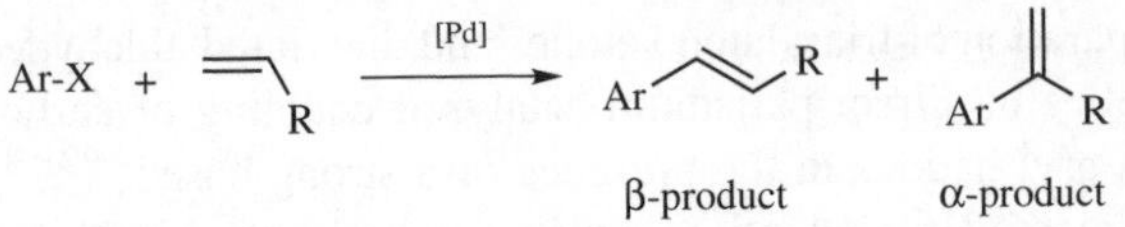

β-product   α-product

Figure 5.1   Formation of regioisomers in a typical Heck reaction.

probably results from the Heck reaction proceeding *via* two competing pathways, one being neutral and the other ionic. Regiocontrol is achievable when the arylating halide is replaced with a triflate or when a stoichiometric quantity of silver or thallium salts is added.[2] The drawback of this approach is that triflates are rarely commercially available and are far more expensive to acquire than halides, and large amounts of toxic inorganic salts are used and generated. A new catalytic system has recently been developed that uses room temperature ionic liquids as solvent and allows for the arylation of electron-rich olefins with aryl halides.[3] The reaction is catalysed by readily available palladium and phosphine species and proceeds with remarkable regioselectivity to give almost exclusively the α-arylated product. By providing an ionic environment, the ionic liquid solvent is believed to play a critical role in promoting the ionic pathway that gives rise to the observed selectivity.

5.2.1   SYNTHESIS OF 1-ACETONAPHTHONE

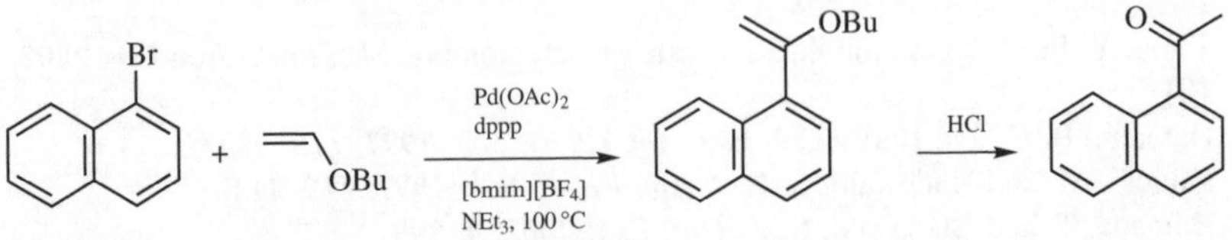

Materials and equipment

- 1-Bromonaphthalene (97%) 414 mg, 2.0 mmol

- n-Butyl vinyl ether (98%) 1.3 ml, 10 mmol
- Palladium(II) acetate (99%) 11 mg, 0.049 mmol
- 1,3-Bis(diphenylphosphino)propane (dppp) (98%), 23 mg, 0.055 mmol
- Dry 1-butyl-3-methylimidazolium tetrafluoroborate ([bmim][BF$_4$]),[4] 2 mL
- Dry triethylamine, 0.36 mL
- Hydrochloric acid (5%), 10 mL
- Dichloromethane
- Ethyl acetate
- Hexane
- Brine
- Anhydrous magnesium sulfate

- Silica gel (Matrex 60A, 37–70 μm), 5 g
- TLC plates, 60F$_{254}$
- One 50-mL two-necked round-bottomed flask with magnetic stirrer bar
- Magnetic stirrer plate
- One glass sintered funnel
- One condenser
- Two 2-mL glass syringes
- One 50-mL separating funnel
- One glass column, diameter 5 cm
- Rotary evaporator

Procedure

1. An oven-dried, two-necked round-bottomed flask containing a stir bar was charged with 1-bromonaphthalene (414 mg, 2.0 mmol), palladium(II) acetate (11 mg, 0.049 mmol), 1,3-bis(diphenylphosphino)propane (23 mg, 0.055 mmol), and the dry ionic liquid 1-butyl-3-methylimidazolium tetrafluoroborate (2 mL) under argon at room temperature. After degassing the flask several times, butyl vinyl ether (1.3 mL, 10 mmol) and NEt$_3$ (0.36 mL, 2 mmol) were injected sequentially. The flask was placed in an oil bath, and the mixture was stirred and heated at 100 °C. After 24 hours the flask was removed from the oil bath and cooled to room temperature. To the mixture hydrochloric acid (5%, 10 mL) was added and, following stirring for 0.5 hours, dichloromethane (10 mL) was added. After separation of the organic layer, the aqueous phase was further extracted with dichloromethane (2 × 10 mL), and the combined organic layer was washed with water until neutral, dried over magnesium sulfate, filtered, and concentrated *in vacuo*.
2. The product 1-acetonaphthone was isolated from the crude reaction mixture by column chromatography (silica gel, hexane/ethyl acetate: 20/1). Yield: 323 mg, 95%.

   ^{1}H NMR (400 MHz, CDCl$_3$) δ 8.78 (d, $J =$ 9.0 Hz, 1H), 7.99 (d, $J =$ 8.0 Hz, 1 H), 7.94 (d, $J =$ 7.0 Hz, 1H), 7.83 (d, $J =$ 8.0 Hz, 1H), 7.62 (m, 1H), 7.55–7.48 (m, 2H), and 2.75 (s, 3H).

^{13}C NMR (100 MHz, CDCl$_3$) δ 202.2, 135.9, 134.4, 133.4, 130.6, 129.0, 128.8, 128.4, 126.8, 126.4, 124.7, 30.3.

Anal. Calcd for C$_{12}$H$_{10}$O: C, 84.68; H, 5.92. Found: C, 84.37; H, 5.93.

5.2.2   SYNTHESIS OF 3-ACETYLBENZONITRILE

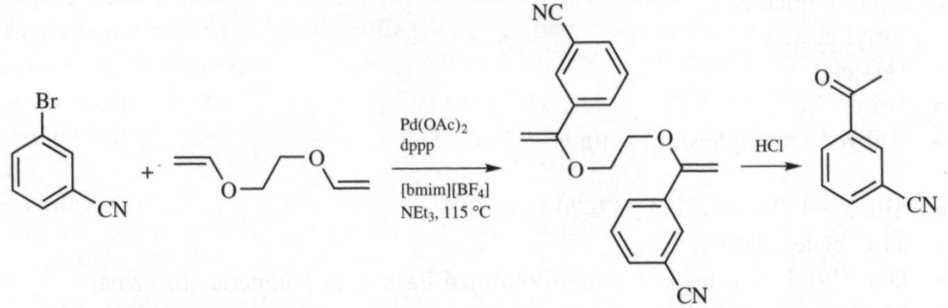

Materials and equipment

- 3-Bromobenzonitrile (99%) 364 mg, 2.0 mmol
- Ethylene glycol divinyl ether (96%) 0.37 ml, 3 mmol
- Palladium(II) acetate (99%) 11 mg, 0.049 mmol
- 1,3-Bis(diphenylphosphino)propane (dppp) (98%), 23 mg, 0.055 mmol
- Dry triethylamine, 0.36 mL
- Dry 1-butyl-3-methylimidazolium tetrafluoroborate ([bmim][BF$_4$]),[4] 2 mL
- Hydrochloric acid (5%), 10 mL
- Dichloromethane
- Ethyl acetate
- Hexane
- Brine
- Anhydrous magnesium sulfate

- Silica gel (Matrex 60A, 37–70 μm), 5 g
- TLC plates, 60F$_{254}$
- One 50-mL two-necked round-bottom flask with magnetic stirrer bar
- Magnetic stirrer plate
- One glass sintered funnel
- One condenser
- Two 1-mL glass syringes
- One 50-mL separatory funnel
- One glass column, diameter 5 cm
- Rotary evaporator

Procedure

1. An oven-dried, two-necked round-bottom flask containing a stir bar was charged with 3-bromobenzonitrile (364 mg, 2.0 mmol), palladium(II) acetate (11 mg,

0.049 mmol), 1,3-bis(diphenylphosphino)propane (23 mg, 0.055 mmol), and the dry ionic liquid 1-butyl-3-methylimidazolium tetrafluoroborate (2 mL) under argon at room temperature. The flask was degassed several times; ethylene glycol divinyl ether (0.37 ml, 3 mmol) and NEt_3 (0.36 mL, 2 mmol) were then injected sequentially. The flask was placed in an oil bath, and the mixture was stirred and heated at 115 °C. After 36 hours reaction time, the flask was removed from the oil bath and cooled to room temperature. To the mixture hydrochloric acid (5%, 10 mL) was added and, following stirring for 1 hour, dichloromethane (10 mL) was added. After separation of the organic layer, the aqueous phase was further extracted with dichloromethane (2 × 10 mL), and the combined organic layer was washed with water until neutral, dried over magnesium sulfate, filtered, and concentrated *in vacuo*.

2. The product 3-acetylbenzonitrile was isolated from the crude reaction mixture by column chromatography (silica gel, hexane/ethyl acetate: 10/1). Yield: 282 mg, 97%.

^{1}H NMR (400 MHz, $CDCl_3$) δ 8.25 (s,1H), 8.19 (d, 1H, $J = 7.9$ Hz), 7.85 (d, 1H, $J = 7.8$ Hz), 7.63 (dd, 1H), 2.65 (s, 3H).

^{13}C NMR (100 MHz, $CDCl_3$) δ 196.3, 138.1, 136.6, 132.6, 132.4, 130.1, 118.3, 113.6, 27.0.

Anal. Calcd for $C_{12}H_{10}O$: C, 74.47; H, 4.86; N 9.65. Found: C, 74.57; H, 4.84; N, 9.70.

CONCLUSION

The chemistry described here allows for the easy α-arylation of vinyl ethers by various aryl halides to give acylated aromatics regardless of the electronic nature of the substituents on the aryl rings. Table 5.4 presents a range of aryl bromides that have been reacted with the vinyl ethers described above. Given that the regiochemistry of the ketone products can be precisely controlled and vinyl ethers are easily available and inexpensive, the protocol provides a powerful supplement to the Friedel-Crafts arylation, which fails with arenes bearing electron-withdrawing substituents.

REFERENCES

1. Larhed, M. and Hallberg, A. In *Handbook of Organopalladium Chemistry for Organic Synthesis,* Negishi, E.-I., Ed.; Wiley: New York, 2002, Vol. 1, pp 1133–1178.
2. Cabri, W. and Candiani, I. *Acc. Chem. Res.* **1995**, 28, 2–7.
3. Xu, L., Chen, W., Ross, J. and Xiao, J. *Org. Lett.* **2001**, 3, 295–297.
4. Holbrey, J. D. and Seddon, K. R. *J. Chem. Soc. Dalton Trans.* **1999**, 2133–2140.

Table 5.4 Heck arylation of vinyl ethers by aryl bromides in [bmim][BF$_4$].

Aryl bromides	Vinyl ethers			
	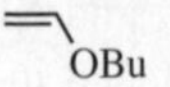 OBu		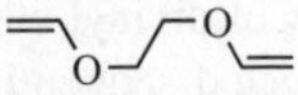 O O	
	Time (h)	Yield (%)	Time (h)	Yield (%)
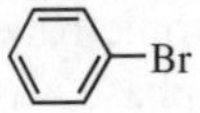 Br	24	97	24	88
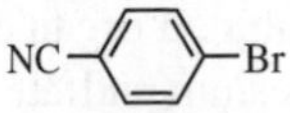 NC—Br	36	94	24	95
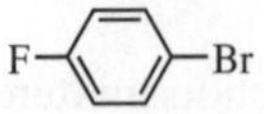 F—Br	24	96	24	95
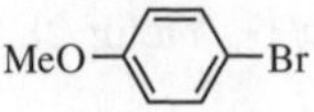 MeO—Br	36	81	36	89
 Ac—Br	24	92	24	94
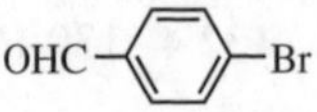 OHC—Br	24	93	24	91
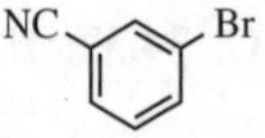 NC Br	36	95	36	97
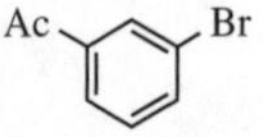 Ac Br	36	85	36	87

5.3   1-[4-(S)-tert-BUTYL-2-OXAZOLIN-2-YL]-2-(S)-(DIPHENYLPHOSPHINO)FERROCENE AS A LIGAND FOR THE PALLADIUM-CATALYSED INTERMOLECULAR ASYMMETRIC HECK REACTION OF 2,3-DIHYDROFURAN

TIM G. KILROY, YVONNE M. MALONE and PATRICK J. GUIRY*

Centre for Synthesis and Chemical Biology, Conway Institute of Biomolecular and Biomedical Research, Department of Chemistry, University College Dublin, Belfield, Dublin 4, Ireland

The intermolecular asymmetric Heck reaction, a palladium-catalysed carbon-carbon bond forming process, is an efficient method for the preparation of optically active cyclic compounds.[1] Very recently, a new catalytic system has been developed based on palladium complexes having 1-[4-(S)-*tert*-butyl-2-oxazolin-2-yl]-2-(S)-(diphenylphosphino)ferrocene (**1**) as the chiral ligand[2] (Figure 5.2), which we have shown to be efficient catalysts for the enantioselective intermolecular Heck reaction of 2,3-dihydrofuran (**2**).[3] In contrast to complexes derived

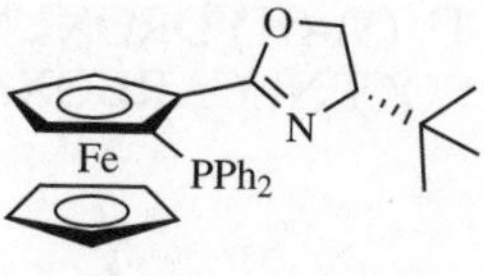

(1)

Figure 5.2

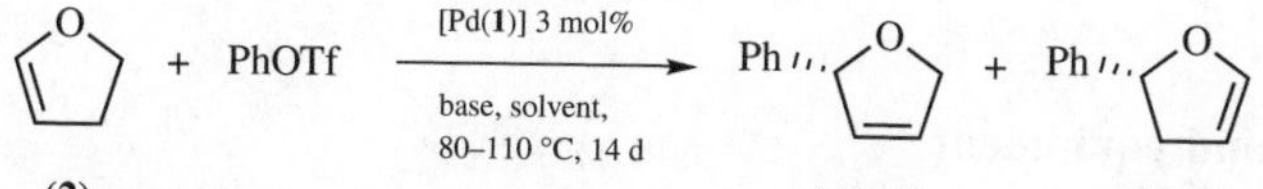

Figure 5.3

from diphosphine complexes the favoured product is the kinetic isomer (**3**) which is formed in good yields and excellent enantiomeric excesses (Figure 5.3).

The ligand, 1-[4-(*S*)-*tert*-butyl-2-oxazolin-2-yl]-2-(*S*)-(diphenylphosphino)ferro-cene (**1**), is prepared in three steps from readily available ferrocenecarboxylic acid chloride (**5**).[4] The steps involved are firstly amide bond formation; secondly cyclisation to the oxazoline and finally diastereoselective lithiation/phosphinylation to afford (**1**) (Figure 5.4).

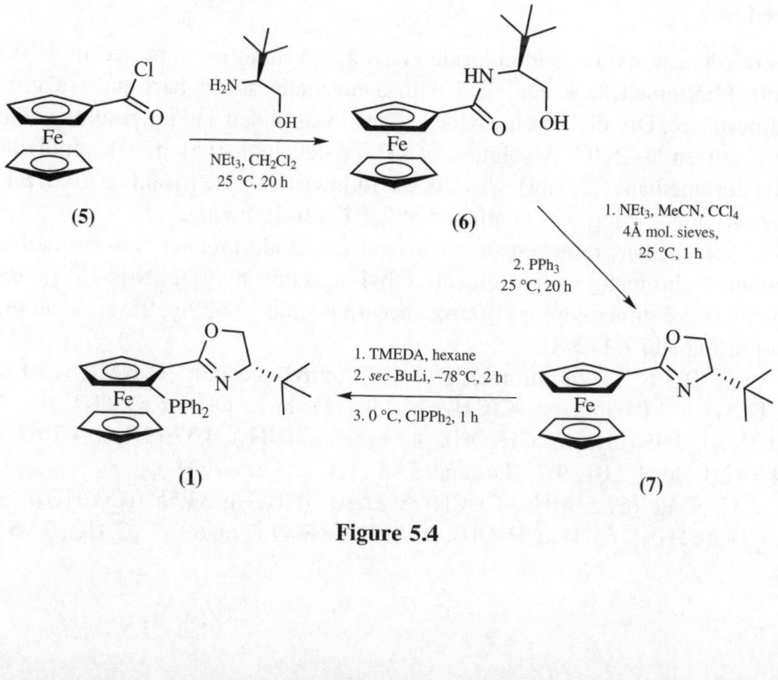

Figure 5.4

5.3.1   SYNTHESIS OF N-[1-(S)-(HYDROXYMETHYL)-2,2-DIMETHYLPROPYL]FERROCENECARBOXAMIDE (6)

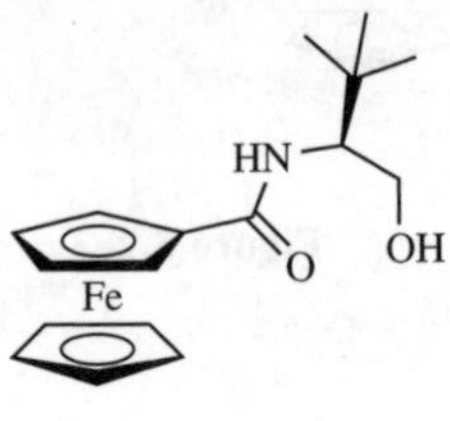

(6)

Materials and equipment

- Ferrocenecarboxylic acid chloride, 1.10 g, 4.4 mmol
- (S)-tert-Leucinol, 0.51 g, 4.4 mmol
- Triethylamine, 0.61 mL, 4.4 mmol
- Dry dichloromethane, 50 mL
- Ethyl acetate, 200 mL

- Two 100-mL round-bottomed flasks, one equipped with a magnetic stirrer bar
- Magnetic stirrer plate
- Silica Merck Kieselgel 60 (Art. 7734), 20 g
- TLC plates, SIL G-60 UV$_{254}$
- 50 Test tubes, diameter 1.5 cm
- Water aspirator
- Rotary evaporator

Procedure

1. Ferrocenecarboxylic acid chloride (1.10 g, 4.4 mmol) was placed in a 100-mL round-bottomed flask equipped with a magnetic stirrer bar, under a nitrogen atmosphere. Dry dichloromethane (25 mL) was added and the resulting solution was stirred at 25 °C. A solution of (S)-tert-leucinol (0.51 g, 4.4 mmol) in dry dichloromethane (25 ml) was added followed by triethylamine (0.61ml, 4.4 mmol) and stirring was continued at 25 °C for 20 hours.

2. The solvent was removed *in vacuo* and the crude product was purified using column chromatography (eluent, ethyl acetate) to give N-[1-(S)-(hydroxymethyl)-2,2-dimethylpropyl]ferrocenecarboxamide, (1.39 g, 96%), as an orange solid; m.p. 170–172 °C

    $[\alpha]_D^{23}$ –9.5 (c 1, chloroform); R$_f$ = 0.53 (ethyl acetate); ^{1}H NMR (270 MHz, CDCl$_3$) δ 1.04 (9H, s, -C(CH$_3$)$_3$), 2.9 (1H, br s, OH), 3.63 (1H, m, -C$\underline{H}$C(CH$_3$)$_3$), 3.95 (2H, m, -CH$_2$OH), 4.23 (5H, s, C$_5$H$_5$), 4.37 (2H, d, J 1.65, -Fc), 4.7 (2H, dd, J 2.01, 9.7, -Fc) and 5.88 (1H, d, J 8.3, -NH-);

    ^{13}C NMR (67.5 MHz, CDCl$_3$) δ 27.10 ((CH$_3$)$_3$), 33.55 ((C(CH$_3$)$_3$), 59.48 (-CHC(CH$_3$)$_3$), 63.41 (CH$_2$OH), 67.8 (Fc), 68.44 (Fc), 69.77 (C$_5$H$_5$), 70.5 (Fc),

70.56 (Fc), 75.91 (*ipso* -Fc) and 171.83 (C=O); m/z (eims, 70eV) 329 (M$^+$, 47%), 213 (100), 185 (26), 121 (18) and 56 (16).

5.3.2 SYNTHESIS OF [4-(*S*)-tert-BUTYL-2-OXAZOLIN-2-YL]FERROCENE (7)

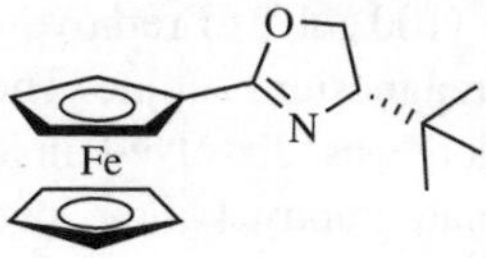

(7)

Materials and equipment

- N-[1-(*S*)-(Hydroxymethyl)-2,2-dimethylpropyl]ferrocenecarboxamide, 1.20 g, 3.6 mmol
- Triethylamine, 23.1 mL, 0.166 mol
- Triphenylphosphine, 3.01 g, 0.011 mol
- Dry acetonitrile, 70 mL
- Carbon tetrachloride, 1.9 mL
- Activated molecular sieves, 4 Å, 0.4–0.8 mm beads, 1 g
- Water, 100 mL
- Saturated solution of sodium chloride
- Anhydrous magnesium sulfate
- Dichloromethane, 3 mL
- Petroleum ether 40–60 °C
- Ethyl acetate

- Two 100-mL round-bottomed flasks, one equipped with a magnetic stirrer bar
- Magnetic stirrer plate
- Silica Merck Kieselgel 60 (Art. 7734), 20 g
- TLC plates, SIL G-60 UV$_{254}$
- 50 Test tubes, diameter 1.5 cm
- One 250-mL Erlenmeyer flask
- One Buchner funnel, diameter 10 cm
- One Buchner flask, 250 mL
- Filter paper
- One 500-mL separatory funnel
- Water aspirator
- Rotary evaporator

Procedure

1. Dry acetonitrile (50 mL) was added to N-[1-(*S*)-(hydroxymethyl)-2,2-dimethyl-propyl]ferrocenecarboxamide, (1.20 g, 3.6 mmol), to form an orange coloured

solution. Triethylamine, (23.1 mL, 0.166 mol), was added to the reaction followed by activated 4 Å molecular sieves and carbon tetrachloride (1.9 mL). The reaction was stirred at 25 °C for 1 hour. Triphenylphosphine, (3.01 g, 0.011 mol), was then added as a solution in acetonitrile (20 mL) and the reaction was stirred for a further 20 hours.

2. After filtration through a Buchner funnel, the filtrate was washed well with water (100 mL) and brine (100 mL) to remove excess triethylamine and was then dried over anhydrous magnesium sulfate. The solvent was then removed to give a brown residue which was dissolved in dichloromethane (3 mL) and purified by column chromatography (silica, petroleum ether 40–60 °C/ethyl acetate 70:30) to give [4-(S)-tert-butyl-2-oxazolin-2-yl]ferrocene, (1.01 g, 90%), as an orange solid, m.p. 137–138 °C.

$[\alpha]_D^{23}$–153.8 (c 1, chloroform); $R_f = 0.45$ (70:30 petroleum ether 40–60 °C/ ethyl acetate); ^{1}H NMR (270 MHz, CDCl$_3$) δ 0.95 (9H, s, C(CH$_3$)$_3$), 3.89 (1H, dd, J$_{1,3}$ 7.33, J$_{1,2}$ 10.0, -NCH), 4.15 (1H, dd, J$_{1,3}$ 7.33, J$_{2,3}$ 8.52, -OCH$_{(3)}$), 4.19 (5H, s, -C$_5$H$_5$), 4.23 (1H, dd, J$_{2,3}$ 8.52, J$_{1,2}$ 10.0, -OCH$_{(2)}$), 4.31–4.32 (2H, m, -Fc), 4.69–4.71 (1H, m, -Fc) and 4.76–4.78 (1H, m, -Fc).

^{13}C NMR (67.5 MHz, CDCl$_3$) δ25.95 ((CH$_3$)$_3$), 33.57 (-C(CH$_3$)$_3$), 68.26 (-OCH$_2$), 68.92, 68.99, 69.46 (C$_5$H$_5$), 70.01, 70.1, 70.77 (ipso-Fc), 76.04 and 165.49 (C=N); m/z (eims, 70 eV) 311 (M$^+$, 57%), 254 (100), 211 (58), 121 (71) and 56 (50).

5.3.3   SYNTHESIS OF 1-[4-(S)-tert-BUTYL-2-OXAZOLIN-2-YL]-2-(S)-(DIPHENYLPHOSPHINO)FERROCENE (1)

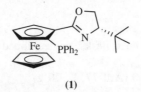

(1)

Materials and equipment

- [4-(S)-tert-Butyl-2-oxazolin-2-yl]ferrocene, 0.10 g, 0.32 mmol
- N,N,N′,N′-Tetramethylethylenediamine, 0.8 mL, 0.42 mmol
- sec-Butyllithium, 0.33 mL, 1.6 M in hexane, 0.42 mmol
- Dry hexane, 9 mL
- Chlorodiphenylphosphine, 0.13 mL, 0.64 mmol
- Aqueous ammonium chloride, 10% solution
- Anhydrous magnesium sulfate
- Dichloromethane, 25 mL
- Petroleum ether 40–60 °C
- Ethyl acetate

- One 5-mL Schlenk tube equipped with a magnetic stirrer bar
- Magnetic stirrer plate
- Silica Merck Kieselgel 60 (Art. 7734), 10 g
- TLC plates, SIL G-60 UV_{254}
- 25 Test tubes, diameter 1.5 cm
- One 100-mL Erlenmeyer flask
- Filter paper
- One 100-mL separating funnel
- Water aspirator
- Rotary evaporator

Procedure

1. Dry hexane (9 mL) was added to [4-(S)-*tert*-butyl-2-oxazolin-2-yl]ferrocene, (0.1 g, 0.32 mmol), in a 5-mL Schlenk tube to give an orange suspension. N,N,N′,N′-Tetramethylethylenediamine, (0.8 mL, 0.42 mmol), was added and the reaction mixture was stirred at ambient temperature for 15 minutes. It was then cooled to –78 °C and *sec*-butyllithium, (0.33 mL, 1.6 M in hexane, 0.42 mmol) was added and stirring was continued at this temperature for 2 hours. After this time the reaction mixture was allowed warm to 0 °C and was stirred at this temperature for 30 minutes to ensure complete lithiation. The reaction was then quenched with chlorodiphenylphosphine, (0.13 mL, 0.64 mmol) and allowed warm to 25 °C and stirred for a further hour.

2. The reaction mixture was poured into a 10% aqueous ammonium chloride solution and dichloromethane (5 × 5 mL) was added. The organic layer was extracted, dried over magnesium sulfate and evaporated to give an orange oil which was purified by column chromatography (silica, petroleum ether 40–60 °C/ethyl acetate: 70/30) to give 1-[4-(S)-*tert*-butyl-2-oxazolin-2-yl]-2-(S)-(diphenylphosphino)ferrocene as a single diastereomer, (0.11 g, 64%), as an orange crystalline solid; m. p. 127 °C.

   $[\alpha]_D^{23}$ –24.2 (*c* 0.50, chloroform); ^{1}H NMR (270 MHz, $CDCl_3$) δ 0.76 (9H, s, $C(CH_3)_3$), 3.6 (1H, m, -Fc), 3.72 (1H, dd, $J_{1,3}$ 7.7, $J_{1,2}$ 9.7, -NCH), 3.85 (1H, app t, $J_{1,3}$ 7.7, $J_{2,3}$ 8.24, $-OCH_{(3)}$), 4.21 (5H, s, $-C_5H_5$), 4.19 (1H, dd, $J_{2,3}$ 8.24, $J_{1,2}$ 9.7, $-OCH_{(2)}$), 4.35 (1H, app t, J 2.38, 2.58, Fc-*o*-PPh_2), 4.95 (1H, m, Fc-*O*-oxazoline) and 7.2–7.5 (10H, m, Ph).

   ^{13}C NMR (67.5 MHz, $CDCl_3$) δ26.23 (($\underline{C}H_3)_3$), 34.08 (-$\underline{C}(CH_3)_3$), 68.97 (-OCH_2), 71.25 (C_5H_5), 71.74, 72.66, 74.37 (d, J 3.2), 76.11 (d, J 16.2), 76.5, 79.18 (d, J 13.9, *ipso*-Fc), 128.35, 128.5 (d, J 6.4, *m*-Ph), 128.68 (d, J 8.0, *m*-Ph), 129.39, 133.0 (d, J 19.3, *o*-Ph), 135.42 (d, J 21.5, *o*-Ph), 138.91 (d, J 14, *ipso*-Ph), 140.16 (d, J 11.8, *ipso*-Ph) and 165.1 (C=N).

   ^{31}P NMR (109.3 MHz, $CDCl_3$) δ-16.77; m/z (eims, 70 eV) 495 (M^+, 100%), 438 (18), 418 (45), 253 (40), 183 (29) and 121 (60).

Notes: This procedure has been scaled up to provide 5 g of 1-[4-(S)-tert-butyl-2-oxazolin-2-yl]-2-(S)-(diphenylphosphino)ferrocene.

Use of n-butyllithium and dry tetrahydrofuran as solvent afforded a 4:1 mixture of diastereomers.

5.3.4   ASYMMETRIC PHENYLATION OF 2,3-DIHYDROFURAN (2)

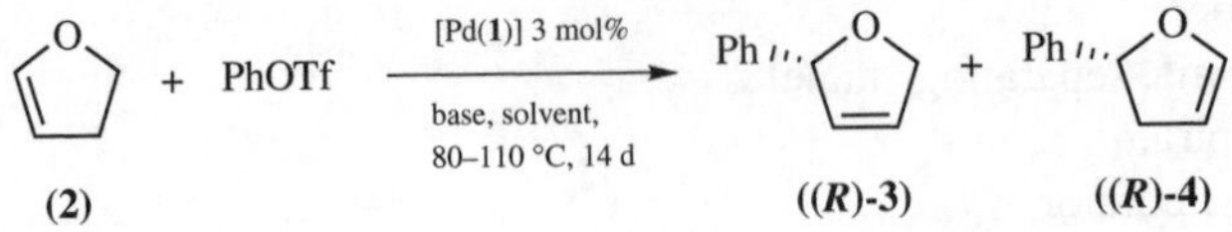

Materials and equipment

- 1-[4-(S)-*tert*-Butyl-2-oxazolin-2-yl]-2-(S)-(diphenylphosphino)ferrocene, 3.96 mg, 0.008 mmol
- Phenyl trifluoromethanesulfonate, 30.0 mg, 0.13 mmol
- $Pd_2(dba)_3$, 2.3 mg, 0.004 mmol
- n-Tridecane, 10.0 mg, 0.054 mmol
- 2,3-Dihydrofuran, 46.2 mg, 0.65 mmol
- Toluene (0.5 mL) or benzene (0.5 mL)
- Bases used:
    - 1,8-bis-(dimethylamino)-naphthalene (proton sponge), 83.58 mg, 0.39 mmol
    - Triethylamine, 39.4 mg, 54 µL, 0.39 mmol
    - Diisopropylethylamine, 50.4 mg, 68 µL, 0.39 mmol
- Pentane, 10 mL
- Diethyl ether, 10 mL

- One 5-mL Schlenk tube equipped with a magnetic stirrer bar
- Magnetic stirrer plate
- Silica Merck Kieselgel 60 (Art. 7734), 10 g
- TLC plates, SIL G-60 UV$_{254}$
- Water aspirator
- Rotary evaporator

Procedure

1. A solution of phenyl trifluoromethanesulfonate (30.0 mg, 0.13 mmol) and n-tridecane (10.0 mg, 0.054 mmol) in either toluene (0.5 mL) or benzene (0.5 mL) was added to a 5-mL Schlenk tube containing $Pd_2(dba)_3$ (2.3 mg, 0.004 mmol) and 1-[4-(S)-*tert*-butyl-2-oxazolin-2-yl]-2-(S)-(diphenylphosphino)ferrocene (3.96 mg, 0.008 mmol) under an atmosphere of nitrogen. To this was added 2,3-dihydrofuran (46.2 mg, 0.65 mmol) and base (0.39 mmol) (Table 5.5). The resulting solution was degassed by three freeze-thaw cycles at 0.01 mbar and left to stir under nitrogen at 80 °C for 14 days giving a red solution with precipitation of Base.HOTf.

Table 5.5 Asymmetric phenylation of 2,3-dihydrofuran using palladium complexes of 1-[4-(S)-*tert*-butyl-2-oxazolin-2-yl]-2-(S)-(diphenylphosphino) ferrocene as catalyst.

Entry	Solvent	Base	Temp. (°C)	Yield (%) (R)-3 ((R)-4)	ee (%) (R)-3
1	Benzene	Proton Sponge	80	38 (12)	99
2	Benzene	i-Pr$_2$NEt	80	19 (-)	99
3	Benzene	Et$_3$N	80	40 (-)	98
4	Toluene	Proton Sponge	110	78 (-)	98
5	Toluene	i-Pr$_2$NEt	110	52 (21)	99
6	Toluene	Et$_3$N	110	57 (6)	98

2. Pentane (10 mL) was added to the reaction mixture and the resulting suspension was filtered through 2 cm of silica with further elution using diethyl ether (10 mL). This solution was concentrated and the yield calculated using GC (Se-30, 11 psi, 50 °C, 4 minutes, 15 °C/min, 170 °C, 10 minutes) by the internal standard method as follows: the burn ratio for the achiral GC column was determined by injecting a known mixture of phenyl trifluoromethanesulfonate and *n*-tridecane (0.5 μL × 5 times) and taking the average integration ratio value and applying the formula:

$$\text{Burn Ratio} = \frac{\text{molar ratio of triflate/tridecane}}{\text{average integration value}}$$

$$\text{To calculate the yield}: \frac{\text{Integration product}}{\text{Integration tridecane}} \times \text{Burn Ratio} = \frac{\text{mmol product}}{\text{mmol triflate}}$$

$$\% \text{ yield} = \frac{\text{mmol product}}{\text{mmol triflate}} \times \frac{100}{1}$$

3. A normal sized TLC plate was run and a strip cut off and visualised with KMnO$_4$. The silica of the remainder of the plate at the same R$_f$ as the product was then scraped off and extracted with dichloromethane to give 2-phenyl-2,5-dihydrofuran. The enantiomeric excess was determined by chiral GC, (γ-CD-TFA, 30 m, 80–90 °C, 0.3 °C min^{-1}, 90–120 °C, 5 °C min^{-1}, 10 minutes, 15 psi, injector temperature 200 °C, detector temperature 220 °C).

Retention times: 2-Phenyl-2,5-dihydrofuran (**3**); R$_T$ (S)-(**3**) 31.8 minutes; (R)-(**3**) 34.0 minutes.

^{1}H NMR (**3**) (270 MHz, CDCl$_3$) δ 4.77 (1H, m, HC(5), 4.86 (1H, m, HC(5)), 5.77–5.82 (1H, m, HC(4), 5.89 (1H, m, HC(2)), 6.04 (1H, m, HC(3)) and 7.24–7.38 (m, 5H, Ph).

^{13}C NMR (67.5 MHz, CDCl$_3$) δ 75.87 (H$_2$C(5)), 87.97 (HC(2)), 126.47 (2 × m-Ph), 126.67 (p-Ph), 127.89 (HC(4)), 128.57 (2 × o-Ph), 130.00 (HC(3)) and 142.10 (ipso-Ph); m/z (eims, 70 eV) 146 (M$^+$, 2%), 145 (7), 115 (9) and 105 (100).

CONCLUSION

The procedure to prepare the ligand is very easy to reproduce. The asymmetric intermolecular phenylation of 2,3-dihydrofuran proceeds in moderate to good yields and with excellent enantioselectivities. The solvent and base employed were varied in an attempt to optimise the reaction, Table 5.5.

REFERENCES

1. (a) Shibasaki, M., Boden, C. D. J. and Kojima, A. *Tetrahedron* **1997**, *53*, 7371. (b) Guiry, P. J., Hennessy, A. J. and Cahill, J. P. *Topics in Catalysis* **1997**, 311. (c) Shibasaki, M. and Vogl, E. M. *J. Organomet. Chem* **1999**, *576*, 1.
2. Malone, M. PhD. Thesis, *National University of Ireland*, 1998.
3. Kilroy, T. G., Hennessy, A. J., Malone, Y. M., Farrell, A. and Guiry, P. J. *J. Mol. Cat. A: Chemical,* **2003**, *196*, 65.
4. (a) Biehl, E. R. and Reeves, P. C. *Synthesis* **1973**, 360. (b) Reeves, P. C. *Organic Syntheses Collective Volumes*, **56**, 28.

6 Sonogashira Coupling Reactions

CONTENTS

6.1 NONPOLAR BIPHASIC CATALYSIS: SUZUKI- AND SONOGASHIRA COUPLING REACTIONS

ANUPAMA DATTA and HERBERT PLENIO*

Institut für Anorganische Chemie, Petersenstr. 18, TU Darmstadt, 64287 Darmstadt, Germany

The palladium-catalysed Suzuki and the Sonogashira couplings are powerful synthetic tools for carbon-carbon bond forming reactions.[1] Developing efficient

Catalysts for Fine Chemical Synthesis, Vol. 3, Metal Catalysed Carbon-Carbon Bond-Forming Reactions
Edited by S. M. Roberts, J. Xiao, J. Whittall, and T. Pickett
© 2004 John Wiley & Sons, Ltd ISBN: 0-470-86199-1

strategies for the separation of catalysts from the products of catalytic reactions is now recognised as an important development in the field of homogeneous catalysis, to enable the recovery of precious metal based catalysts and to avoid contamination of the product by such metals. In this context, the modification of catalysts by attaching groups that determine the solubility of the respective catalysts (phase tags) is one of the several strategies that allows separation of such catalysts in biphasic[2,3], nanofiltration[4] or precipitation catalysis.[5] Soluble phase-tagged catalysts with linear polymers, which govern the solubility properties, have been used in polar and nonpolar biphasic catalysis. In this work, a nonpolar phase tag, poly(4-methylstyrene), has been attached to 1-diadamantyl phosphine.[4,6] The palladium complex of the phosphinated polymer is an efficient catalyst for Suzuki and Sonogashira reactions.[7] Catalyst separation from the products can be done by a simple phase separation of the cyclohexane (phase catalyst) and the DMSO or nitromethane solutions (phase product).

6.1.1 NONPOLAR BIPHASIC SONOGASHIRA REACTION OF 4-BROMOACETOPHENONE AND PHENYLACETYLENE TO 1-(4-PHENYLETHYNYL-PHENYL)-ETHANONE

Materials and equipment

- Pd(PhCN)$_2$Cl$_2$ (5.7 mg, 0.015 mmol), CuI (5.7 mg, 0.03 mmol)
- Phosphinated polymer (15% loading, 150 mg, 0.03 mmol)
- Dried and deoxygenated cyclohexane (7 ml), DMSO (4 ml) and HN^{i}Pr$_2$ (0.1 ml)
- Phenyl acetylene (198 µl, 1.8 mmol), 4-bromoacetophenone (299 mg, 1.5 mmol)
- Silica gel MN60 (63–200 µm)

- Schlenk tube
- Magnetic stirrer, magnetic stirrer bars
- Contact thermometer

Procedure

1. In a Schlenk tube Pd(PhCN)$_2$Cl$_2$ and phosphinated polymer were added to cyclohexane. After addition of CuI, HN^{i}Pr$_2$, PhC≡CH, 4-bromoacetophenone and DMSO the catalyst is formed and the reaction was stirred at 60 °C till completion (2 hours) under nitrogen.
2. After cooling to room temperature, the DMSO layer was separated, evaporated and the remaining crude product purified by column chromatography, (silica, cyclohexane/ethylacetate) to obtain pure 1-(4-phenylethynyl-phenyl)-ethanone (yield 95%). When the reaction is carried out on a larger scale, evaporation of the DMSO is impractical, instead addition of water to the DMSO solution followed by extraction of the product with ether is easier.

3. The catalyst can be recycled at least four times which involves the addition of fresh DMSO, HN^{i}Pr$_2$ and the two substrates to the catalyst phase for the next reaction cycle, with the cumulative product yield of 95%.

Note: The method was also applied for the coupling of bromobenzene (yield: 92%, reaction time: 5 hours), 4-bromoanisole (89%, 10 hours) and 4-bromoethylbenzoate (96%, 2 hours) with phenylacetylene.

6.1.2 NONPOLAR BIPHASIC SUZUKI REACTION FOR THE SYNTHESIS OF 1-BIPHENYL-4-YL-ETHANONE

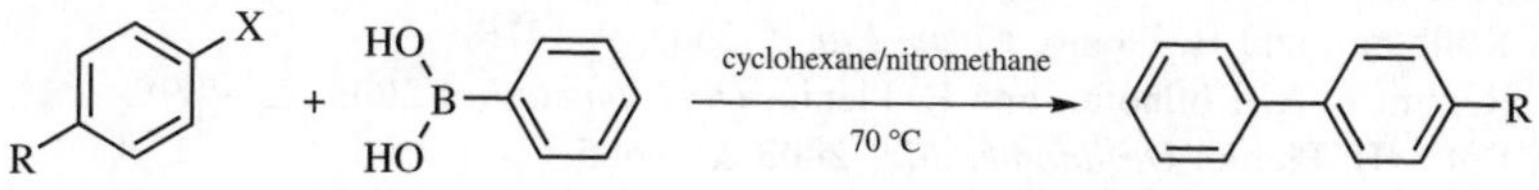

Materials and equipment

- Pd(OAc)$_2$ (3.3 mg, 0.015 mmol)
- Phosphinated polymer (15% loading, 150 mg, 0.030 mmol)
- Dried and deoxygenated cyclohexane (7 ml), nitromethane (4 ml)
- Phenyl boronic acid (183 mg, 1.5 mmol), 4-bromoacetophenone (199 mg, 1.0 mmol)
- K$_3$PO$_4$ (water free, 440 mg)
- Silica gel MN60 (63–200 µm)

- Schlenk tube
- Magnetic stirrer, magnetic stirrer bars
- Contact thermometer

Procedure

1. In a Schlenk tube Pd(OAc)$_2$ and phosphinated polymer were added to cyclohexane. After addition of K$_3$PO$_4$ (potassium phosphate), phenyl boronic acid, 4-bromoacetophenone and nitromethane the catalyst is formed and the reaction was stirred at 70 °C till completion (6 hours) under nitrogen.
2. After cooling to room temperature, the nitromethane solvent was separated, evaporated and the remaining crude product purified by column chromatography (silica, cyclohexane/ethyl acetate), to obtain pure 1-biphenyl-4-yl-ethanone (yield 94%).
3. The catalyst can be recycled at least five times by the addition of fresh nitromethane, K$_3$PO$_4$ and the two substrates to the catalyst phase and stirring the reaction at 70 °C for 6 hours in each cycle. The overall product yield was 95%.

Note: The method was also used for the coupling of bromobenzene (92%, 14 hours), 4-bromoanisole (90%, 14 hours), 4-chloroacetophenone (83%, 24 hours) with phenyl boronic acid.

CONCLUSION

Palladium catalyst supported on poly(4-methylstyrene) can be used for biphasic nonpolar Suzuki and Sonogashira coupling of aryl bromides and chlorides with good product yields. The catalyst can be recycled five times with negligible decrease in the activity.

REFERENCES

1. Applied *Homogeneous Catalysis with Organometallic Compounds*, (Cornils, B.; Herrmann, W. A. eds.), Wiley-VCH, Weinheim **2001**.
2. A. Köllhofer, and H. Plenio, *Chem. Eur. J.* **2003**, *9*, 1416.
3. H. Remmele, A. Köllhofer, and H. Plenio, *Organometallics* **2003**, *22*, 4098.
4. A. Datta, H. Plenio, *Organometallics* **2003** *22*, 4685.
5. Thematic Issue: Recoverable Catalysts and Reagents, *Chem. Rev.* **2002**, 102, (ed. J. A. Gladysz).
6. The copolymer of (4-methylstyrene) and (4-bromomethylstyrene) (15 mol% Br loaded, 7 g, 55 mmol) and (1-Ad) $_2$PH (11.5 mmol, 3.5 g) were dissolved in xylene (30 ml) and stirred

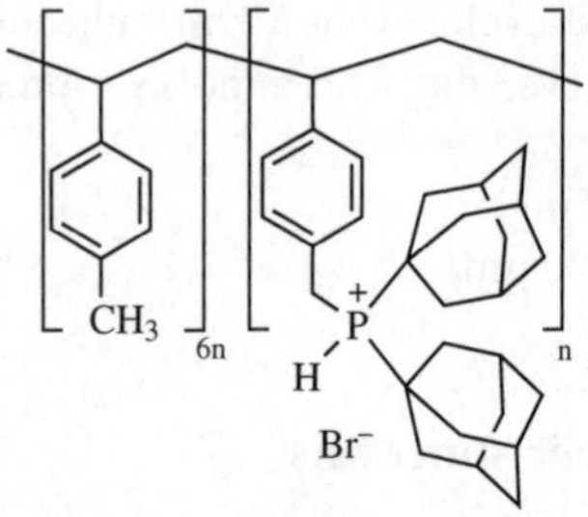

at 125 °C. After 20 hours, the reaction mixture was cooled to room temperature and poured into diethylether (120 ml). The precipitated polymer was filtered, washed with diethylether and dried under vacuum. Yield: 8.5 g (90%).[4]
7. A. Datta, and H. Plenio, *Chem. Commun.* **2003**, 1504.

6.2 POLYSTYRENE-SUPPORTED SOLUBLE PALLADACYCLE CATALYST AS RECYCLABLE CATALYST FOR HECK, SUZUKI AND SONOGASHIRA REACTIONS

CHIH-AN LIN and FEN-TAIR LUO*

Institute of Chemistry, Academia Sinica, Taiwan, ROC

A new type of soluble polystyrene-supported palladium complex was synthesised (Figure 6.1) as an excellent and recyclable palladacycle catalyst for carbon-carbon bond formation in Heck, Suzuki and Sonogashira reactions to give high yields of the desired products.

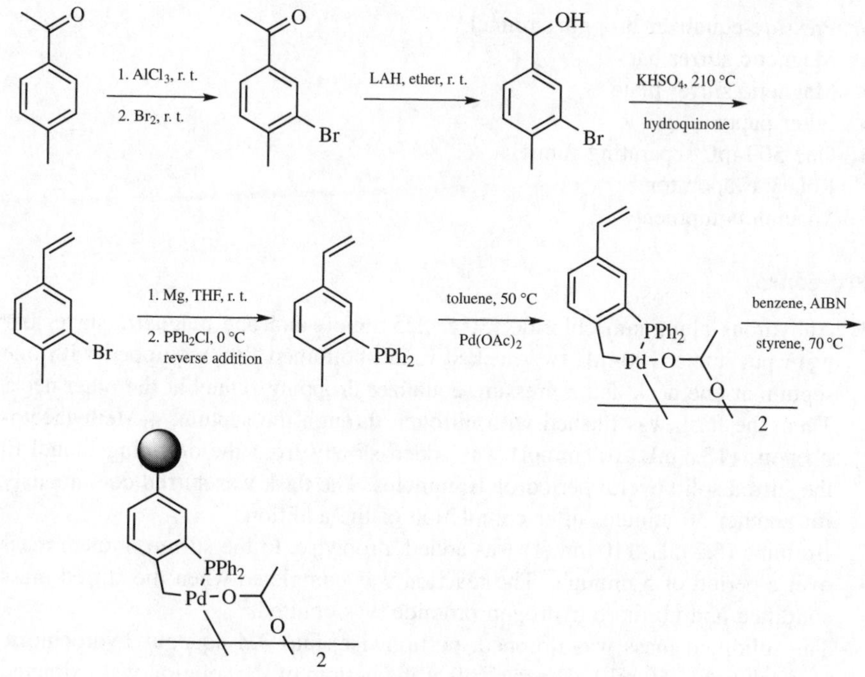

Figure 6.1 Synthesis of a polystyrene-supported palladacycle.

6.2.1 SYNTHESIS OF 3-BROMO-4-METHYLACETOPHENONE

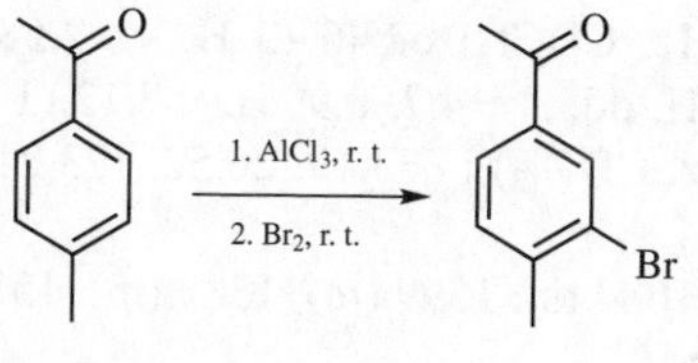

Materials and equipment

- Anhydrous aluminum chloride, 30 g, 225 mmol
- 4-Methylacetophenone, 13.3 mL, 100 mmol
- Bromine, 5.7 mL, 110 mmol
- 3 M hydrochloric acid solution, 250 mL
- Ether, 90 mL
- Saturated ammonium chloride, 50 mL
- Anhydrous magnesium sulfate powder

- One 100-mL two-necked round-bottomed flask
- Thermometer
- Distillation equipment

- Pressure-equalizer dropping funnel
- Magnetic stirrer bar
- Magnetic stirrer plate
- Filter paper
- One 500-mL separating funnel
- Rotary evaporator
- Vacuum equipment

Procedure

1. Anhydrous aluminum chloride (30 g, 225 mmol) and one magnetic stirrer bar were put into a 100-mL two-necked round-bottomed flask equipped with one septum at one neck and a pressure-equalizer dropping funnel at the other neck. Then, the flask was flushed with nitrogen through the septum. 4-Methylaceto-phenone (13.3 mL, 100 mmol) was added slowly from the dropping funnel to the stirred solid over a period of 10 minutes. The flask was stirred continuously for another 30 minutes after completion of the addition.
2. Bromine (5.7 mL, 110 mmol) was added, dropwise, to the stirred, molten mass over a period of 5 minutes. The reaction was completed when the stirred mass solidified and no more hydrogen bromide was emitted.
3. The solidified mass was dropped, portionwise, into 3M aqueous hydrochloric acid solution (250 mL). The dark oil at the bottom of the solution was extracted by ether (3 × 30 mL). The organic layer was then washed by saturated ammonium chloride aqueous solution (50 mL), dried with magnesium sulfate, and concentrated to get the crude product.
4. The crude product was further purified by distillation at low pressure 118 °C (3 mm Hg) to give 18.53 g (87% yield) of the desired product; m.p. 94–95 °C.

    ^{1}H NMR (300 MHz, CDCl$_3$) δ2.46 (3 H, s), 2.58 (3 H, s), 7.33 (1 H, d, $J = 4.7$ Hz), 7.79 (1 H, dd, $J = 4.7$, 0.99 Hz), 8.12 (1 H, d, $J = 0.99$ Hz).

    ^{13}C NMR (75 MHz, CDCl$_3$) : δ23.19, 26.52, 123.21, 127.08, 130.87, 132.32, 136.46, 143.53, 196.4.

    IR(neat) 3024 (s), 1684 (s), 1599 (m), 1383 (m), 1521 (s), 1039 (m), 958 (w), 907 (m), 831 (w) cm^{-1}.

    MS m/z 212 (M$^+$), 199, 197, 171, 169, 89; HRMS calcd for C$_9$H$_9$OBr, 211.9837; found 211.9840.

6.2.2   SYNTHESIS OF 1-(3-BROMO-4-METHYL-PHENYL)-ETHANOL

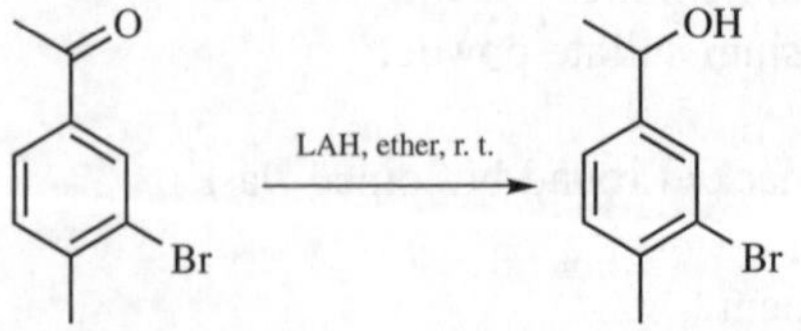

Material and equipment

- Lithium aluminum hydride (LAH), 6.45 g, 170 mmol
- 3-Bromo-4-methylacetophenone, 38 g, 180 mmol
- Dry ether, 50 mL
- 3 M hydrochloric acid solution, 240 mL
- Ethyl acetate (EAC), 120 mL
- Saturated ammonium chloride aqueous solution, 50 mL
- Anhydrous magnesium sulfate powder

- One 250-mL round-bottomed flask
- One 500-mL separating funnel
- Filter paper
- Distillation equipment
- Vacuum equipment
- Rotary evaporator
- Magnetic stirrer bar
- Magnetic stirrer plate
- Thermometer

Procedure

1. Lithium aluminium hydride (LAH) (6.45 g, 170 mmol) and one magnetic stirrer bar were added into a 250-mL round-bottomed flask, which was then closed with a septum and dried under vacuum followed by filling with nitrogen. One part of dry ether (50 mL) was first injected into the stirred LAH. The mixture of 3-bromo-4-methylacetophenone (38 g, 180 mmol) in dry ether (50 mL) was injected into the flask dropwise. Stirring was continued after the completion of addition of 3-bromo-4-methylacetophenone until no more gas was generated.
2. The mixture was poured into dilute hydrochloric acid aqueous solution (240 mL), extracted with ethyl acetate (4 × 30 mL), washed over saturated aqueous ammonium chloride solution (50 mL), dried over anhydrous magnesium sulfate, and concentrated at low pressure.
3. The crude product was further purified by distillation at low pressure 130 °C (4 mm Hg) to give 31.35 g (81% yield) of the desired product.
   ^{1}H NMR (300 MHz, CDCl$_3$) δ1.47 (3 H, t, $J = 6.4$ Hz), 1.76 (1 H, s), 2.38 (3 H, s), 4.84 (1 H, q, $J = 6.4$ Hz), 7.20 (2 H, s), 7.55 (1 H, s).
   ^{13}C NMR (75 MHz, CDCl$_3$) δ22.54, 25.16, 69.33, 124.30, 124.93, 129.34, 130.82, 136.88, 145.27.
   IR (neat) 3024 (s), 2997 (s), 2927 (m), 1605 (w), 1562 (w), 1494 (m), 1452 (m), 1381 (m), 1331 (w), 1255 (m), 1091 (m), 1038 (m), 1009 (m), 908 (m), 822 (m), 733 (s) cm^{-1}.

MS m/z 214 (M^+), 199, 197, 171, 169, 119, 91; HRMS calcd for $C_9H_{11}OBr$, 213.9993; found 213.9995.

6.2.3   SYNTHESIS OF 3-BROMO-4-METHYL-STYRENE

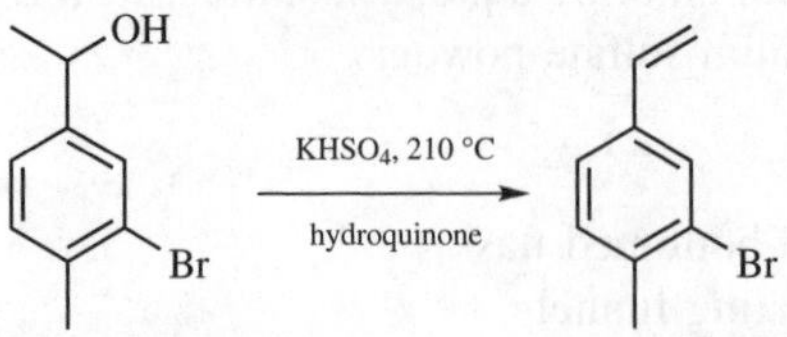

Materials and equipment

- Potassium hydrogen sulfate, 0.55 g, 34 mmol
- Hydroquinone, 0.142 g, 1.3 mmol
- 1-(3-Bromo-4-methyl-phenyl)-ethanol, 17.2 g, 80 mmol

- One 25-mL round-bottomed flask
- Magnetic stirrer plate
- Magnetic stirrer bar
- Distillation equipment
- Thermometer

Procedure

1. Potassium hydrogen sulfate (0.55 g, 34 mmol), hydroquinone (0.142 g, 1.3 mmol), and 1-(3-bromo-4-methyl-phenyl)-ethanol (17.2 g, 80 mmol) were delivered into a 25-mL round bottomed flask, equipped with a distillation equipment. The mixture was kept stirring.
2. The system was evacuated to 3 mm Hg. The dehydration and distillation processes 102 °C (3 mm Hg) proceeded at the same time. The isolated product was obtained in 13.6 g (86% yield).

   ^{1}H NMR (300 MHz,CDCl$_3$) δ2.37 (3 H, s), 5.23 (1 H, d, J = 10.9 Hz), 5.69 (1 H, d, J = 17.6 Hz), 6.60 (1 H, dd, J = 10.9, 17.6 Hz), 7.16 (1 H, d, J = 7.8 Hz), 7.21 (1 H, d, J = 7.8 Hz), 7.57 (1 H, s).

   ^{13}C NMR (75 MHz, CDCl$_3$) δ22.63, 114.32, 125.04, 125.09, 129.92, 130.76, 135.30, 137.11, 137.22 ppm.

   IR (neat) 3091 (m), 3063 (m), 3016 (s), 2985 (s), 2964 (m), 1899 (w), 1833 (w), 1701 (w), 1629 (m), 1602 (m), 1554 (s), 1492 (s), 1450 (s), 1380 (s), 1304 (w), 1278 (m), 1205 (m), 1037 (s), 989 (s), 914 (s), 833 (s), 854 (s), 826 (s) cm^{-1}.

   MS m/z 197(M^+ + 1) 171, 169, 117, 115, 91, 89; HRMS calcd for C_9H_9Br, 195.9888; found 195.9887.

6.2.4 SYNTHESIS OF 3-(DIPHENYLPHOSPHINO)-4-METHYL-STYRENE

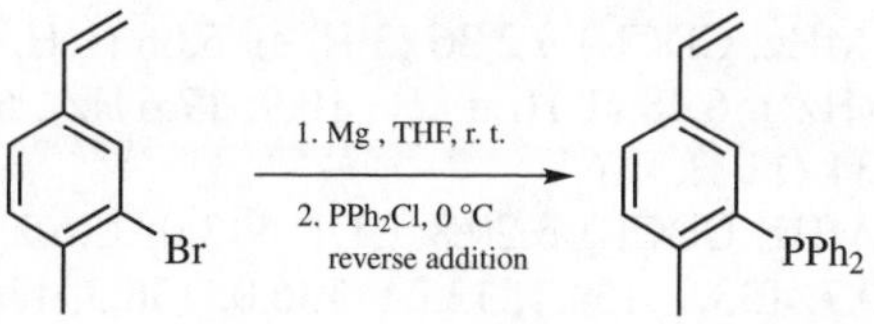

Materials and equipment

- Magnesium, 0.72 g, 30 mmol
- Dry tetrahydrofuran, 70 mL
- 3-Bromo-4-methyl-styrene, 3.94 g, 20 mmol
- Chlorodiphenylphosphine, 5.25 g, 25 mmol
- Saturated aqueous ammonium chloride solution, 10 mL
- Anhydrous magnesium sulfate powder
- Hexane
- Ethyl acetate

- Silica gel (Merck, 60 Å)
- Two 50-mL round-bottomed flasks
- Syringe
- Magnetic stirrer plate
- Magnetic stirrer bar
- One 500-mL separatory funnel
- Glass column, diameter 2 cm
- Ice bath

Procedure

1. Magnesium (0.72 g, 30 mmol) was added into a 50-mL round-bottomed flask, which was dried and filled with nitrogen. Dry tetrahydrofuran (THF) (10 mL) was injected first and then stirred with the magnesium. Half the solution of 3-bromo-4-methyl-styrene (3.94 g, 20 mmol) in dry THF (10 mL) was injected dropwise. After the exothermic reaction started, the rest of the solution was injected.
2. The Grignard reagent was added into the mixture of chlorodiphenylphosphine (5.25 g, 25 mmol) in dry THF (10 mL) at 0 °C. After completion of the addition, the temperature was raised to the room temperature.
3. After stirring for another 20 hours, the mixture was poured into a saturated aqueous ammonium chloride solution (10 mL) at 0 °C and extracted with dry THF (3 × 20 mL). The organic layer was dried over anhydrous magnesium sulfate and concentrated to 5 mL. Hexane was added into the organic solution to deposit the by-products.

4. After filtration and concentration, the residue was purified by a flash column chromatography (hexane/EAC = 4/1). The isolated product was 2.42 g (39% yield).

^{1}H NMR (300 MHz, CDCl$_3$) δ 2.36 (3 H, s), 5.06 (1 H, d, J = 10.9 Hz), 5.42 (1 H, d, J = 17.6 Hz), 6.48 (1 H, q, J = 10.9, 17.6 Hz), 6.79 (1 H, q, J = 2.1, 5.8 Hz,), 7.20–7.34 (11 H, m).

^{13}C NMR (75 MHz, CDCl$_3$) δ 20.8, 21.1, 113.0, 126.2, 128.5, 128.6, 128.8, 130.2, 130.3, 130.7, 133.9, 134.1, 135.1, 136.0, 136.3, 136.5, 141.7 ppm.

^{31}P NMR (121 MHz, CDCl$_3$) δ −12.34 ppm.

IR (neat) 3057 (m), 3016 (s), 2974 (m), 1957 (w), 1900 (w), 1819 (w), 1629 (m), 1589 (m), 1479 (s), 1435 (s), 1380 (m), 1306 (w), 1262 (m), 1179 (m), 1152 (m), 1093 (m), 1028 (m), 992 (m), 911 (s), 830 (s), 699 (s).

MS m/z 302 (M$^+$) 223, 183, 165, 152, 115, 78; HRMS calcd for C$_{21}$H$_{19}$P, 302.1124; found 302.1227.

6.2.5   SYNTHESIS OF TRANS-DI(μ-ACETATO)-BIS[3-(DIPHENYL-PHOSPHINO)-4-STYRYL]DIPALLADIUM(II)

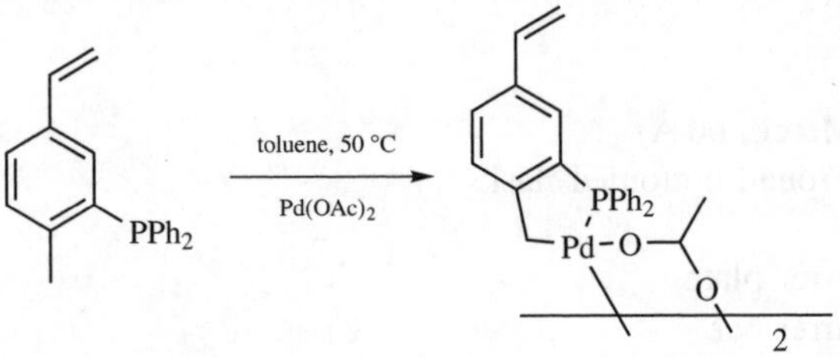

Materials and equipment

- Pd(OAc)$_2$, 0.225 g, 1 mmol
- Dry toluene, 24 mL
- 3-(Diphenylphosphino)-4-methyl-styrene, 0.333 g, 1.1 mmol
- Hexane
- Toluene

- One 50-mL round-bottomed flask
- Magnetic stirrer bar
- Magnetic stirrer plate
- Several test tubes

Procedure

1. Pd(OAc)$_2$ (0.225 g, 1 mmol) was dissolved in stirring dry toluene (20 mL) in a 50-mL round bottomed flask. The mixture of 3-(diphenylphosphino)-4-methyl-styrene (0.333 g, 1.1 mmol) and dry toluene (4 mL) was injected into the flask. After that, the mixture was heated at 50 °C for 5 minutes, and then cooled slowly to the room temperature.

2. The mixture was concentrated to one third of the original volumn. Hexane (25 mL) was added to the mixture for precipitation of the product. Recrystallisation was undertaken repeatedly in the system of toluene/hexane or dichloromethane/hexane. The isolated product was 0.34 g (74% yield).
   ^{31}P NMR (121 MHz, CDCl$_3$) δ 19 ppm.

6.2.6   SYNTHESIS OF POLYMER-SUPPORTED PALLADACYCLE CATALYST

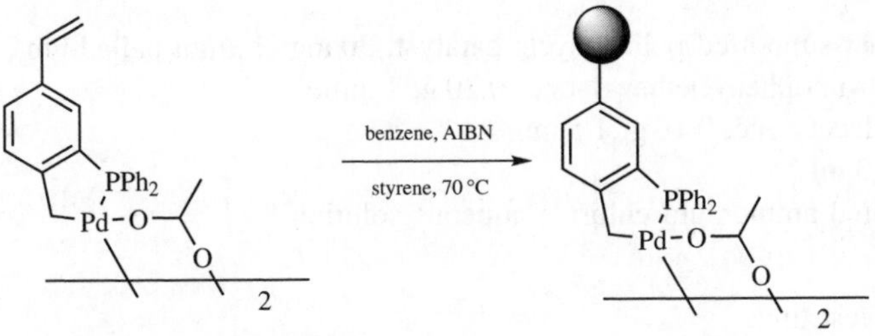

Materials and equipment

- Trans-di(μ-acetato)-bis[3-(diphenylphosphino)-4-styryl]dipalladium(II), 93 mg, 0.1 mmol
- Styrene, 62 mg, 0.6 mmol
- Benzene, 3 mL
- AIBN, 18 mg, 0.01 mmol
- Tetrahydrofuran
- Hexane

- Pyrex test tube
- Magnetic stirrer bar
- Magnetic stirrer plate

Procedure

1. *Trans*-di(μ-acetato)-bis[3-(diphenylphosphino)-4-styryl]dipalladium(II) (93 mg, 0.1 mmol) was added to a test tube and flushed with nitrogen. A mixture of styrene (62 mg, 0.6 mmol) and benzene (2 mL) was injected into the test tube. Then, a mixture of AIBN (azobisisobutyronitrile) (18 mg, 0.01 mmol) and benzene (1 mL) was injected into the stirring test tube. The reaction mixture was then heated at 70 °C for 40 hours.
2. The mixture was concentrated to dryness. The dry solid was repeatedly washed by the solvent (THF/hexane = 1:15). The isolated polymer was 0.11 g (66% yield).
   ^{31}P NMR (121 MHz, CDCl$_3$) δ 19 ppm.

6.2.7 SYNTHESIS OF 1-[4-(2-PHENYLETHYNYL)PHENYL]ETHAN-1-ONE *VIA* SONOGASHIRA REACTION BY THE USE OF POLYMER-SUPPORTED PALLADACYCLE CATALYST

98%

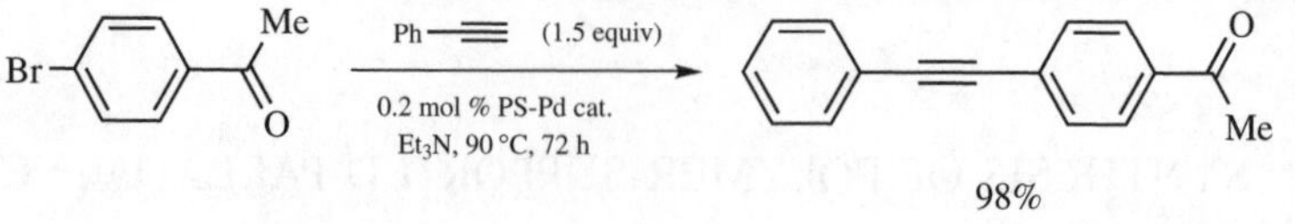

Materials and equipment

- Polymer-supported palladacycle catalyst, 30 mg, 2 μmol palladium
- 1-(4-Bromophenyl)ethan-1-one, 0.20 g, 1 mmol
- Phenylacetylene, 0.16 g, 1.5 mmol
- Et$_3$N, 3 mL
- Saturated ammonium chloride aqueous solution

- Pyrex test tube
- Magnetic stirrer bar
- Magnetic stirrer plate
- Centrifuge

Procedure

1. Polystyrene-supported palladacycle catalyst (30 mg, 2 μmol Pd), 1-(4-bromo-phenyl)ethan-1-one (0.20 g, 1 mmol), phenylacetylene (0.16 g, 1.5 mmol), and triethylamine (3 mL) were sequentially added into a 15 mL septum-sealed test tube under nitrogen atmosphere. The mixture was then heated at 90 °C for 72 hours.
2. After cooling to the room temperature, to the mixture was added 8 mL of dry ether to precipitate the polystyrene-supported palladacycle catalyst. The catalyst was further removed by centrifugation; the upper liquid layer of the reaction mixture was transferred *via* a syringe into another 20-mL round-bottom flask.
3. The above procedure (2) was repeated one more time.
4. The combined liquid layers were concentrated under reduced pressure. Saturated ammonium chloride solution (5 mL) was then added to the oily mixture. The organic product was extracted three times into ethyl acetate (10 mL × 3), dried over magnesium sulfate, filtrated, and concentrated to give the crude product.
5. The crude product was further purified by column chromatography (silica gel, hexane/ethyl acetate = 4/1) to give 0.21 g (98% yield) of the desired product.

CONCLUSION

The newly invented polystyrene-supported palladacycle catalyst was prepared in six steps with high yields. The simple precipitation and filtration process to recycle the catalyst after model reactions for Heck, Suzuki, and Sonogashira reactions is noteworthy.[1]

REFERENCE

1. Chih-An Lin and Fen-Tair Luo, *Tetrahedron Lett.*, **2003**, *44*, 7565–7568.

7 Cross-Coupling Reactions

CONTENTS

7.1   CROSS-COUPLING REACTION OF ALKYL HALIDES WITH GRIGNARD REAGENTS IN THE PRESENCE OF 1,3-BUTADIENE CATALYSED BY NICKEL, PALLADIUM, OR COPPER

JUN TERAO and NOBUAKI KAMBE*

Department of Molecular Chemistry & Science and Technology Center for Atoms, Molecules and Ions Control, Graduate School of Engineering, Osaka University, Yamadaoka 2-1, Suita, Osaka 565-0871, Japan

Transition metal-catalysed cross-coupling reaction between organic halides and organometallic reagents constitutes one of the most straightforward methods for the formation of carbon-carbon bonds.[1] As for the scope of the substrates, there have been employed a variety of organometallic reagents containing boron, magnesium, lithium, tin, aluminium and zinc connecting to alkyl, alkenyl, aryl, alkynyl, allyl, and benzyl groups, whereas the scope of the coupling partner had long been limited to aryl and alkenyl halides. The use of alkyl halides usually gave unsatisfactory results due mainly to the slow oxidative addition to transition metal catalysts and the fast β-elimination from the alkylmetal intermediates. However, cross-coupling reaction of alkyl halides with organometallic reagents has extensively been studied in the past several years[2,3] and alkyl halides as well as sulfonates have now become promising candidates as the reagents in transition metal-catalysed cross-coupling reaction. Progress of this field has been aided by providing a novel catalytic system where nickel,[4] palladium,[5] or copper[4b] catalyses cross-coupling reaction of alkyl chlorides, bromides and tosylates with Grignard reagents in the presence of a 1,3-butadiene as an additive (Figure 7.1). These results include the first examples of cross-coupling reaction using non-activated alkyl chlorides[4a] and fluorides.[4b]

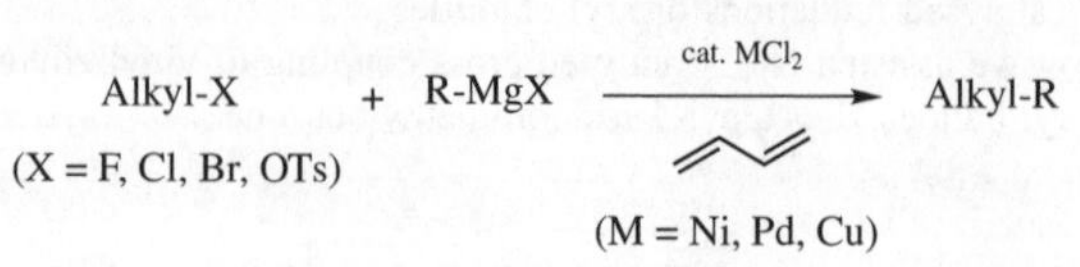

$$\text{Alkyl-X} \quad + \quad \text{R-MgX} \quad \xrightarrow{\text{cat. MCl}_2} \quad \text{Alkyl-R}$$

(X = F, Cl, Br, OTs)

(M = Ni, Pd, Cu)

Figure 7.1   Cross-coupling reactions involving Grignard reagents.

7.1.1   SYNTHESIS OF NONYLCYCLOPROPANE

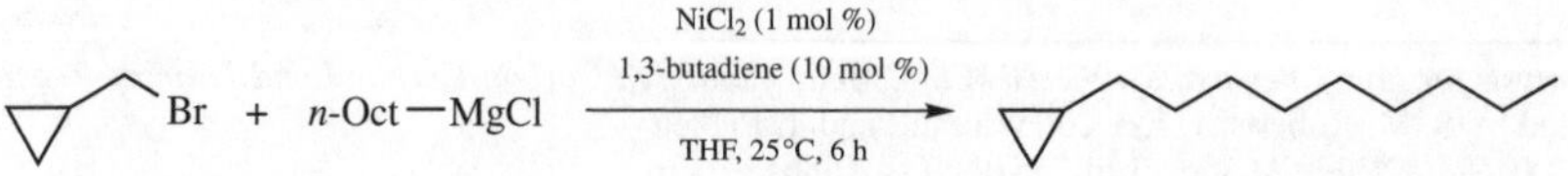

Materials and equipment

- Cyclopropylmethyl bromide (99%), 270 mg, 2 mmol
- ^{n}OctMgCl, of 1.0 M in THF, 2.6 mL, 2.6 mmol
- 1,3-Butadiene (>98%), 4.48 mL at 20 °C under 1 atm, 0.2 mmol
- Nickel(II) chloride (>98%), 2.6 mg, 0.02 mmol
- Saturated solution of ammonium chloride, 10 mL
- Diethyl ether, 10 mL
- Anhydrous magnesium sulfate

- Septum
- 20-mL Schlenk tube with a magnetic stirring bar
- Magnetic stirrer
- One 200-mL Erlenmeyer flask
- One funnel, diameter 10 cm
- One Kjéldahl flask, 100 mL
- Filter paper
- One 200-mL separating funnel
- Rotary evaporator
- Kugelrohr distillation equipment

Procedure

1. Cyclopropylmethyl bromide (270 mg, 2 mmol) and ^{n}OctMgCl (1.0 M in THF, 2.6 mL, 2.6 mmol) were placed in a 20-mL Schlenk tube equipped with a magnetic stirring bar under nitrogen. The system was closed and cooled to −78 °C to get slightly reduced pressure. Into the vessel was introduced 1,3-butadiene (4.48 mL at 20 °C under 1 atmosphere) through a septum using a syringe. The system was opened to an atmospheric pressure comprising a gentle nitrogen stream and a catalytic amount of nickel(II) chloride (2.6 mg, 0.02 mmol) was added. The system was closed again, warmed to 25 °C and stirred for 6 hours. A saturated aqueous ammonium chloride solution (10 mL) was added, and the product was extracted with ether (2 ×10 mL). The organic layer was dried over magnesium sulfate and the solvent was evaporated on a rotary evaporator to afford a colourless crude product (87% GC yield).
2. The corresponding coupling product was purified by vacuum distillation in Kugelrohr [90–100 °C, 0.030–0.035 mmHg, cooling the recipient flask with dry ice (79% yield)]. (NMR data).

   IR (neat): 2924, 2853, 1654, 1458, 1320, 1013 cm^{-1}.
   ^{1}H NMR (400 MHz, CDCl$_3$): δ = 1.39–1.31 (m, 2H), 1.29–1.22 (m, 12H), 1.19 (q, J = 7.1 Hz, 2H), 0.88 (t, J = 6.8 Hz, 3H), 0.70–0.59 (m, 1H), 0.40–0.32 (m, 2H), −0.02 (m, 2H).
   ^{13}C NMR (100 MHz, CDCl$_3$): δ = 34.9, 32.1, 29.9, 29.8 (two carbons), 29.7, 29.5, 22.9, 14.3, 11.1, 4.6.

7.1.2   SYNTHESIS OF 4-BROMO-HEXYLBENZENE

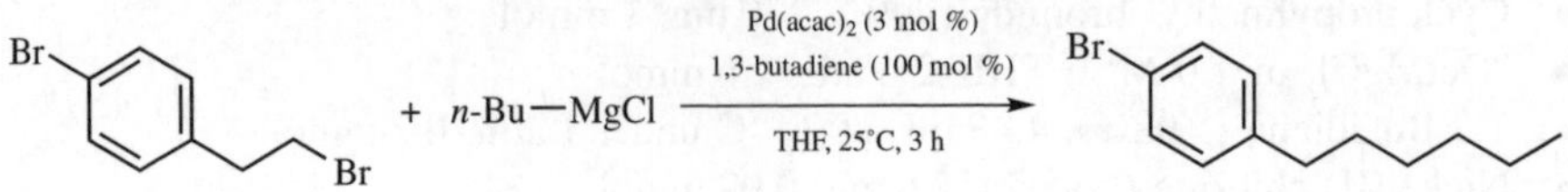

Materials and equipment

- Toluene-4-sulfonic acid 2-(4-bromo-phenyl)-ethyl ester, 264 mg, 1 mmol
- ^{n}BuMgCl, 0.9 M in THF, 1.67 mL, 1.5 mmol
- 1,3-Butadiene (>98%), 22.4 mL at 20 °C under 1 atm, 1 mmol
- Pd(acac)$_2$, 9.1 mg, 0.03 mmol
- Saturated solution of ammonium chloride, 10 mL
- Diethyl ether, 10 mL
- Chloroform, 200 mL
- Anhydrous magnesium sulfate

- Septum
- TLC plates, SIL G-60 UV$_{254}$
- 20-mL Schlenk tube with a magnetic stirring bar
- Magnetic stirrer
- One 200-mL Erlenmeyer flask
- One funnel, diameter 10 cm
- One Kjéldahl flask, 100 mL
- Filter paper
- One 200-mL separating funnel
- Rotary evaporator

Procedure

1. Toluene-4-sulfonic acid 2-(4-bromophenyl)ethyl ester (254 mg, 1 mmol) and ^{n}BuMgCl (0.9 M in THF, 1.67 mL, 1.5 mmol) were placed in a 20-mL Schlenk tube equipped with a magnetic stirring bar under nitrogen. The system was closed and cooled to −78 °C to get slightly reduced pressure. Into the vessel was introduced 1,3-butadiene (22.4 mL at 20 °C under 1 atmosphere, 1 mmol) through a septum using a syringe. The system was opened with a gentle nitrogen stream and a catalytic amount of Pd(acac)$_2$ (9.1 mg, 0.03 mmol) was added. The system was closed again, warmed to 25 °C and stirred for 3 hours. A saturated aqueous ammonium chloride solution (10 mL) was added, and the product was extracted with ether (2 × 10 mL). The organic layer was dried over magnesium sulfate and the solvent was evaporated on a rotary evaporator to afford a colourless crude product (80% GC yield).
2. The corresponding coupling product was purified by HPLC with chloroform as an eluent to afford 171 mg (71%) of 4-bromo-hexylbenzene.

    IR (neat): 2956, 2928, 2856, 1488, 1466, 1072, 1012 cm^{-1}.

^{1}H NMR (400 MHz, CDCl$_3$): δ = 7.38 (d, J = 8.0, 2H), 7.04 (d, J = 8.0, 2H), 2.55 (t, J = 7.6 Hz, 2H), 1.59–1.56 (m, 2H), 1.31–1.28 (m, 6H), 0.88 (t, J = 6.8, 3H).

^{13}C NMR (100 MHz, CDCl$_3$) δ = 141.7, 131.1, 130.0, 119.1, 35.5, 31.8, 31.4, 29.0, 22.7, 14.2.

Anal. calcd for C$_{12}$H$_{17}$Br: C, 59.76; H, 7.11. found: C, 60.02; H, 7.15.

7.1.3 SYNTHESIS OF 1,1-DIPHENYL-1-NONENE

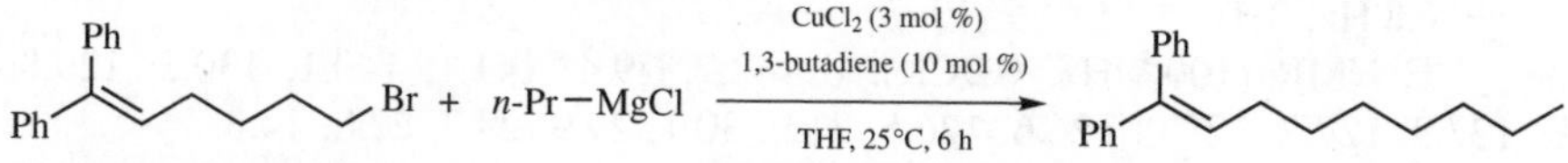

Materials and equipment

- 1,1-Diphenyl-6-fluoro-1-hexene (99%), 254 mg, 1 mmol
- ^{n}PrMgBr, 0.93 M in THF, 1.4 mL, 1.3 mmol
- 1,3-Butadiene (>98%), 2.24 mL at 20 °C under 1 atmosphere, 0.1 mmol
- Copper (II) chloride (>95%), 4.0 mg, 0.03 mmol
- Saturated solution of ammonium chloride, 10 mL
- Diethyl ether, 10 mL
- Anhydrous magnesium sulfate

- Septum
- Silica gel (Fuji-Davison silicagel, 100–250 mesh), 100 g
- TLC plates, SIL G-60 UV$_{254}$
- 20-mL Schlenk tube with a magnetic stirring bar
- Magnetic stirrer
- One 200-mL Erlenmeyer flask
- One funnel, diameter 10 cm
- One Kjéldahl flask, 100 mL
- Filter paper
- One 200-mL separating funnel
- One glass column, diameter 5 cm
- Rotary evaporator

Procedure

1. 1,1-Diphenyl-6-fluoro-1-hexene (254 mg, 1 mmol) and ^{n}PrMgBr (0.93 M in THF, 1.4 mL, 1.3 mmol) were placed in a 20-mL Schlenk tube equipped with a magnetic stirring bar under nitrogen. The system was closed and cooled to −78 °C to get slightly reduced pressure. Into the vessel was introduced 1,3-butadiene (2.24 mL at 20 °C under 1 atmosphere, 0.1 mmol) through a septum using a syringe. The system was readjusted to atmospheric pressure with a gentle nitrogen stream and a catalytic amount of copper (II) chloride (4.0 mg, 0.03 mmol) was added. The system was closed again, warmed to 25 °C and

stirred for 6 hours. A saturated aqueous ammonium chloride solution (10 mL) was added, and the product was extracted with ether (2 × 10 mL). The organic layer was dried over magnesium sulfate and the solvent was evaporated on a rotary evaporator to afford a colourless crude product (98% NMR yield).

2. The coupling product was isolated easily in pure form by column chromatography (silica gel, hexane), (250 mg, 90% yield).

IR (neat): 2924, 2853, 1598, 1493, 1443, 1073, 1029, 761, 700 cm^{-1}.

^{1}H NMR (400 MHz, CDCl$_3$): δ 7.38–7.16 (m, 10H), 6.08 (t, $J = 7.5$ Hz, 1H), 2.10 (dt, $J = 7.3$, 7.5 Hz, 2H), 1.52–1.39 (m, 2H), 1.28–1.23 (m, 8H), 0.86 (t, $J = 6.8$ Hz, 3H).

^{13}C NMR (100 MHz, CDCl$_3$): δ 142.7, 29.4, 141.1, 140.1, 130.2, 129.8, 127.9, 127.8, 127.0, 126.6, 126.5, 31.9, 30.1, 29.9, 29.3, 22.8, 14.3.

MS (EI) m/z (relative intensity, %) 278 (M$^+$, 73), 193 (100), 180 (57), 115 (36), 91 (17).

HRMS calcd for C$_{21}$H$_{26}$ 278.2034, found 278.2030; Anal. Calcd for: C, 90.59; H, 9.41. Found: C, 90.37; H, 9.32.

CONCLUSION

This reaction has wide generalities in terms of the scope of reagents, i.e. primary and secondary alkyl and aryl Grignard reagents and primary alkyl fluorides, chlorides, bromides and tosylates. Table 7.1 summarises the results of the cross-coupling reaction using different substrates and catalysts. The present catalytic system has several advantages over the conventional systems for large scale production since the reaction proceeds efficiently under mild conditions using less expensive nickel, palladium and copper salts as the catalysts and 1,3-butadiene as the additive instead of phosphine ligands.

Table 7.1 Transition-metal catalysed cross-coupling reaction of alkyl halides with grignard reagents in the presence of 1,3-butadiene.

Entry	Alkyl-X	R	Catalyst	Yield (%)
1	n-C$_{10}$H$_{21}$-Br	n-Butyl	NiCl$_2$	100
2	n-C$_8$H$_{17}$-Br	Ph	NiCl$_2$	90
3	n-C$_8$H$_{17}$-Cl	n-Butyl	NiCl$_2$	96
4	n-C$_8$H$_{17}$-F	i-Propyl	CuCl$_2$	81
5	n-C$_8$H$_{17}$-F	t-Butyl	CuCl$_2$	99
6	n-C$_7$H$_{15}$-OTs	s-Butyl	Pd(acac)$_2$	71

REFERENCES

1. a) For the history and recent development of transition metal-catalysed cross-coupling reactions, see *"International Symposium on 30 years of the Cross-coupling Reaction"* in *J. Organomet. Chem.*, **2002**, *653*, 1. b) *Metal-catalyzed Cross-coupling Reactions*, Diederich, F. and Stang, P. J. (Eds), Wiley-VCH, New York, 1998.

2. For a recent review and articles of transition metal-catalysed cross-coupling reactions using alkyl halides, see: a) Luh, T. -Y., Leung, M. and Wong, K. -T. *Chem. Rev.*, **2000**, *100*, 3187. b) Cárdenas, D. J. *Angew. Chem., Int. Ed. Engl.*, **1999**, *38*, 3018. c) Cárdenas, D. J. *Angew. Chem., Int. Ed. Engl.*, **2003**, *42*, 384.
3. After the above publications, the following studies were reported: Zhou, J. and Fu, G. C. *J. Am. Chem. Soc.*, **2003**, *125*, 12527. M. Eckhardt, G. C. Fu, *J. Am. Chem. Soc.* **2003**, *125*, 13642–13643: J. Zhou, G. C. Fu, *J. Am. Chem. Soc.* **2003**, *125*, 14726–14727: S. L. Wiskur, A. Korte, G. C. Fu, *J. Am. Chem. Soc.* **2004**, *126*, 82–83: J. Zhou, G. C. Fu, *J. Am. Chem. Soc.* **2004**, *126*, 1340–1341: M. Nakamura, K. Matsuo, S. Ito, E. Nakamura, *J. Am. Chem. Soc.* **2004**, *126*, 3686–3687: T. Nagano, T. Hayashi, *Org. Lett.* **2004**, *6*, 1297–1299.
4. a) Terao, J., Watanabe, H., Ikumi, A., Kuniyasu, H. and Kambe, N. *J. Am. Chem. Soc.* **2002**, *124*, 4222. b) Terao, J., Ikumi, A., Kuniyasu, H. and Kambe, N. *J. Am. Chem. Soc.*, **2003**, *125*, 5646.
5. Terao, J., Naitoh, Y., Kuniyasu, H. and Kambe, N. *Chem. Lett.*, **2003**, *32*, 890.

7.2 TRIORGANOINDIUM COMPOUNDS AS EFFICIENT REAGENTS FOR PALLADIUM-CATALYSED CROSS-COUPLING REACTIONS WITH ARYL AND VINYL ELECTROPHILES

LUIS A. SARANDESES and JOSÉ PÉREZ SESTELO

Departamento de Química Fundamental, Universidade da Coruña, E-15071 A Coruña, Spain

Transition metal-catalysed cross-coupling reactions of organometallic compounds with organic electrophiles represent one of the most powerful methods for the construction of carbon-carbon bonds, especially between unsaturated carbons.[1] Recently, an efficient methodology to perform these types of reactions was developed using triorganoindium compounds (R_3In). The cross-coupling reaction of R_3In with aryl or vinyl halides and pseudohalides, under palladium catalysis, proceeds in excellent yields and with high chemoselectivity.[2] In addition, when the reaction is performed in the presence of carbon monoxide atmosphere, unsymmetrical ketones were obtained in good yields (Figure 7.2).[3] In these coupling reactions, the triorganoindium compounds transfer, in a clear example of atom economy, all three of the organic groups attached to the metal. These novel features reveal indium organometallics as useful alternatives in cross-coupling reactions.

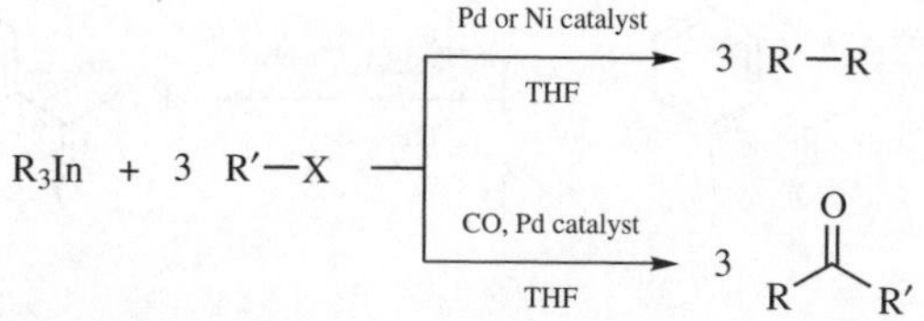

Figure 7.2 Cross-coupling reactions involving organoindium compounds.

7.2.1   PREPARATION OF TRIPHENYLINDIUM

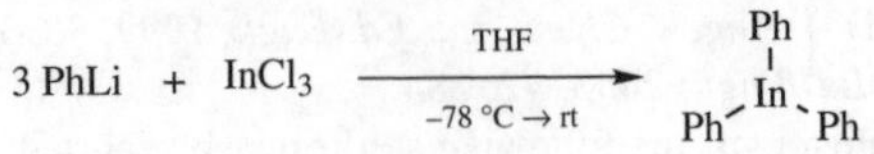

$$3\ PhLi\ +\ InCl_3\ \xrightarrow[-78\ °C\ \to\ rt]{THF}\ \underset{Ph}{\overset{Ph}{\underset{\diagdown}{\overset{|}{In}}}}\ Ph$$

Materials and equipment

- Indium trichloride (Aldrich, 99.9%), 0.41 g, 1.85 mmol
- Phenyllithium, 3.05 mL, 5.51 mmol, 1.8 M in cyclohexane:Et$_2$O (7:3) (Aldrich)
- Dry tetrahydrofuran, 7 mL

- Two 25-mL round-bottomed flasks with magnetic stirrer bars
- Magnetic stirrer plate
- Vacuum line
- Inert gas (argon) line
- Syringe-cannula equipment
- Heat-gun
- Cooling bath (−78 °C)

Procedure

1. Indium trichloride (0.41 g, 1.85 mmol) was placed in a 25-mL round-bottomed flask equipped with a magnetic stirrer bar and was dried under vacuum, heating externally with a heat gun. After the mixture was cooled and a positive argon pressure was established, dry tetrahydrofuran (7 mL) was added and the resulting clear solution was cooled to −78 °C. To this solution was slowly added a solution of phenyllithium [3.05 mL, 5.51 mmol, 1.8 M in cyclohexane:Et$_2$O (7:3)] over 15 minutes. After was stirring for 30 minutes, the cooling bath was removed and the reaction mixture was warmed to room temperature. The resulting solution of triphenylindium (ca. 0.18 M in THF) was directly used in the cross-coupling reaction.

7.2.2   SYNTHESIS OF 4-ACETYLBIPHENYL

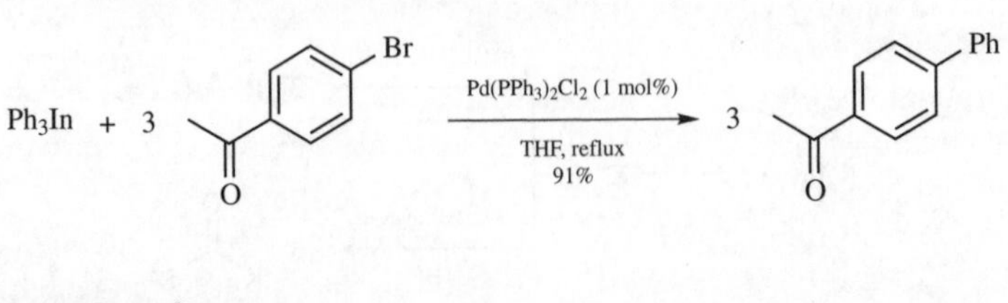

Materials and equipment

- Triphenylindium, 10.3 mL, 1.85 mmol, 0.18 M in dry tetrahydrofuran
- 4′-Bromoacetophenone (Aldrich, 98%), 1.00 g, 5.01 mmol

- Pd(PPh$_3$)$_2$Cl$_2$ (Aldrich, 98%), 35 mg, 0.05 mmol
- Dry tetrahydrofuran, 20 mL
- Diethyl ether, 40 mL
- HCl aqueous solution, 5 wt%, 20 mL
- Saturated aqueous solution sodium hydrogencarbonate, 25 mL
- Saturated aqueous sodium chloride solution, 25 mL
- Anhydrous sodium sulfate
- TLC plates (Merck), silica gel 60 F$_{254}$
- Silica gel 60 (Merck, 0.040–0.063 mm, 230–400 mesh ASTM), 35 g
- 2% EtOAc/hexanes, ca. 200 mL

- 50-mL round-bottomed flask with magnetic stirrer bar
- Magnetic stirrer plate with heating
- Silicone oil bath
- Inert gas (argon) line
- Vacuum line
- Syringe-cannula equipments
- Separation funnel
- 100-mL round-bottomed flask
- Chromatographic glass column, diameter 2.5 cm
- Rotary evaporator

Procedure

1. 4′-Bromoacetophenone (1.00 g, 5.01 mmol) and Pd(PPh$_3$)$_2$Cl$_2$ (35 mg, 1 mol%) were placed in a 50-mL round-bottomed flask equipped with a magnetic stirrer bar and a Liebig cooler under argon atmosphere. Dry tetrahydrofuran (20 mL) was then added and the resulting mixture was heated at reflux in a silicone oil bath. To this mixture was added, *via* cannula through the Liebig condenser, a solution of triphenylindium (10.3 mL, 1.85 mmol, 0.18 M in dry tetrahydrofuran). The resulting solution was stirred under reflux for 1.5 hours until the starting material has been consumed (TLC or GC test), and the reaction was then quenched by addition of few drops of methanol.
2. The crude product was isolated by concentration of the reaction mixture and addition of diethyl ether (40 mL). The organic phase was successively washed with aqueous hydrochloric acid (5%, 20 mL), saturated aqueous sodium hydrogencarbonate (25 mL) and saturated aqueous sodium chloride (25 mL), dried with anhydrous sodium sulfate, filtered and concentrated in a rotary evaporator at reduced pressure.
3. The crude 4-acetylbiphenyl was purified by column chromatography (silica gel 60, 35 g, 2% EtOAc/hexanes) followed by concentration and high vacuum drying (895 mg, 91% yield, as a white solid).[4]

    ^{1}H NMR (200 MHz, CDCl$_3$) δ 2.65 (s, 3H), 7.47 (m, 3H), 7.67 (m, 4H), 8.05 (d, $J = 8.4$ Hz, 2H).

^{13}C NMR (50 MHz, CDCl$_3$) δ 26.7 (CH$_3$), 127.2 (2 × CH), 127.3 (2 × CH), 128.2 (CH), 128.9 (2 × CH), 129.0 (2 × CH), 135.8 (C), 139.9 (C), 145.8 (C), 197.8 (C).

MS (EI, 70 eV) m/z (%) 196 (M$^+$, 68), 181 (M$^+$ − CH$_3$, 100), 152 (45).

7.2.3   SYNTHESIS OF 1,3-DIPHENYL-2-PROPEN-1-ONE

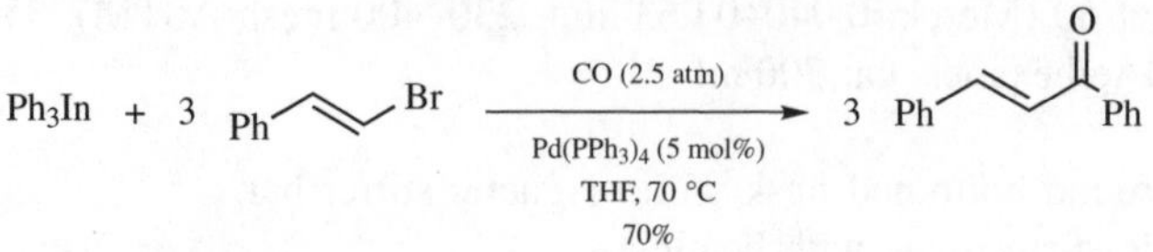

Ph$_3$In + 3 Ph⌇⌇Br $\xrightarrow[\substack{\text{Pd(PPh}_3)_4 \text{ (5 mol%)} \\ \text{THF, 70 °C} \\ \text{70%}}]{\text{CO (2.5 atm)}}$ 3 Ph⌇⌇C(O)Ph

Materials and equipment

- Triphenylindium, 11.2 mL, 2.02 mmol, 0.18 M in dry tetrahydrofuran
- β-Bromostyrene (Aldrich, 99%), 1.00 g, 5.46 mmol
- Pd(PPh$_3$)$_4$ (Aldrich, 99%), 316 mg, 0.27 mmol
- Carbon monoxide (Air liquide, N37)
- Dry tetrahydrofuran, 25 mL
- Diethyl ether, 40 mL
- Saturated aqueous ammonium chloride solution, 25 mL
- Anhydrous sodium sulfate
- TLC plates (Merck), silica gel 60 F$_{254}$
- Silica gel 60 (Merck, 0.040–0.063 mm, 230–400 mesh ASTM), 35 g
- 5% EtOAc/hexanes, ca. 250 mL

- 50-mL Schlenk tube with magnetic stirrer bar
- Magnetic stirrer plate with heating
- Silicone oil bath
- Vacuum line
- Syringe-cannula equipments
- Separating funnel
- 100-mL round-bottomed flask
- Chromatographic glass column, diameter 2.5 cm
- Rotary evaporator

Procedure

1. β-Bromostyrene (1.00 g, 5.46 mmol), Pd(PPh$_3$)$_4$ (316 g, 5 mol%) and dry tetra-hydrofuran (25 mL) were placed in a 50-mL Schlenk tube equipped with a magnetic stirrer bar. A solution of triphenylindium (11.2 mL, 2.02 mmol, 0.18 M in dry tetrahydrofuran) was then added and the resulting mixture was flushed with a carbon monoxide atmosphere and evacuated under vacuum

successively three times. The tube was finally charged with carbon monoxide (2.5 atm) and heated at 70 °C for 3 hours. The mixture was cooled, the pressure was released and the reaction was quenched by the addition of few drops of MeOH.

2. The mixture was concentrated in vacuo and diethyl ether (40 mL) was added. The organic phase was washed with saturated aqueous ammonium chloride (25 mL), dried (anhydrous sodium sulfate), filtered, and concentrated under vacuum.

3. The crude 1,3-diphenyl-2-propen-1-one was purified by column chromatography (5% EtOAc/hexanes) followed by concentration and high-vacuum drying (0.79 g, 70% yield, as a yellowish oil).[5]

^{1}H NMR (200 MHz, CDCl$_3$) δ 7.39–7.68 (m, 9 H), 7.84 (d, J = 15.6 Hz, 1 H), 8.02–8.08 (m, 2 H).

^{13}C NMR (50 MHz, CDCl$_3$) δ 122.1 (CH), 128.4 (2 × CH), 128.5 (2 × CH), 128.6 (2 × CH), 129.0 (2 × CH), 130.5 (CH), 132.8 (CH), 134.9 (C), 138.2 (C), 144.8 (CH), 190.5 (C).

MS (EI, 70 eV) m/z (%) 208 (M$^+$, 78), 207 (M$^+$ – H, 100), 131 (40), 77 (96).

CONCLUSION

In conclusion, the use of R$_3$In is a new and generally high yielding cross-coupling methodology. This reaction allows to triorganoindium compounds containing alkyl, vinyl, alkynyl and aryl groups to transfer, with high atom efficiency, the three organic groups to aryl and vinyl electrophiles under palladium catalysis (Table 7.2).

Table 7.2 Cross-coupling reactions between triorganoindium compounds and aryl and vinyl electrophiles.

Entry	R	R'—X	Conditions	Time (h)	Product yield (%)
1	Ph	4-MeC$_6$H$_4$—I	Pd(PPh$_3$)$_2$Cl$_2$	1	96
2	CH$_2$=CH		Pd(PPh$_3$)$_2$Cl$_2$	0.5	89
3	Me$_3$SiC≡C		Pd(PPh$_3$)$_2$Cl$_2$	1	93
4	n-Bu		Pd(PPh$_3$)$_2$Cl$_2$	4	82
5	Ph		CO, Pd(PPh$_3$)$_4$	3	83
6	PhC≡C		CO, Pd(PPh$_3$)$_4$	3	90
7	Me		CO, Pd(PPh$_3$)$_4$	3	70
8	CH$_2$=CH	4-AcC$_6$H$_4$—Br	Pd(dppf)Cl$_2$	2.5	94
9	Me$_3$SiC≡C		Pd(dppf)Cl$_2$	1	91
10	n-Bu		Pd(PPh$_3$)$_2$Cl$_2$	1.5	91
11	Ph		CO, Pd(PPh$_3$)$_4$	3	70
12	Ph	4-AcC$_6$H$_4$—OTf	Pd(PPh$_3$)$_2$Cl$_2$	3	95
13	CH$_2$=CH		Pd(dppf)Cl$_2$	3	89
14	Me$_3$SiC≡C		Pd(dppf)Cl$_2$	1	93
15	n-Bu		Pd(PPh$_3$)$_2$Cl$_2$	1.5	91
16	PhC≡C	PhCH=CHBr	CO, Pd(PPh$_3$)$_4$	3	75

In addition, the reaction in the presence of a carbon monoxide atmosphere constitutes an efficient synthesis of unsymmetrical ketones in good yields. These novel features make indium organometallics promising reagents for organic synthesis.

REFERENCES

1. (a) Diederich, F. and Stang, P. J., Eds. *Metal-Catalyzed Cross-Coupling Reactions*, Wiley-VCH, Weinheim, 1998. (b) Geissler, H. In *Transition Metals for Organic Synthesis*, Beller, M., Bolm, C., Eds.; Wiley-VCH, Weinheim, 1998; Chapter 2.10, pp 158–183. (c) Tsuji, J. *Transition Metal Reagents and Catalysts*, Wiley, Chichester, U.K., 2000; Chapter 3, pp 27–108.
2. (a) Pérez, I., Pérez Sostelo, J. and Sarandeses, L. A. *Org. Lett.* **1999**, *1*, 1267–1269. (b) Pérez, I., Pérez Sostelo, J. and Sarandeses, L. A. *J. Am. Chem. Soc.* **2001**, *123*, 4155–4160. (c) Pena, M. A., Pérez, I., Pérez Sostelo, J. and Sarandeses, L. A. *Chem. Commun.* **2002**, 2246–2247.
3. Pena, M. A., Pérez Sostelo, J. and Sarandeses, L. A. *Synthesis* **2003**, 780–784.
4. Echavarren, A. M. and Stille, J. K. *J. Am. Chem. Soc.* **1987**, *109*, 5478–5486.
5. Kang, S.-K., Lim, K.-H., Ho, P.-S. and Kim, W.-Y. *Synthesis* **1997**, 874–876.

7.3   CROSS-COUPLING REACTIONS CATALYSED BY HETEROGENEOUS NICKEL-ON-CHARCOAL

BRYAN A. FRIEMAN and BRUCE H. LIPSHUTZ*

Department of Chemistry & Biochemistry, University of California, Santa Barbara, CA 93106, USA

Nickel-on-Charcoal (Ni/C) is an environmentally benign, inexpensive and recyclable heterogeneous catalyst applicable to formation of carbon-carbon, carbon-nitrogen, and carbon-hydrogen bonds. Use of nickel to mediate cross-couplings as an alternative to both palladium(0) in solution and Pd/C provides an easy entry to aryl chlorides as substrates. Thus far, Ni/C has been documented for use in Negishi-,[1a] Kumada-,[1b] and Suzuki-type couplings[1c], as well as in aminations,[1d] reductions of aryl halides,[1e] and vinyl alane-benzylic chloride couplings.[1f] Most recently, microwave-assisted vinyl and alkyl zirconocene-aryl halide bond constructions have been shown to be rapid under the influence of Ni/C.[1g] The active catalyst has been extensively studied[2] in order to assess the nature of the distribution of Ni atoms within the charcoal matrix using sophisticated spectroscopic analyses (Transmission Electron Microscopy; TEM, energy dispersion X-ray analysis, inductively coupled plasma-atomic emission spectroscopy; ICP, and React-IR). ICP-AES studies on samples from each reaction type were performed to determine the amount of nickel present in solution. The data indicated that less than three percent of the (usually *ca.* 5%) nickel on the solid support present in the reaction mixture had leached. Control experiments strongly implicate catalysis occurring largely, if not totally, in solution but within the charcoal matrix.

7.3.1 PREPARATION OF THE HETEROGENEOUS CATALYST: NICKEL-ON-CHARCOAL

$$Ni(NO_3)_2 \cdot 6H_2O \ + \ C_c \ + \ H_2O \xrightarrow[\substack{\text{2) } \Delta, \ H_2O \ \text{removal} \\ \text{3) dry under vacuum}}]{\text{1) sonication (30 min)}} Ni(II)/C_c$$

Materials and equipment

- $Ni(NO_3)_2 \cdot 6H_2O$ (92% by ICP), 727 mg, 2.30 mmol
- Activated Carbon (Darco® KB-100 mesh, 25% H_2O) 5.00 g
- Water, 75 mL

- Ultrasonic bath
- 100-mL round-bottomed flask equipped with stir bar
- 250-mL round-bottomed flask
- 50-mL round-bottomed flask
- Shortpath distillation head
- Thermometer
- Magnetic stir plate

- Argon needle
- Heating mantle with sand
- Rheostat temperature controller
- 100-mL graduated cylinder
- Micro spatula
- 150-mL course fritted funnel containing a male joint on both ends and a Vacuum output below the frit
- High-pressure vacuum line

Procedure

1. Darco® KB (5.00 g, 100 mesh) activated carbon (25% water content) was added to a 100-mL round-bottomed flask containing a stir bar. A solution of 727 mg $Ni(NO_3)_2 \cdot 6H_2O$ (Aldrich, 24,407-4, Ni content by ICP determination: 92%; 2.30 mmol) in 35 mL of deionized water was added to the activated carbon using 40 mL of deionized water to wash down the sides of the flask. The flask was purged under argon and stirred vigorously for 1 minute. The flask was then submerged in an ultrasonic bath under a positive argon flow for 30 minutes.
2. The flask was attached to an argon purged distillation setup and placed in a pre-heated 175–180 °C sand bath above a stir plate. Once the distillation ended, the flask temperature rises automatically but should be held below 220 °C for an additional 15 minutes.
3. Upon cooling to room temperature, the black solid was washed with water (2 × 50 mL) under argon into a pre-dried *in vacuo* 150-mL course fritted funnel. The 100 mL of water used to wash the Ni/C was removed *via* rotary evaporation

and examined for any remaining nickel. The fritted funnel was turned upside down under vacuum for 3 hours until the Ni/C falls from the frit into the collection flask. The collection flask is then dried *in vacuo* at 100 °C for 18 hours. Using these specific amounts, all of the nickel is impregnated on the support, which corresponds to 0.552 mmol Ni(II)/g catalyst, or 3.2% Ni/catalyst by weight.

7.3.2 Ni/C-CATALYSED SUZUKI COUPLINGS: 2-CYANOBIPHENYL

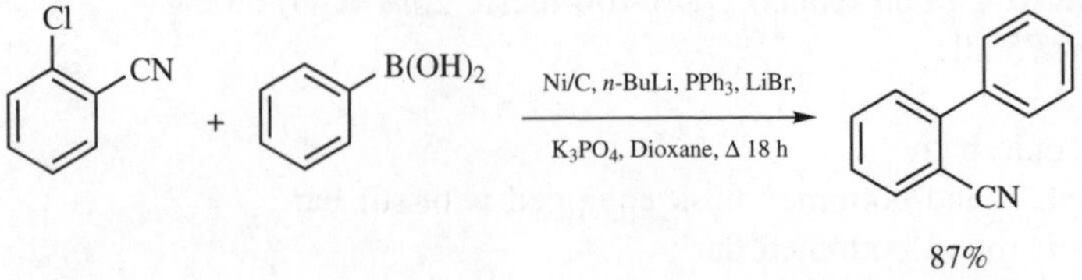

87%

Materials and equipment

- Ni/C, 96 mg, 0.75 mmol/g, 0.07 mmol
- Triphenylphosphine; recrystallized from hexanes (75 mg, 0.28 mmol)
- *n*-Butyllithium (2.55 M in hexanes, 120 μL, 0.28 mmol)
- Phenylboronic acid; 98% (135 mg, 1.08 mmol)
- 2-Chlorobenzonitrile; 99% (99 mg, 0.72 mmol)
- Dry dioxane (2.5 mL)
- Potassium phosphate; 97% (547 mg, 2.58 mmol)
- Lithium bromide; 99% (150 mg, 1.73 mmol)
- Methanol (40 mL)

- Two 10-mL round-bottomed flasks equipped with a stir bar and purged under argon
- 100-mL round-bottomed flask
- 5-mL dried glass syringe
- 250-μL dried glass syringe
- Ice water bath
- Sand bath and rheastat
- Celite®
- 60-mL course fritted funnel
- Magnetic stir plate
- Argon-purged reflux condenser
- Rotary evaporator
- Micro spatula

Procedure

1. In a 10-mL round-bottomed flask under argon triphenylphosphine (75 mg, 0.28 mmol, 0.4 equiv.) and 0.75 mmol/g Ni/C (96 mg, 0.07 mmol, 0.1 equiv.)

were combined. Dioxane (2.5 mL) was added *via* syringe and the mixture allowed to stir for 20 minutes.

2. In a second flask K_3PO_4 (547 mg, 2.58 mmol, 3.6 equiv.), lithium bromide (150 mg, 1.73 mmol), phenylboronic acid (135 mg, 1.08 mmol, 1.5 equiv.) and 2-chlorobenzonitrile (99 mg, 0.72 mmol) were combined.

3. To the flask containing the Ni/C mixture *n*-butyllithium (2.55 M, 120 μL, 0.28 mmol, 0.4 equiv.) was introduced dropwise *via* syringe to generate the active Ni(0)/C complex. The mixture was allowed to stir for 5 minutes and the flask was next placed in an ice water bath until the contents solidified.

4. The contents of the second flask were added to the frozen mixture and the flask was fitted with an argon purged condenser. The mixture was allowed to warm to room temperature and then placed in a sand bath preheated to 135 °C, where it was refluxed for 18 hours.

5. Upon cooling, the mixture was poured onto a fritted funnel containing a pad of Celite® and washed with methanol (40 mL). The filtrate was concentrated using a rotary evaporator, and collected and absorbed onto silica gel. The mixture was then subjected to column chromatography. The product was eluted with 3% EtOAc/hexanes and isolated as a white solid (112 mg, 87%).

7.3.3 Ni/C-CATALYSED AROMATIC AMINATIONS: N-(4-CYANOPHENYL)-MORPHOLINE

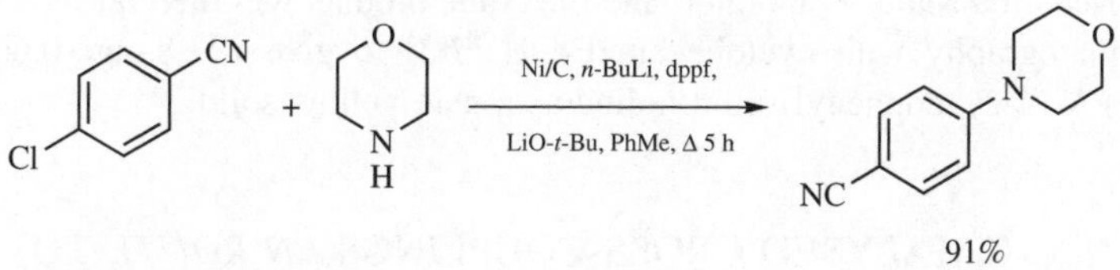

91%

Materials and equipment

- Ni/C (50.5 mg, 0.74 mmol/g, 0.038 mmol)
- dppf (10.7 mg, 0.019 mmol)
- *n*-Butyllithium (2.42 M in hexanes, 31 μL, 0.075 mmol)
- Morpholine, distilled (132 μL, 1.50 mmol)
- 4-Chlorobenzonitrile, 99% (104.2 mg, 0.75 mmol)
- Dry toluene (2.5 mL)
- Lithium *tert*-butoxide; 97% (74.3 mg, 0.90 mmol)
- Dichloromethane (40 mL)
- Methanol (40 mL)

- Two 5-mL round-bottomed flasks equipped with stir bar and purged under argon
- 100-mL round-bottomed flask
- 5-mL dried glass syringe
- Two 250-μL dried glass syringe
- Cannula (18 gauge)
- Oil bath

- Celite®
- 60-mL course fritted funnel
- Magnetic stir plate
- Argon-purged reflux condenser
- Rotary evaporator
- Micro spatula

Procedure

1. Ni(II)/C (50.5 mg, 0.038 mmol, 0.74 mmol/g), dppf (10.7 mg, 0.019 mmol), and lithium *tert*-butoxide (74.3 mg, 0.90 mmol) were added to a flame-dried 5-mL round-bottomed flask under a blanket of argon at room temperature. Dry toluene (0.5 mL) was added by syringe and the slurry allowed to stir for 90 minutes.
2. *n*-Butyllithium (31 µL, 2.42 M in hexanes, 0.075 mmol) was added dropwise with stirring.
3. After 30 minutes, 4-chlorobenzonitrile (104.2 mg, 0.75 mmol), which was dissolved in dry toluene (0.5 mL), and morpholine (132 µL, 1.50 mmol) were added *via* cannula from an argon-purged 5-mL round-bottomed flask, followed by heating to reflux for 2.5 hours (the oil bath temperature was set to 130 °C).
4. After cooling to room temperature, the crude reaction mixture was then filtered through a 60-mL course fritted funnel and the filter cake further washed with methanol and dichloromethane. The filtrate was collected, solvents were removed on a rotary evaporator, and the crude product was then purified by flash chromatography with cyclohexane/EtOH (7/3) to give 123.8 mg (0.66 mmol; 88%) *N*-(4-cyanophenyl)-morpholine as a pale yellow solid.

7.3.4 Ni/C-CATALYSED CROSS-COUPLINGS *EN ROUTE* TO ALLYLATED AROMATICS: TOLUENE-4-SULFONIC ACID 2-(3,7,11,15,19,23,27,31,35,39-DECAMETHYLTETRACONTA-2,6,10,14,18,22,26,-30,34,38-DECAENYL)-5,6-DIMETHOXY-3-METHYLPHENYL ESTER (COENZYME Q$_{10}$ PRECURSOR)

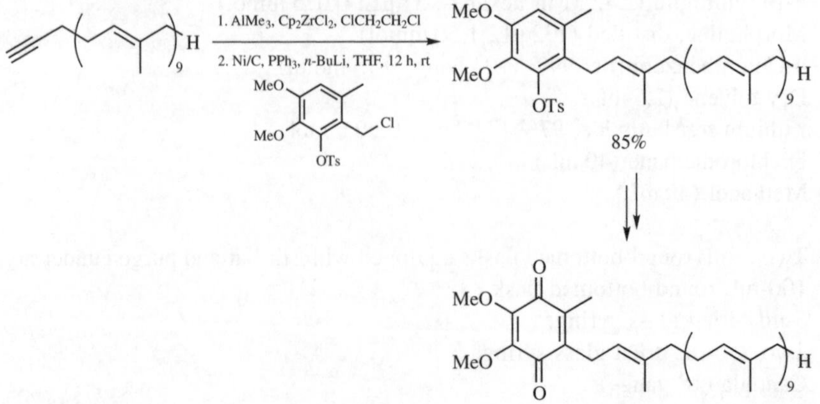

Materials and equipment

- Ni/C, 25.7 mg, 0.552 mmol/g, 0.017 mmol
- Triphenylphosphine (8.9 mg, 0.034 mmol)
- n-Butyllithium (2.45 M in hexanes, 10 μL, 0.034 mmol)
- Alkyne (200 mg, 0.31 mmol)
- Tosylated benzyl chloride (90.8 mg, 0.25 mmol)
- Dry dichloroethane (0.25 mL)
- Trimethylaluminum (2 M in hexanes, 0.31 μL, 0.61 mmol)
- Bis(cyclopentadienyl)zirconium dichloride; 97% (45 mg, 0.15 mmol)
- Dry tetrahydrofuran (1.3 mL)
- Dry pentane (2 mL)
- Dichloromethane (15 mL)
- Diethyl ether (20 mL)

- Three 10-mL round-bottomed flasks each equipped with stir bar and purged under argon
- 100-mL round-bottomed flask
- Three 2-mL dried glass syringes
- Two 250-μL dried glass syringes
- Three cannulas (18 gauge)
- Ice water bath
- Celite®
- 60-mL course fritted funnel
- Magnetic stir plate
- Rotary evaporator
- Micro spatula

Procedure

Carboalumination

1. To a 10-mL round-bottomed flask under argon, Cp_2ZrCl_2 (45 mg, 0.15 mmol) was added. A solution of Me_3Al (0.31 mL, 2.0 M in hexanes, 0.61 mmol) was added at 0 °C *via* syringe and the solution stirred under reduced pressure until the hexanes were removed.
2. Dichloroethane was added (0.25 mL) and the solution was allowed to stir and warm to room temperature over 10 minutes. To this solution the alkyne (200 mg, 0.31 mmol) was added and the mixture was stirred at 0 °C and then warmed to room temperature over 5 hours, after which time carboalumination was complete (as determined by TLC).
3. The solvent was removed *in vacuo* and freshly distilled pentane (1 mL) was added which was then also removed *in vacuo*. Additional pentane (1 mL) was then added to the flask so as to precipitate zirconium salts.
4. The vinylalane in pentane was then transferred to a second flask *via* cannula *(with great care taken to avoid contamination by the zirconium salts)*. The

yellow pentane solution was concentrated under reduced pressure and the residue dissolved in THF (0.4 mL).

Nickel-on-charcoal catalysed coupling

5. To a 10-mL round-bottomed flask, Ni/C (25.7 mg, 0.017 mmol) and Ph$_3$P (8.9 mg, 0.034 mmol) were added under argon at room temperature. Tetrahydrofuran (0.9 mL) was added followed by n-BuLi (10 μL, 2.45 M in hexanes, 0.034 mmol). The solution was allowed to stir at room temperature for 5 minutes.

6. The vinylalane was then transferred via cannula to the Ni/C catalyst at room temperature.

7. The tosylated benzyl chloride (90.8 mg, 0.25 mmol) was dissolved in THF (0.25 mL) and added via cannula at room temperature and the cross-coupling reaction followed by TLC analysis.

8. When the reaction was complete (12 hours), the solution was diluted with diethyl ether (5 mL) and quenched at 0 °C by carefully adding 1 M hydrochloric acid dropwise (1.5 mL). The mixture was filtered and the aqueous layer extracted with diethyl ether (3 × 5 mL) and dichloromethane (3 × 5 mL).

9. The combined organic layers were dried (anhydrous sodium sulfate) and concentrated in vacuo. Silica gel column chromatography of the residue (5% dichloromethane-petroleum ether) afforded 208.1 mg of the title compound (85%) as a clear oil; R$_f$ 0.28 (5% CH$_2$Cl$_2$-petroleum ether).

7.3.5   Ni/C-CATALYSED REDUCTIONS OF ARYL CHLORIDES

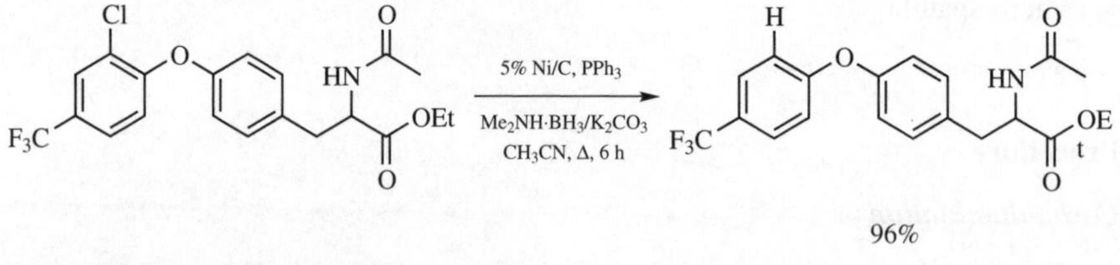

96%

Materials and equipment

- Ni/C, 77.6 mg, 0.640 mmol/g, 0.05 mmol
- Triphenylphosphine (39 mg, 0.15 mmol)
- Dimethylamine-borane complex (98%) 66 mg, 1.1 mmol
- Potassium carbonate (99%) 152 mg, 1.1 mmol
- Dry acetonitrile (2 mL)
- Aryl chloride (429 mg, 1.0 mmol)
- Dry pentane (2 mL)
- Ethyl acetate (20 mL)

- 10-mL round-bottomed flask equipped with a stir bar and purged under argon
- 100-mL round-bottomed flask

- 2-mL dried glass syringes
- Buchner funnel with filter paper
- Magnetic stir plate
- Rotary evaporator
- Micro spatula
- Argon purged reflux condenser
- Sand bath

Procedure

1. To a flame-dried, 10-mL round-bottomed flask at room temperature Ni(II)/C (77.6 mg, 0.05 mmol, 0.64 mmol/g catalyst), triphenylphosphine (39 mg, 0.15 mmol), 98% dimethylamine-borane complex (66 mg, 1.1 mmol) and potassium carbonate (152 mg, 1.1 mmol) were added, all under an argon atmosphere.
2. Dry, deoxygenated acetonitrile (2 mL) was added *via* syringe and the slurry allowed to stir for 2 hours.
3. The aryl chloride (1.0 mmol, 429 mg) was added with stirring and the mixture was then heated to reflux in a 100 °C sand bath for 6 hours.
4. The crude mixture was filtered through filter paper and the filter cake further washed with EtOAc. The filtrate was concentrated under reduced pressure and the residue purified on a silica gel column eluting with hexane-EtOAc (1 : 1) to afford the product as white crystals (380.7 mg, 96%).

7.3.6 MICROWAVE ASSISTED Ni/C-CATALYSED CROSS COUPLING OF VINYL ZIRCONOCENES AND ARYL HALIDES: 1-OCTENYL-4-TRIFLUOROMETHYLBENZENE

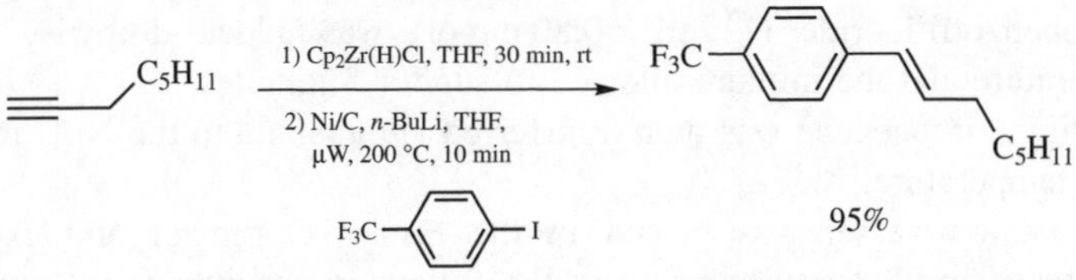

Materials and equipment

- Ni/C, 67.3 mg, 0.552 mmol/g, 0.04 mmol
- *n*-Butyllithium (2.55 M in hexanes, 31 μL, 0.08 mmol)
- 4-Iodobenzotrifluoride; distilled (117 μL, 0.80 mmol)
- 1-Octyne; distilled (149 μL, 1.00 mmol)
- $Cp_2Zr(H)Cl$ (99%) 259 mg, 1.00 mmol
- Dry tetrahydrofuran (3.0 mL)

- Diethyl ether (30 mL)
- Hexanes (30 mL)

- 10-mL round-bottomed flask equipped with a stir bar and purged under argon
- Emrys Optimizer 2-5 mL pyrex reaction vessel equipped with a stir bar and purged under argon
- 5-mL dried glass syringe
- 250-μL dried glass syringe
- Cannula (18 gauge)
- Fullers Earth
- 60-mL course fritted funnel
- Magnetic stir plate
- Rotary evaporator
- Micro spatula
- Emrys Optimizer

Procedure

Hydrozirconation

1. To a 10-mL round-bottomed flask wrapped in aluminum foil and under argon Cp$_2$Zr(H)Cl (259 mg, 1.00 mmol) was added. Tetrahydrofuran (2.0 mL) was added followed by 1-octyne (149 μL, 1.00 mmol) *via* syringe.
2. After 30 minutes, the hydrozirconation was complete by GC.

Nickel-on-charcoal catalysed coupling

3. To an Emrys Optimizer 2-5 mL pyrex reaction vessel Ni/C (67.3 mg, 0.04 mmol) was added under argon at room temperature. Tetrahydrofuran (1 mL) was added followed by *n*-BuLi (31 μL, 2.55 M in hexanes, 0.08 mmol).
4. The solution was allowed to stir at room temperature for 5 minutes after which 4-iodobenzotrifluoride (117 μL, 0.80 mmol) was added dropwise at room temperature and the mixture allowed to stir for 5 minutes.
5. The vinyl zirconocene was then transferred *via* cannula to the Ni/C mixture at room temperature.
6. The reaction vessel was placed in the Emrys Optimizer and exposed to microwave irradiation according to the following specifications: temperature: 200 °C, time: 600 seconds, fixed hold time: on, sample absorption: normal, pre-stirring: 30 seconds.
7. After cooling to room temperature, the crude reaction mixture was filtered through a glass frit containing Fullers Earth to remove the Ni/C and zirconium salts, and the filter cake was further washed with ether and hexanes.
8. The filtrate was collected, solvents were removed on a rotary evaporator, and the crude mixture was purified by flash chromatography on silica gel with pet ether. The title compound was obtained (244 mg; 95%) as a clear, viscous oil; R$_f$ 0.80 (pet ether).

CONCLUSION

Nickel-on-charcoal can be made in multi-gram quantities *via* a straightforward and reproducible procedure using readily available and inexpensive materials. The initially formed Ni(II)/C is smoothly reduced by *n*-BuLi to active Ni(0)/C, a catalyst that mediates many types of coupling reactions. It can be easily removed from reaction mixtures and recycled with little loss of catalytic activity and with similar isolated yields.[1e,f] Reaction rates involving this catalyst may be increased significantly when these couplings are performed under microwave irradiation.

REFERENCES

1. (a) Lipshutz, B. H. and Blomgren, P. A. *J. Am. Chem. Soc.* **1999**, *121*, 5819. (b) Lipshutz, B. H., Tomioka, T., Blomgren, P. A. and Sclafani, J. A. *Inorg. Chim. Acta* **1999**, *296*, 164. (c) Lipshutz, B. H., Sclafani, J. A. and Blomgren, P. A. *Tetrahedron* **2000**, *56*, 2139. (d) Lipshutz, B. H. and Ueda, H. *Angew. Chem. Int Ed.* **2000**, *39*: 4492; *Angew. Chem.* **2000**, *112*, 4666. (e) Lipshutz, B. H., Tomioka, T. and Sato, K. *Synlett* **2001**, 970. (f) Lipshutz, B. H., Frieman, B. and Pfeiffer, S. S. *Synthesis* **2002**, 2110. (g) Lipshutz, B. H. and Frieman, B. *Tetrahedron* **2004**, *60*, 1309.
2. Lipshutz, B. H., Tasler, S., Chrisman, W., Spliethoff, B. and Tesche, B. *J. Org. Chem.* **2003**, *68*, 1177.

7.4 CARBON-CARBON BOND FORMATION USING ARYLBORON REAGENTS WITH RHODIUM(I) CATALYSTS IN AQUEOUS MEDIA

JOHN MANCUSO, MASAHIRO YOSHIDA and MARK LAUTENS*

Department of Chemistry, University of Toronto, 80 St. George Street, Toronto, Ontario M5S 3H6, Canada

Rhodium catalysts solubilised in an aqueous phase through the use of water-soluble ligands (1) and (2) (Figure 7.3) allow for the addition of arylboron reagents to

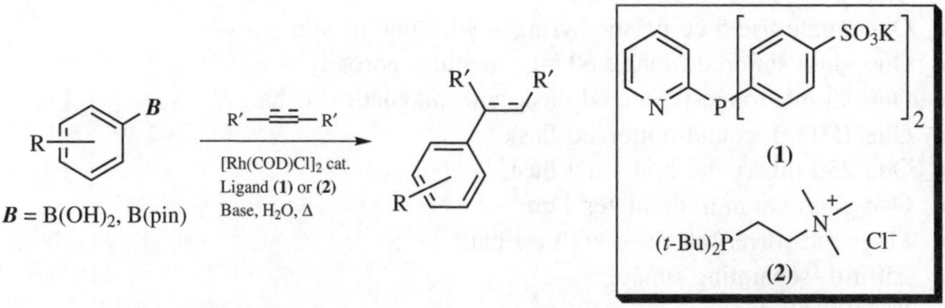

Figure 7.3 Addition of arylboron reagents to alkenes and alkynes.

carbon-carbon π-bonds. Intermolecular cross-coupling reactions using unactivated alkenes have been achieved[1] and more recently, through the use of ligand (1), additions to 2-vinyl- and 2-alkynyl-(azaheteroaromatics) have been accomplished.[2] An intramolecular cyclization protocol using a bifunctional arylboronate ester with [Rh(COD)Cl]₂/ligand (2) and a strained olefin yields functionalized indanes with high yield and diastereoselectivity.[3] Examples of both inter- and intramolecular processes using ligands (1) and (2) are described here.[2,3]

7.4.1   SYNTHESIS OF (E)-2-[2-(2-METHYLPHENYL)-1-HEXENYL]PYRIDINE

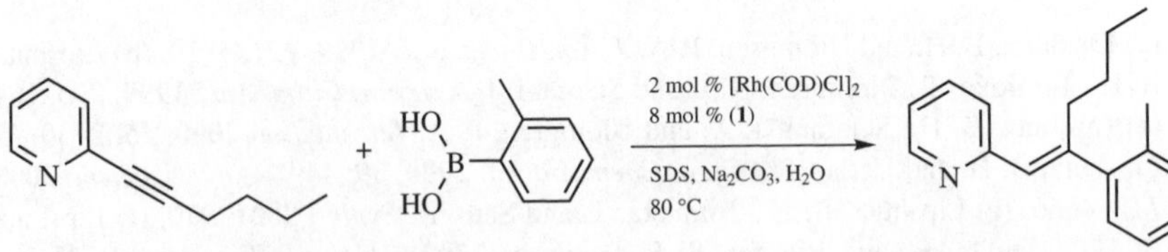

Materials and equipment

- 2-(1-Hexynyl)pyridine,[4] 47.9 mg, 0.321 mmol
- 2-Methylphenylboronic acid, 109 mg, 0.802 mmol
- [Rh(COD)Cl]₂ (98%), 3.2 mg, 6.42 µmol
- Ligand (1),[5] 12.8 mg, 25.7 µmol
- Sodium carbonate (99.5+%), 68 mg, 0.642 mmol
- Sodium dodecyl sulphate (SDS) (≥99%), 52 mg, 0.160 mmol
- Degassed distilled water, 2 mL
- Diethyl ether (technical grade)
- Saturated aqueous sodium chloride solution
- 15% Diethyl ether in hexanes
- Anhydrous magnesium sulfate

- Silica gel (Silicycle 60Å, 40–63 µm, 230–400 mesh)
- TLC plates, SIL G-60 UV$_{254}$
- One single-use 5-cc plastic syringes with one needle
- One glass sintered funnel, 60 mL, medium porosity
- One 10-mL round-bottomed flask with magnetic stir bar
- One 100-mL round-bottomed flask
- One 250-mL round-bottomed flask
- One glass column, diameter 1 cm
- Magnetic stirrer/hot plate with oil bath
- 250-mL separating funnel
- Water aspirator
- Rotary evaporator

Procedure

1. To a 10-mL round-bottomed flask, containing a stir bar, 2-(1-hexynyl)pyridine (1.0 equiv, 47.9 mg, 0.321 mmol), [Rh(COD)Cl]$_2$ (0.02 equiv, 3.2 mg, 6.42 μmol), ligand (1) (0.08 equiv, 12.8 mg, 25.7 μmol) and degassed distilled water (2 mL) were added and the suspension was left to stir under nitrogen for 5 minutes at room temperature (the rhodium complex remained as an insoluble powder). Sodium carbonate (2 equiv, 68 mg, 0.642 mmol,), sodium dodecyl sulfate (0.5 equiv, 52 mg, 0.160 mmol) and 2-methylphenylboronic acid (2.5 equiv, 109 mg, 0.802 mmol) were then added. An opaque yellow emulsion formed and was heated at 80 °C under nitrogen until all the starting material was consumed (the rhodium complex was completely dissolved in the solution at 80 °C). The progress of the reaction was monitored by TLC and was complete after 1 hour.

2. The reaction mixture was transferred to a 100-mL round-bottomed flask with stir bar through a glass funnel, diluted with 25 mL diethyl ether, and vigorously stirred for 1 hour at room temperature. The suspension was transferred to a 250-mL separating funnel and the water layer was removed. The organic layer was washed with water (10 mL) and saturated sodium chloride solution (10 mL). The organic layer was dried (magnesium sulfate), filtered through a sintered glass funnel and the filtrate concentrated under reduced pressure. The residue was purified by silica gel flash column chromatography (10 mm column, 6-inch silica height, 15% ether in hexanes) to give 61.0 mg (81%) of the title compound as a colorless oil and as a single diastereomer.

IR (neat) $\nu = 2956, 2870, 1634, 1585, 1558$ cm^{-1}.

^{1}H-NMR (300 MHz, CDCl$_3$) $\delta = 8.65$–8.63 (1H, pyridyl H, m), 7.64 (1H, pyridyl H, ddd, $J = 7.8, 7.5$ and 1.5 Hz), 7.26–7.18 (5H, pyridyl and aryl H, m), 7.13–7.09 (1H, pyridyl H, m), 6.35 (1H, alkenyl H, s), 2.90 (2H, allyl H, t, $J = 7.5$ Hz), 2.37 (3H, benzylic methyl, s), 1.44–1.23 (4H, alkyl methylene, m), 0.83 (3H, alkyl methyl, t, $J = 6.9$ Hz).

^{13}C-NMR (75 MHz, CDCl$_3$) $\delta = 156.9, 149.2, 148.4, 144.1, 135.9, 134.9, 130.1, 128.4, 128.3, 126.8, 125.3, 124.0, 120.9, 32.5, 30.1, 22.9, 20.0, 13.8$. MS m/z 251 (M$^+$).

HRMS m/z calcd for C$_{18}$H$_{21}$N$_2$ [M$^+$]: 251.1674; found 251.1679.

7.4.2 SYNTHESIS OF METHYL (2EZ)-3-[2-(4,4,5,5-TETRAMETHYL-1,3,2-DIOXABOROLAN-2-YL)PHENYL]ACRYLATE

Materials and equipment

- 2-Formylphenylboronic acid, 1.93 g, 12.9 mmol
- Pinacol (98%), 1.67 g, 14.1 mmol
- Methyl (triphenylphosphoranylidene) acetate, (98%), 5.6 g, 16.8 mmol
- Benzene, 65 mL
- Hexanes (technical grade), diethyl ether (technical grade)
- 30% Ethyl acetate in hexanes

- Silica gel (Silicycle 60Å, 40–63 μm, 230–400 mesh)
- TLC plates, SIL G-60 UV$_{254}$
- 100-mL round-bottomed flask with magnetic stir bar
- 250-mL round-bottomed flask
- Magnetic stirrer/hot plate with oil bath
- One glass sintered funnel, 60 mL, medium porosity
- Teflon® tape
- 15 mL Dean-Stark Trap (DST)
- One glass column, diameter 3 cm
- Water aspirator
- Rotary evaporator

Procedure

1. 2-Formylphenylboronic acid (1.0 equiv, 1.93 g, 12.9 mmol) was suspended in 50 mL benzene in a 100-mL round-bottomed flask with stir bar and pinacol (1.1 equiv, 1.67 g, 14.1 mmol) was added. A Dean-Stark trap (DST) was attached and filled with 15 mL benzene. The joints were sealed with Teflon® tape and the mixture was vigorously refluxed (90 °C) with stirring under air until all the calculated amount of water present had collected within the DST (ca. 4–5 hours). After cooling to room temperature, the DST was drained and removed. Methyl (triphenylphosphoranylidene) acetate (1.3 equiv, 5.6 g, 16.8 mmol) was then added in one portion, the round-bottomed flask was sealed with a rubber septum, and the mixture stirred at room temperature for 16 hours.

2. The solvent was completely removed *in vacuo* and the residue suspended in 20 mL diethyl ether with stirring for 15 minutes. The suspension was filtered through a sintered glass funnel to remove Ph$_3$P=O, washing the filter cake with diethyl ether (3 × 5 mL). The filtrate was concentrated *in vacuo* and the residue immediately purified by silica gel flash column chromatography (3.0 cm column diameter, 7-inch silica height, 30% ethyl acetate in hexanes) to give 3.24 g (87%) of the title compound as a light yellow solid which was determined to be a mixture of *trans/cis* (85 : 15) isomers. The purity is sufficient for use in the cross-coupling step. An analytically pure sample of the *trans*-isomer of the product was obtained by recrystallization from hexanes.

   IR (neat) ν = 2979, 1720, 1634, 1443, 1381, 1349, 1319, 1273, 1171, 1144, 861, 770, 656 cm^{-1}.

^{1}H-NMR (400 MHz, CDCl$_3$) $\delta = 8.58$ (1H, vinyl H, d, $J = 16.0$ Hz), 7.86 (1H, aryl H, dd, $J = 7.6$ Hz, 1.6 Hz), 7.70 (1H, aryl H, d, $J = 8.0$ Hz), 7.47 (1H, aryl H, t, $J = 7.2$ Hz), 7.39 (1H, aryl H, t, $J = 7.6$ Hz), 6.41 (1H, vinyl H, d, $J = 16.0$ Hz), 3.84 (3H, -OCH$_3$, s), 1.41 (12H, -CH$_3$, s).

^{13}C-NMR (100 MHz, CDCl$_3$) $\delta = 167.6$, 145.9, 140.2, 136.1, 131.0, 129.0, 125.6, 118.5, 84.1, 51.5, 24.8.

HRMS: m/z calcd for C$_{16}$H$_{21}$BO$_4$ [M$^+$]: 288.1541; found: 288.1532.

Note: *The product should be stored at 4 °C when not in use.*

7.4.3 SYNTHESIS OF METHYL (1S^*,4R^*,4aS^*,9S^*,9aS^*)-2,3,4,4a,9,9a-HEXAHYDRO-1H-1,4-METHANO-FLUOREN-9-YLACETATE

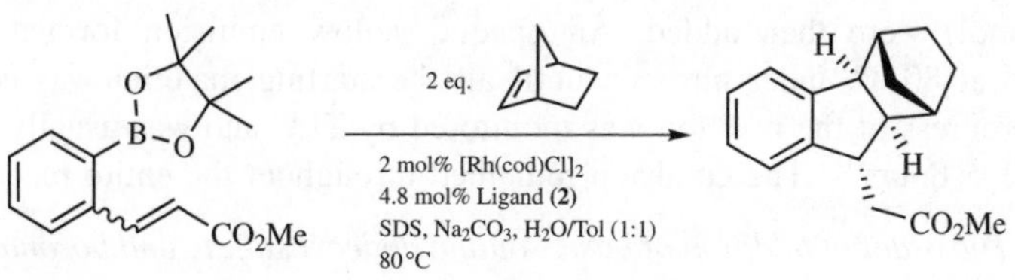

Materials and equipment

- Methyl (2EZ)-3-[2-(4,4,5,5-tetramethyl-1,3,2-dioxaborolan-2-yl)phenyl]acrylate, 115.2 mg, 0.4 mmol
- Norbornylene (99%), 75.3 mg, 0.8 mmol
- [Rh(COD)Cl]$_2$ (98%), 3.9 mg, 0.008 mmol
- Ligand (2) (*tert*-butyl AMPHOS Cl),[6] 5.1 mg, 0.019 mmol
- Sodium carbonate (99.5+%), 42 mg, 0.4 mmol
- Sodium dodecyl sulphate ($\geq$99%), 69 mg, 0.24 mmol
- Degassed ACS-grade toluene
- Degassed doubly distilled water
- Diethyl ether (technical grade)
- Saturated aqueous sodium chloride solution
- 5% Diethyl ether in hexanes
- Anhydrous magnesium sulfate

- Silica gel (Silicycle 60Å, 40–63 μm, 230–400 mesh)
- TLC plates, SIL G-60 UV$_{254}$
- Two single-use 1-cc plastic syringes with needles
- One disposable 3 dram glass vial with stir bar
- One glass sintered funnel, 60 mL, medium porosity
- One 100-mL round-bottomed flask with magnetic stir bar
- One 250-mL round-bottomed flask

- Magnetic stirrer/hot plate
- 250-mL separating funnel
- Water aspirator
- Rotary evaporator

Procedure

1. A 3 dram screwcap vial, containing a stir bar, [Rh(COD)Cl]$_2$ (0.02 equiv, 3.9 mg, 0.008 mmol) and ligand (**2**) (0.048 equiv, 5.1 mg, 0.0192 mmol, 1 : 1.2 Rh to ligand) was sealed with a rubber septum and flushed under nitrogen for 5 minutes. Degassed water (1 mL) and degassed toluene (1 mL) were injected and norbornylene (2 equiv, 75.3 mg, 0.8 mmol)[7] was added and the biphasic system left to stir under nitrogen for 5 minutes at room temperature. In order, sodium carbonate (1 equiv, 42 mg, 0.4 mmol,), sodium dodecyl sulfate (0.6 equiv, 69 mg, 0.24 mmol) and the boronate ester (1 equiv, 115.2 mg, 0.4 mmol) were then added. An opaque yellow emulsion formed and was heated at 80 °C under nitrogen until all the starting material was consumed. The progress of the reaction was monitored by TLC and was usually complete after 2–5 hours.[8] The emulsion remained throughout the entire reaction.

   Note: The order of addition of base, sodium dodecyl sulfate and boronate ester is important. This allows for the generation of the active hydroxorhodium(I) catalyst prior to the addition of boron substrate to the emulsion.

2. The reaction mixture was transferred to a 100-mL round-bottomed flask with stir bar, diluted with 50 mL diethyl ether, and stirred for 1 hour. The suspension was transferred to a 250-mL separating funnel and the water layer was removed. The organic layer was washed with water (10 mL) and saturated sodium chloride solution (10 mL). The organic layer was dried (magnesium sulfate), filtered through a sintered glass funnel and the filtrate concentrated under reduced pressure. The residue was purified by silica gel flash column chromatography (10 mm column, 6-inch silica height, 5% ether in hexanes) to give 97 mg (95%) of the title compound as a colorless oil and as a single diastereomer.

   IR (neat) ν = 3068, 3020, 2950, 2870, 1739, 1481, 1455, 1436, 1363, 1337, 1253, 1163, 1102, 1024, 987, 848, 751 cm^{-1}.

   ^{1}H-NMR (400 MHz, CDCl$_3$): δ = 7.08–7.18 (4H, aryl H, m), 3.73 (3H, —OCH$_3$, s), 3.24 (1H, methine CH, t, $J = 5.2$ Hz), 3.14 (1H, methine CH, d, $J = 7.6$ Hz), 2.73 (1H, -CH$_2$-, dd, $J = 5.2$ Hz, 15.6 Hz), 2.47 (1H, -CH$_2$-, dd, $J = 9.6$ Hz, 15.6 Hz), 2.25 (2H, bridgehead CH, d, $J = 8.4$ Hz), 2.02 (1H, -CH$_2$-, dd, $J = 3.6$, 7.2 Hz), 1.51–1.60 (2H, -CH$_2$-, m), 1.37 (1H, -CH$_2$-, t, $J = 10.4$ Hz), 1.25 (1H, -CH$_2$-, t, $J = 9.2$ Hz), 1.02 (2H, apical CH, AB system, dd, $J = 10.4$ Hz, 25.4 Hz).

   ^{13}C-NMR (100 MHz, CDCl$_3$): δ = 173.0, 146.4, 146.1, 127.1, 126.5, 124.6, 123.3, 54.4, 53.1, 51.5, 47.6, 43.2, 43.1, 42.1, 32.7, 28.9, 28.8.

   HRMS: *m/z* calcd for C$_{17}$H$_{20}$O$_2$ [M$^+$]: 256.1463; found: 256.1459.

REFERENCES

1. Lautens, M., Roy, A., Fukuoka, K., Fagnou, K., and Martín-Matute, B. *J. Am. Chem. Soc.* **2001**, *123*, 5358.
2. (a) Lautens, M., and Yoshida, M. *Org. Lett.* **2002**, *4*, 123. (b) Lautens, M., and Yoshida, M. *J. Org. Chem.* **2003**, *68*, 762.
3. (a) Lautens, M. and Mancuso, J. *Org. Lett.* **2002**, *4*, 2105: Lautens, M. and Mancuso, J. *J. Org. Chem.* **2004**, *69*, 3478.
4. For the preparation of 2-(1-hexynyl)pyridine, see: Sato, N., Hayakawa, A., and Takeuchi, R. *J. Heterocyl. Chem.* **1990**, *27*, 503.
5. For the preparation of ligand (**1**), see: Herd, O., Langhans, P. K., and Stelzer, O. *Angew. Chem. Int. Ed.* **1993**, *32*, 1058.
6. For the preparation of ligand (**2**), see: Shaughnessy, K. H., and Booth, R. S. *Org. Lett.* **2001**, *3*, 2757.
7. Two equivalents of norbornylene were used to ensure reaction completion, however the loading can be reduced to 1.05 equivalents with a slight decline in yield (ca. 2–5%).
8. The reaction can be performed at room temperature, but requires longer time (12–16 hours).

8 Regioselective or Asymmetric 1,2-Addition to Aldehydes

CONTENTS

Catalysts for Fine Chemical Synthesis, Vol. 3, Metal Catalysed Carbon-Carbon Bond-Forming Reactions
Edited by S. M. Roberts, J. Xiao, J. Whittall, and T. Pickett
© 2004 John Wiley & Sons, Ltd   ISBN: 0-470-86199-1

8.1   DEVELOPMENT OF A HIGHLY REGIOSELECTIVE METAL-MEDIATED ALLYLATION REACTION IN AQUEOUS MEDIA

KUI-THONG TAN, SHU-SIN CHNG, HIN-SOON CHENG AND TECK-PENG LOH

Department of Chemistry, National University of Singapore, 3 Science Drive 3, Singapore 117543

Being important building blocks and versatile synthons, homoallylic alcohols are featured in the organic syntheses of many biological active molecules such as macrolides, polyhydroxylated natural products and polyether antibiotics. Among the existing means to construct these synthetically and biologically important molecules, metal-mediated Barbier-type allylation is one of the easiest and most convenient methods.

Such reactions have provided an easy access to homoallylic alcohols by adding reagents and metal together at room temperature using environmentally benign solvents like water and/or ethanol. Metals like indium, zinc, and tin are always used in typical metal-mediated Barbier-type allylation with carbonyl compounds and allylic halides. Although this Barbier-type allylation has been found to be highly regio- and stereoselective, one severe limitation inherent in this reaction is the difficulty in obtaining the linear adduct when allylic metals are employed.[1]

Herein, a general strategy to obtain homoallylic alcohols using three different metals, indium, tin and zinc, in aqueous medium is reported.[2]

8.1.1   SYNTHESIS OF 1-CYCLOHEXYLPENT-3-EN-1-OL USING INDIUM-MEDIATED ALLYLATION

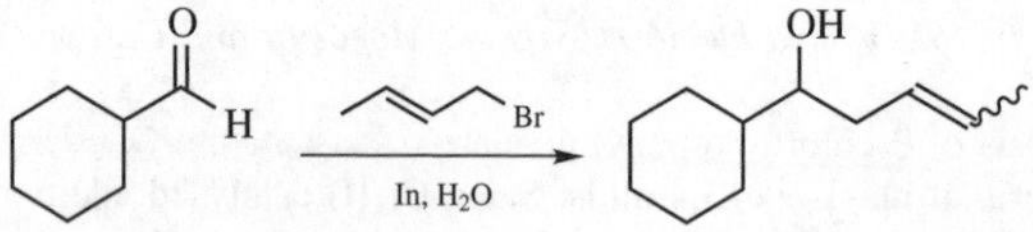

Materials and equipment

- Cyclohexane carboxaldehyde (98% Purity from Sigma-Aldrich)
- Indium powder (100 mesh, from Sigma-Aldrich)
- Crotyl bromide (85% Purity, from Sigma-Aldrich)
- Distilled water
- Anhydrous ether
- 1M Hydrochloric acid
- Saturated sodium chloride solution
- Anhydrous magnesium sulfate
- Ceric molybdate stain

- 5-mL round-bottomed flask
- 50-mL water bath
- Magnetic stirring bar
- Magnetic stirring plate
- Filter Paper
- 100-mL separating funnel
- Merck 60 F_{254} precoated silica gel plate (0.2 mm thickness)
- UV radiator Spectroline Model ENF-24061/F 254 nm
- Merck silica gel 60
- Infrared spectra, BIO-RAD FTS 165 FTIR spectrometer
- ^{1}H NMR, Bruker ACF 300 and Bruker Avance DPX 300
- ^{13}C NMR, Bruker Avance DPX 300
- Mass spectra, VG 7035 micromass mass spectrometer

Procedure

A mixture of cyclohexane carboxaldehyde (0.11 g, 1 mmol) and indium powder (0.17 g, 100 mesh, 1.5 mmol) in water (0.11 ml, 6 mmol) was added crotyl bromide (0.16 g, 1.2 mmol) slowly at room temperature in 5-mL round-bottomed flask. The reaction mixture was stirred for 12 hours followed by heated at 40 °C and monitored by TLC[3] until all the adduct was converted into its linear regioisomer. Ether was added to dilute the reaction mixture followed by 1 M hydrochloric acid to quench the reaction. The reaction mixture was extracted with ether. The combined organic layer was washed with brine, and dried over anhydrous magnesium sulfate, filtered and the solvent was removed *in vacuo*.

The crude product was purified by column chromatography (100 mL of hexane followed by 300 mL of Hexane/Ether: 20/1, $Rf = 0.45$, 4:1 hexane/ethyl acetate), to afford 85% (143 mg) yield of homoallylic alcohol (99% regioselectivity) (Entry 2, Table 8.1) as a colorless oil.

^{1}H NMR (300 MHz, CDCl$_3$) E isomer: δ 5.61–5.38 (m, 1H), 3.39–3.29 (m, 1H), 2.24 (t, $J = 8.4$ Hz, 1H), 2.05 (t, $J = 8.4$ Hz, 1H), 1.69 (dd, $J = 5.6$, 0.7 Hz, 3H), 1.87–0.94 (m, 11H). Z isomer: δ 1.64 (dd, $J = 5.9$, 0.7 Hz, 3H).

^{13}C NMR (75.4 MHz, CDCl$_3$) E isomer: δ 128.7, 127.6, 74.9, 42.9, 37.4, 29.0, 28.1, 26.5, 26.2, 26.1, 18.0; Z isomer: δ 127.0, 126.6, 75.5, 42.59, 37.46, 31.7, 29.0, 28.1, 26.5, 26.2, 26.1.

FTIR (film) 3591, 3422, 3020, 2925, 2854, 1625, 1448, 1261, 1086, 1065, 1030, 968, 892 cm^{-1}.

HRMS Calcd for C$_{11}$H$_{20}$O [M$^+$]: 168.1514. Found: 168.1520.

8.1.2 SYNTHESIS OF 1-CYCLOHEXYLPENT-3-EN-1-OL USING TIN-MEDIATED ALLYLATION

Materials and equipment

- Cyclohexane carboxaldehyde (98% Purity from Sigma-Aldrich)
- Tin powder (100% purity, from Fisher Scientific)
- Crotyl bromide (85% Purity, from Sigma-Aldrich)
- Distilled water
- Anhydrous dichloromethane distilled over calcium hydride
- Anhydrous ether
- Ethyl acetate
- Saturated sodium chloride solution
- Anhydrous magnesium sulfate
- Ceric molybdate stain

- 10-mL round-bottomed flask
- Glass septum
- 50-mL water bath
- Magnetic stirring bar
- Magnetic stirring plate
- Filter paper
- 100-mL separating funnel
- Merck 60 F_{254} precoated silica gel plate (0.2 mm thickness)
- UV radiator Spectroline Model ENF-24061/F 254 nm
- Merck silica gel 60
- Infrared spectra, BIO-RAD FTS 165 FTIR spectrometer
- ^{1}H NMR, Bruker ACF 300 and Bruker Avance DPX 300
- ^{13}C NMR, Bruker Avance DPX 300
- Mass spectra, VG 7035 micromass mass spectrometer

Procedure

A mixture of cyclohexanecarboxaldehyde (0.120 mL, 1.0 mmol, 1 equiv.), tin powder (0.178 g, 1.5 mmol, 1.5 equiv.) and crotyl bromide (0.125 mL, 1.2 mmol, 1.2 equiv.) in dichloromethane (0.385 mL, 6 equiv.) and water (0.108 mL, 6 equiv.) was stirred in a 10-mL round-bottomed flask equipped with a glass septum at room temperature for 48 hours. 50-mL water was added and the reaction mixture extracted with 50 mL portions of ethyl acetate. The combined organic layer was washed with 50 mL brine and 50 mL water and dried over anhydrous magnesium sulfate. The mixture was then filtered and the solvent removed *in vacuo*. The crude product was purified by column chromatography to afford a 78% (131 mg) yield mixture of (95%) and (5%) homoallylic alcohols (Entry 13, Table 8.1) as a colorless oil.

8.1.3   SYNTHESIS OF 1-CYCLOHEXYLPENT-3-EN-1-OL USING ZINC-MEDIATED ALLYLATION

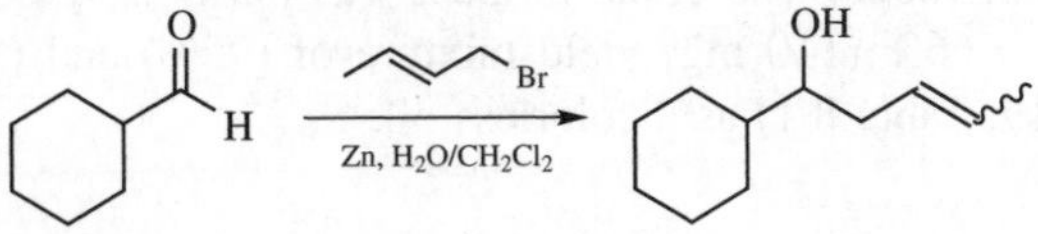

Materials and equipment

- Cyclohexane carboxaldehyde (98% Purity from Sigma-Aldrich)
- Zinc powder (>95% Purity, <45 mesh from Merck)
- Crotyl bromide (85% Purity, from Sigma-Aldrich)
- Distilled water
- 5% Hydrochloric acid
- Anhydrous dichloromethane distilled over calcium hydride
- Anhydrous ether
- Saturated sodium bicarbonate
- Saturated sodium chloride solution
- Anhydrous magnesium sulfate
- Ceric molybdate stain

- 5-mL round-bottomed flask
- Glass septum
- 50-mL water bath
- Magnetic stirring bar
- Magnetic stirring plate
- Filter paper
- 100-mL separating funnel
- Merck 60 F_{254} precoated silica gel plate (0.2 mm thickness)
- UV radiator Spectroline Model ENF-24061/F 254 nm
- Merck silica gel 60
- Infrared spectra, BIO-RAD FTS 165 FTIR spectrometer
- ^{1}H NMR, Bruker ACF 300 and Bruker Avance DPX 300
- ^{13}C NMR, Bruker Avance DPX 300
- Mass spectra, VG 7035 micromass mass spectrometer

Procedure

A mixture of cyclohexanecarboxaldehyde (0.120 mL, 1.0 mmol, 1 equiv.), zinc powder (0.098 g, 1.5 mmol, 1.5 equiv.) and crotyl bromide (0.125 mL, 1.2 mmol, 1.2 equiv.) in dichloromethane (0.128 mL, 2 equiv.) and water (0.036 mL, 2 equiv.) was stirred in a 5-mL round-bottomed flask equipped with a glass septum at room temperature for 120 hours (monitored by TLC). The reaction mixture was diluted with 50 mL diethyl ether and then extracted with 10 mL of 5% hydrochloric acid.

The ether layer was washed with 20 mL sodium bicarbonate, followed with 20 mL brine and dried over anhydrous magnesium sulfate. The mixture was filtered and the solvent removed *in vacuo*. The crude product was purified by column chromatography to afford a 66% (110 mg) yield mixture of (95%) and (5%) homoallylic alcohols (Entry 19, Table 8.1) as a colorless oil.

CONCLUSION

In conclusion, a general regioselective metal-mediated allylation reaction has been developed. It has been demonstrated that indium, tin and zinc can be used to carry out this reaction, giving moderate to good yields and selectivities. Table 8.1 gives a brief summary of the different aldehydes and allyl bromides which are used to synthesize homoallylic alcohols in this procedure.

Table 8.1　Metal-mediated allylation of aldehydes with bromides.

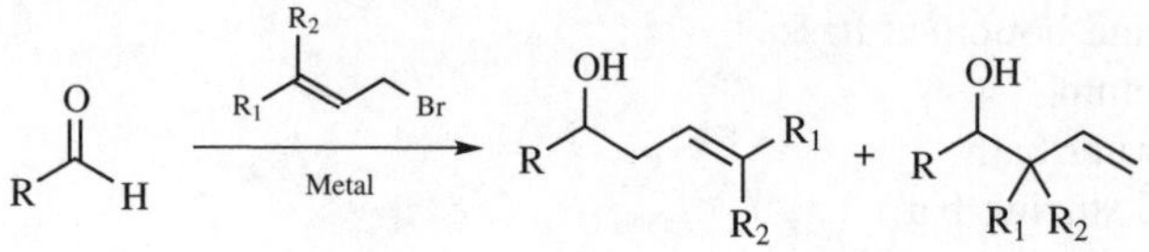

Entry	R	R_1	R_2	Metal	Yield % (1:2) (E/Z)
1	Ph	Me	H	In	60 (99:1) (55/45)
2	c-C_6H_{11}	Me	H	In	85 (99:1) (70/30)
3	n-C_5H_{11}	Me	H	In	75 (98:2) (65/35)
4	$PhCH_2CH_2$	Me	H	In	67 (97:3) (55/45)
5	Ph	Ph	H	In	66 (98:2) (E)
6	c-C_6H_{11}	Ph	H	In	73 (96:4) (98/2)
7	n-C_5H_{11}	Ph	H	In	71 (99:1) (90/10)
8	$PhCH_2CH_2$	Ph	H	In	50 (99:1) (95/5)
9	c-C_6H_{11}	CO_2Et	H	In	90 (80:20) (80/20)
10	Me	CO_2Et	H	In	56 (85:15) (90/10)
11	c-C_6H_6	Me	Me	In	30 (95:5)
12	Ph	Me	H	Sn	83 (55:45) (75:25)
13	c-C_5H_{11}	Me	H	Sn	78 (95:5) (67/33)
14	Ph CH_2CH_2	Me	H	Sn	82 (71:29) (53/47)
15	t-Bu	Me	H	Sn	42 (99:1) (83/17)
16	c-C_5H_{11}	Ph	H	Sn	54 (99:1) (100:0)
17	Ph	Ph	H	Sn	80 (99:1) (100:0)
18	t-Bu	Me	H	Zn	34 (45:55) (62/38)
19	c-C_6H_{11}	Me	H	Zn	66 (95:5) (60/40)
20	Ph	Me	H	Zn	50 (97:3) (77/23)

REFERENCES

1. (a) Yamamoto, Y., and Asao, N. *Chem. Rev.* **1993**, *93*, 2207. (b) Li, C. J., and Chan, T. H. *Organic Reactions in Aqueous Media, John Wiley & Sons, New York*, **1997**.
2. (a) Tan, K. T., Chng, S. S., Cheng, H. S., and Loh, T. P. *J. Am. Chem. Soc.* **2003**, *125*, 2958. (b) Loh, T. P., Tan, K. T., Yang, J. Y., and Xiang, C. L. *Tetrahedron Lett.* **2001**, *42*, 8701. (c) Loh, T. P., Tan, K. T., and Hu, Q. Y. *Tetrahedron Lett.* **2001**, *42*, 8705.
3. Reaction must be closely monitored by TLC, in order to obtain high regioselectivity of homoallylic alcohols.

8.2 BORONIC ACIDS AS ARYL SOURCE FOR THE CATALYSED ENANTIOSELECTIVE ARYL TRANSFER TO ALDEHYDES

JENS RUDOLPH AND CARSTEN BOLM*

Department of Organic Chemistry RWTH Aachen, Professor Pirlet Str. 1, 52074 Aachen, Germany

Enantiopure diaryl methanols (**1**) are important intermediates for the synthesis of biologically active compounds.[1] Here, a synthetic approach which employs aromatic aldehydes and readily accessible and commercially available aryl boronic acids is reported.[2,3] The catalysis with planar chiral ferrocene (**2**)[4] yields a broad range of optically active diaryl methanols with excellent enantioselectivities. Upon addition of DiMPEG (dimethoxy polyethyleneglycol) to the reaction mixture the enantioselectivity is increased.

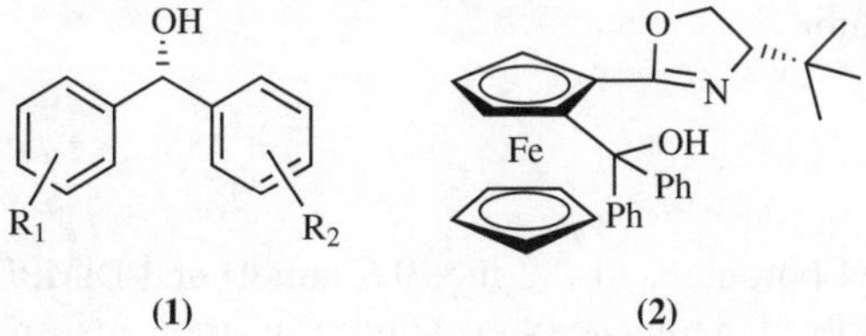

(1)   (2)

8.2.1 PREPARATION OF (*S*)-4-TOLYL-PHENYL METHANOL

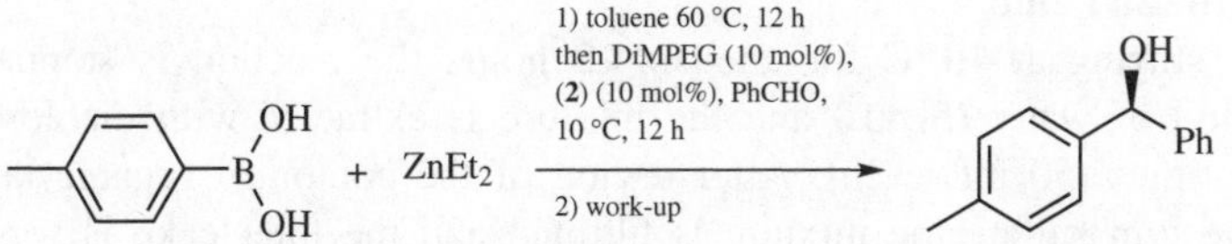

91% yield, 96% ee

Materials and equipment

- 4-Methylphenylboronic acid, 82 mg, 0.6 mmol
- Diethyl zinc, 184 μL, 1.8 mmol
- Dry toluene, 5 mL
- (S, R_p)-2-(α-Diphenylhydroxymethyl)-ferrocenyl-(4-*tert*-butyl)-oxazoline, 12 mg, 0.025 mmol
- DiMPEG, 50 mg, 0.025 mmol, M = 2000 g/mol
- Benzaldehyde, 25 μL, 0.25 mmol
- Dichloromethane, 200 mL
- Water, 5 mL
- Pentane, 300 mL
- Diethyl ether, 60 mL
- Magnesium sulfate, 30 g

- One Glass vial, 2 cm diameter, 5 cm high
- One Teflon® plug for the vial
- One Metal seal ring
- One Stirring bar 0.3 cm diameter, 1 cm
- Two Hamilton syringes (100 μL and 250 μL)
- Hamilton needle (10 cm)
- Silica gel (Merck 40–63 μm), 25 g
- TLC plates, Merck F_{254}
- Magnetic stirrer plate
- One 250-mL Erlenmeyer flask
- Filter paper
- One 100-mL separating funnel
- One glass column, diameter 2 cm
- Rotary evaporator
- Cryostat

Procedure

1. 4-Methylphenyl boronic acid (82 mg, 0.6 mmol) and DiMPEG (50 mg, 0.025 mmol) are dissolved in toluene (4 mL) under an argon atmosphere. Then diethyl zinc (184 μL, 1.8 mmol) is added carefully using a Hamilton syringe. The resulting mixture is stirred for 12 hours at 60 °C. After cooling to room temperature ferrocene (**2**) is added as dry toluene solution (1 mL) and stirred for 15 minutes. Then the reaction mixture is cooled to 10 °C and benzaldehyde (25 μL, 0.25 mmol) is added in one portion. **Ensure careful handling of the neat diethyl zinc!**

2. After stirring at 10 °C for another 12 hours the reaction is stopped by the addition of water (5 mL) and the mixture is extracted with dichloromethane three times (50 mL each). After drying of the combined organic layers with magnesium sulfate the mixture is filtrated and the filter cake is washed with dichloromethane (50 mL). After removal of the organic solvents the residue is

purified by column chromatography with pentane/diethyl ether (85:15) and the desired product is obtained as a slowly crystallizing white solid in 45 mg (0.22 mmol, 91% yield).

The yields and the enantioselectivities depend on the quality of the aryl boronic acid. Rigorously dried boronic acids do not seem to be the reagents of choice.

CONCLUSION

The reported procedure allows the transfer of aryl moieties to aromatic aldehydes affording diaryl methanols with excellent enantiomeric excesses in high yields (Table 8.2). It relys on the use of a readily available ferrocene catalyst and

Table 8.2   Aryl transfer to benzaldehyde using DiMPEG (10 mol%), $ZnEt_2$ and ferrocene (**2**), as well as various aryl boronic acids as aryl source.

$$ArB(OH)_2 \ + \ ZnEt_2 \xrightarrow[\text{2) work-up .}]{\substack{\text{1) toluene 60 °C, 12 h,} \\ \text{then DiMPEG (10 mol%),} \\ \text{(2) (10 mol%), PhCHO,} \\ \text{10 °C, 12 h}}} \underset{Ph \quad Ar}{\overset{OH}{\diagup}}$$

Entry	ArB(OH)$_2$ with Ar =	Yield (%)	ee (%)
1	4-chlorophenyl	93	97 (*S*)
2	4-biphenyl	75	97
3	4-methylphenyl	91	96
4	4-methoxyphenyl	86	90
5	1-naphthyl	91	85
6	2-bromophenyl	58	88
7	4-bromophenyl	48	96

avoids the expensive and difficult to handle diphenyl zinc. Interestingly, by the appropriate choice of the boronic acid and the aldehyde both enantiomers of the diaryl methanols are available with the same catalyst.

REFERENCES

1. (a) Meguro, K., Aizawa, M., Sohda, T., Kawamatsu, Y., and Nagaoka, A. *Chem. Pharm. Bull.* **1985**, *33*, 3787. (b) Toda, F., Tanaka, K., and Koshiro, K. *Tetrahedron: Asymmetry* **1991**, *2*, 873. (c) Stanev, S., Rakovska, R., Berova, N., and Snatzke, G. *Tetrahedron: Asymmetry* **1995**, *6*, 183. (d) Botta, M., Summa, V., Corelli, F., Pietro, G. Di, and Lombardi, P. *Tetrahedron: Asymmetry* **1996**, *7*, 1263.
2. Bolm, C. and Rudolph, J. *J. Am. Chem. Soc.* **2002**, *124*, 14850.
3. For previous work on catalysed aryl transfer reactions with ferrocene (**2**) and related systems, see: (a) Bolm, C. and Muñiz, K. *Chem. Commun.* **1999**, 1295. (b) Bolm, C., Hermanns, N., Hildebrand, J. P., and Muñiz, K. *Angew. Chem. Int. Ed.* **2000**, *39*, 3465.

(c) Bolm, C., Kesselgruber, M., Hermanns, N., and Hildebrand, J. P. *Angew. Chem. Int. Ed.* **2001**, *39*, 1488. (d) Bolm, C., Kesselgruber, M., Grenz, A., Hermanns, N., and Hildebrand, J. P. *New J. Chem.* **2001**, *25*, 13. (e) Bolm, C., Hermanns, N., Kesselgruber, M., and Hildebrand, J. P. *J. Organomet. Chem.* **2001**, *624*, 157. (f) Review: Bolm, C., Hildebrand, J. P., Muñiz, K., and Hermanns, N. *Angew. Chem. Int. Ed.* **2001**, *40*, 3284.
4. For the synthesis of ferrocene (**2**), see: (a) Bolm, C., Muñiz Fernandez, K., Seger, A., and Raabe, G. *Synlett* **1997**, 1051. (b) Bolm, C., Muñiz-Fernandez, K., Seger, A., Raabe, G., and Günther, K. *J. Org. Chem.* **1998**, *63*, 7860.

8.3 JACOBSEN'S SALEN AS A CHIRAL LIGAND FOR THE CHROMIUM-CATALYSED ADDITION OF 3-CHLORO-PROPENYL PIVALATE TO ALDEHYDES: A CATALYTIC ASYMMETRIC ENTRY TO SYN-ALK-1-ENE-3,4-DIOLS

MARCO LOMBARDO*, SEBASTIANO LICCIULLI, STEFANO MORGANTI AND CLAUDIO TROMBINI*

Dipartimento di Chimica "G. Ciamician", Università di Bologna, via Selmi 2, I-40126, Bologna, Italy

The Nozaki-Hiyama-Kishi (NHK) reaction involves the mild addition of chromium(II) organometallics to aldehydes to give homoallylic alcohols in a regio- and stereo-controlled fashion.[1] A very significant achievement in the chromium organometallic chemistry was accomplished by Fürstner who developed a catalytic version of the NHK reaction based on the coupled use of the redox Mn(0)/Cr(III) couple and trimethylsilyl chloride (TMSCl).[2] Moreover, the integration of the Fürstner protocol with the addition of the Jacobsen's Salen [*N,N'*-bis(3,5-di-*t*-butylsalicylidene)-1,2-cyclohexanediamine] and triethylamine allowed Cozzi, Umani-Ronchi, *et al.* to develop a catalytic enantioselective route to homoallylic alcohols.[3]

3-Halopropenyl esters (**1**) (Figure 8.1),[4] have recently been brought to the attention of synthetic chemists as precursors of heterofunctionalised allylic

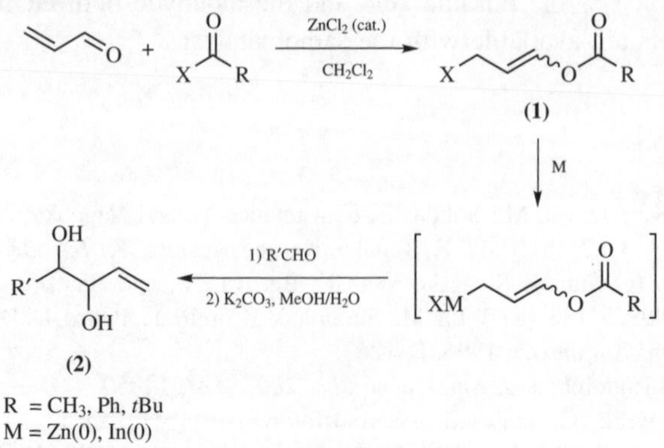

(1)

(2)

R = CH₃, Ph, *t*Bu
M = Zn(0), In(0)

Figure 8.1   Synthesis of alk-1-ene-3,4-diols.

organometallic compounds under typical Barbier or Grignard conditions. Using indium(0) or zinc(0), the resulting organometallic species were found to add efficiently to carbonyl compounds to give racemic *syn-* or *anti*-alk-1-ene-3,4-diols (**2**), in both aqueous and organic solvents.[5]

When the Salen-Cr(II) catalysed system is applied to the reaction of 3-chloropropenyl pivalate (**1a**) (R = *t*-Bu, X = Cl) with aldehydes, (S,S)-*syn*-alk-1-ene-3,4-diols (**2**) are produced in 50–70% d.e. and in high ee's, particularly when aliphatic aldehydes are used (ee >90%). Up to now, this represents the first catalytic asymmetric entry to *syn*-alk-1-ene-3,4-diols (Figure 8.2).[6]

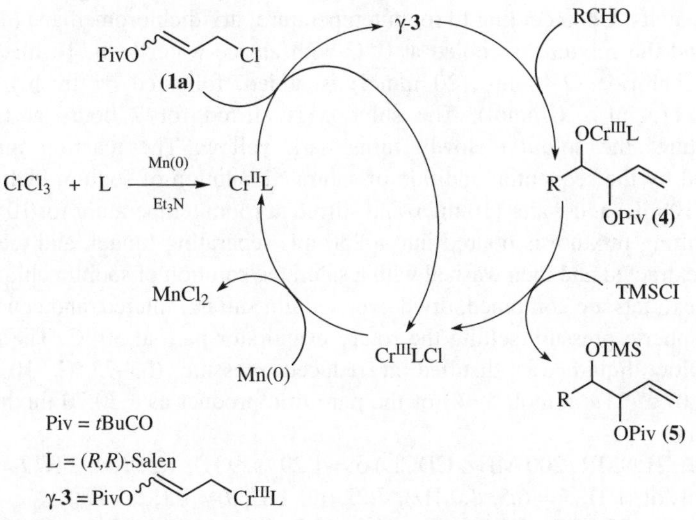

Figure 8.2 Catalytic cycle for the conversion of 3-chloropropenyl pivalate to a protected alk-1-ene-3,4-diol (**5**).

8.3.1   SYNTHESIS OF 3-CHLORO-PROPENYL PIVALATE

Materials and equipment

- Pivaloyl chloride (Fluka, 98%), 2.46 mL, 20 mmol
- Zinc chloride (Aldrich), 0.027 g, 0.2 mmol
- Acrolein (Fluka, ~95%), 1.4 mL, 20 mmol
- Dry dichloromethane (Fluka, H_2O < 0.005%), 40 mL
- Saturated solution of sodium hydrogencarbonate
- Saturated solution of sodium chloride
- Anhydrous sodium sulfate

- Two-necked 100-mL round-bottomed flask equipped with a magnetic stirrer bar
- Magnetic stirrer plate

- One 250-mL Erlenmeyer flask
- One 250-mL separating funnel
- Rotary evaporator
- Distillation equipment

Procedure

An oven dried, two-necked 100-mL round-bottomed flask equipped with a magnetic stirrer bar, bearing an argon inlet/outlet vented through a mineral oil bubbler, is placed under an argon atmosphere and charged with zinc chloride (0.027 g, 0.2 mmol). The flask is heated with a heat-gun under argon until all zinc chloride melts. Upon cooling to room temperature, dry dichloromethane (40 mL) is added and the mixture is cooled at 0 °C with an ice-water bath. To this solution pivaloyl chloride (2.46 mL, 20 mmol) is added, followed by freshly distilled acrolein (1.4 mL, 20 mmol). The solution is stirred for 2 hours at the same temperature; the solution slowly turns dark yellow. The reaction mixture is quenched by the sequential addition of saturated solution of sodium hydrogencarbonate (10 mL) and water (10 mL), and stirred at room temperature for 10 minutes. The resulting mixture is poured into a 250-mL separating funnel, and the organic layer is extracted and then washed with a saturated solution of sodium chloride. The organic extracts are combined, dried over sodium sulfate, filtered, and concentrated at atmospheric pressure setting the rotary evaporator bath at 50 °C. The resulting pale yellow liquid was distilled at reduced pressure (65–75 °C, 10 torr) to afford 2.07 g (11.8 mmol, 59%) of the pure title product as a 30:70 mixture of E/Z isomers.

(Z)-**1a**: ^{1}H NMR (200 MHz, CDCl$_3$) $\delta = 1.29$ (s, 9 H), 4.21 (dd, 2 H, $J = 1.1$, 8.0 Hz), 5.18 (dt, 1 H, $J = 6.5$, 8.0 Hz), 7.22 (dt, 1 H, $J = 1.1$, 6.5 Hz).
^{13}C NMR (75 MHz, CDCl$_3$) $\delta = 36.2$, 59.3, 69.0, 110.1, 135.8, 166.6.

(E)-**1a**: ^{1}H NMR (200 MHz, CDCl$_3$) $\delta = 1.26$ (s, 9 H), 4.11 (dd, 2 H, $J = 1.1$, 8.0 Hz), 5.64(dt, 1 H, $J = 8.0$, 12.7 Hz), 7.22 (dt, 1 H, $J = 1.1$, 12.7 Hz).
^{13}C NMR (75 MHz, CDCl$_3$) $\delta = 40.7$, 59.2, 69.0, 111.7, 138.2, 166.9.

8.3.2   SYNTHESIS OF ALK-1-ENE–3,4-DIOLS: SALEN-Cr(II) CATALYSED ADDITION OF 3-CHLORO-PROPENYL PIVALATE TO CYCLOHEXANECARBOXALDEHYDE (TABLE 1, ENTRY 1)

Materials and equipment

- Chromium(III) chloride (Aldrich, 99%), 0.016 g, 0.1 mmol
- Manganese chips (Aldrich, 99.98%), 0.110 g, 2 mmol
- (R,R)-Salen (Fluka, >98%), 0.110 g, 0.2 mmol
- 3-Chloro-propenyl pivalate **1a**, 0.245 mL, 1.5 mmol
- Trimethylsilyl chloride (Fluka, 99%), 0.152 mL, 1.2 mmol
- Triethylamine, 0.056 mL, 0.4 mmol

- Cyclohexanecarboxaldehyde (Fluka), 0.120 mL, 1.0 mmol
- Dry acetonitrile (Aldrich, 99.8%), 2.5 mL
- Celite® 521 (Aldrich)
- Saturated solution of sodium hydrogencarbonate
- Anhydrous sodium sulfate
- Silica gel (Aldrich, Merck grade 9385, 230–400 mesh, 60A)
- Alumina (Aldrich, Brockmann I, 150 mesh, 58A)

- Two-necked 10-mL round-bottomed flask equipped with a magnetic stirrer bar
- Magnetic stirrer plate
- One 50-mL Erlenmeyer flask
- One 25-mL separatory funnel
- One glass sintered filter funnel, porosity 3
- Rotary evaporator
- One glass column, diameter 2 cm

Procedure

An oven dried, two-necked 10-mL round-bottomed flask equipped with a magnetic stirrer bar, bearing an argon inlet/outlet vented through a mineral oil bubbler, is placed under an argon atmosphere and charged with chromium(III) chloride (0.016 g, 0.1 mmol). The flask is heated with a heat-gun under argon at about 300 °C for 5 minutes. Upon cooling to room temperature, dry acetonitrile (2.5 mL) and freshly crushed manganese (0.110 mg, 2 mmol) are added. The mixture is stirred for about 1 minute, and is then left undisturbed for 5 minutes and again vigorously stirred for 2 hours at 20–25 °C. The following chemicals are added at well defined time intervals, while maintaining an efficient stirring at 20–25 °C: i) (R,R)-Salen (0.110 mg, 0.2 mmol) and triethylamine (0.056 mL, 0.4 mmol), 1 hour (the mixture turns dark red-brown); ii) 3-chloro-propenyl pivalate (**1a**) (0.245 mL, 1.5 mmol), 1 hour; iii) cyclohexanecarboxaldehyde (0.120 mL, 1.0 mmol) and trimethylsilyl chloride (0.152 mL, 1.2 mmol), 18 hours (the solution slowly turns yellow-brown and becomes thicker). The reaction mixture is quenched with saturated solution of sodium hydrogencarbonate (0.5 mL), filtered over a small pad of Celite®, extracted with ether and flash-chromatographed on silica. Two fractions are collected, a major one corresponding to $syn/anti$-(**5**) and a minor one corresponding to desilylated (**5**). In order to unambiguously establish both diastereo and diastereofacial selectivity, pivalate and trimethylsilyl protective groups are removed using lithium aluminium hydride (tetrahydrofuran, 0 °C) to afford corresponding diols (**2**) in quantitative yields.

The following procedures need to be followed to achieve good results: i) crushing manganese chips (Aldrich, 99.98%) in a mortar invariably affords a much more reactive metal with respect to the use of commercial manganese powder of the same purity, and ii) immediately before use, trimethylsilyl chloride is purified by elution through a short column packed with basic alumina.

CONCLUSION

A constantly increasing pressure is applied to organic chemists by efficiency, economy and environmental concerns; developing new synthons, exploitable in asymmetric catalytic transformations able to produce high value intermediates in a short and sustainable way, is a main challenge in the area of organic synthesis.

An example, developed by our group, has been presented: 3-chloro-propenyl pivalate, the new synthon we propose as a formal α-hydroxy allyl anion, and chromium(II) chemistry allow to achieve the formal diastereo- and enantio-selective α-hydroxyallylation of carbonyl compounds by applying a catalytic cycle based on an *in situ* produced Salen-chromium complex (Table 8.3).

Table 8.3   (*R,R*)-Salen-Cr(II) catalysed reaction of 3-chloropropenyl pivalate with a set of representative aldehydes.

Entry	R of RCHO	Yield (%)	Syn:Anti	Syn ee (%)	Anti ee (%)
1	Cyclohexyl	68	82:18	94 (*S,S*)	67
2	Pentyl	50	83:17	93 (*S,S*)	65
3	2-Phenylethyl	78	83:17	99 (*S,S*)	85
4	Isopropyl	42	85:15	92 (*S,S*)	77
5	2-Methylpropyl	55	72:28	92 (*S,S*)	60
6	Phenyl	77	80:20	64 (*S,S*)	43
7	4-Metoxyphenyl	82	74:26	65 (*S,S*)	51
8	2-Naphthyl	60	78:22	73 (*S,S*)	23

Chiral *syn*-alk-1-ene-3,4-diols produced in this way embody, besides two stereodefined alcoholic centres, a carbon-carbon double bond amenable to further stereoselective functionalizations, thus opening entries to densely functionalized carbon chains, typically present in a plethora of naturally occurring compounds.

REFERENCES

1. (a) Okude, Y., Hirano, S., Hiyama, T., and Nozaki, H. *J. Am. Chem. Soc.* **1977**, *99*, 3179. (b) Takai, K., Kimura, K., Kuroda, T., Hiyama, T., and Nozaki, H. *Tetrahedron Letters* **1983**, *24*, 5281. (c) Jin, H., Uenishi, J., Christ, W. J., Kishi, Y. *J. Am. Chem. Soc.* **1986**, *108*, 5644.
2. Fürstner, A. *Chem. Rev.* **1999**, *99*, 991.
3. (a) Bandini, M., Cozzi, P. G., Melchiorre, P., and Umani-Ronchi, A. *Angew. Chem. Int. Ed. Eng.* **1999**, *38*, 3357. (b) Bandini, M., Cozzi, P. G., and Umani-Ronchi, A. *Angew. Chem. Int. Ed. Eng.* **2000**, *39*, 2327.
4. (a) Ulich, L. H. and Adams, R. *J. Am. Chem. Soc.* **1921**, *43*, 660. (b) Neuenschwander, M., Bigler, P., Christen, K., Iseli, R., Kyburz, R., and Mühle, H. *Helv. Chim. Acta* **1978**, *61*, 2047.
5. (a) Lombardo, M., Girotti, R., Morganti, S., and Trombini, C. *Org. Lett.* **2001**, *3*, 2981. (b) Lombardo, M., Girotti, R., Morganti, S., and Trombini, C. *Chem. Comm.* **2001**, 2310. (c) Lombardo, M., D'Ambrosio, F., R., Morganti, S., and Trombini, C. *Tetrahedron Letters*, **2003**, *44*, 2823. (d) Lombardo, M., Morganti, S., and Trombini, C. *Org. Chem.* **2003**, *68*, 997.
6. Lombardo, M., Licciulli, S., Morganti S., and Trombini, C. *Chem. Comm.* **2003**, 1762.

9 Olefin Metathesis Reactions

CONTENTS

9.1 HIGHLY ACTIVE RUTHENIUM (PRE)CATALYSTS FOR METATHESIS REACTIONS

SYUZANNA HARUTYUNYAN, ANNA MICHROWSKA and KAROL GRELA*

Institute of Organic Chemistry, Polish Academy of Sciences. Kasprzaka 44/52. POBox 58, 01-224 Warsaw, Poland.

The development of stable and active well-defined ruthenium carbene complex $Cl_2(PCy_3)_2Ru = CHPh$ and its "second-generation" successors (**1a**) and (**1b**) (Figure 9.1) has meant that olefin metathesis reactions have become valuable tools in synthetic organic chemistry.[1] Among them the ring-closing (RCM), en-yne and cross-metathesis (CM) reactions have received much attention as they

Catalysts for Fine Chemical Synthesis, Vol. 3, Metal Catalysed Carbon-Carbon Bond-Forming Reactions
Edited by S. M. Roberts, J. Xiao, J. Whittall, and T. Pickett
© 2004 John Wiley & Sons, Ltd   ISBN: 0-470-86199-1

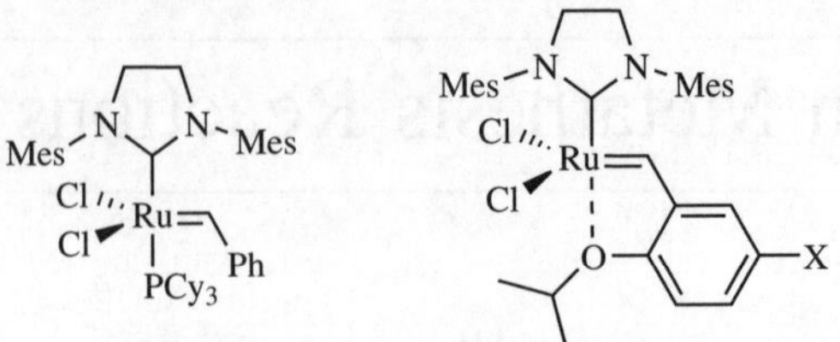

(1a)

(1b) (X = H)
(1c) (X = NO₂)

Figure 9.1   Second generation ruthenium carbene complexes for metathesis reactions.

offer great potential for an intra- and intermolecular C=C bond construction.[1] Recently, the stable and easy-to-use nitro-substituted catalyst (**1c**) has been developed; it possesses a dramatically enhanced reactivity, e.g. promoting metathesis even at 0 C.[2a-b]

9.1.1   SYNTHESIS OF THE RUTHENIUM (PRE)CATALYST (1C)

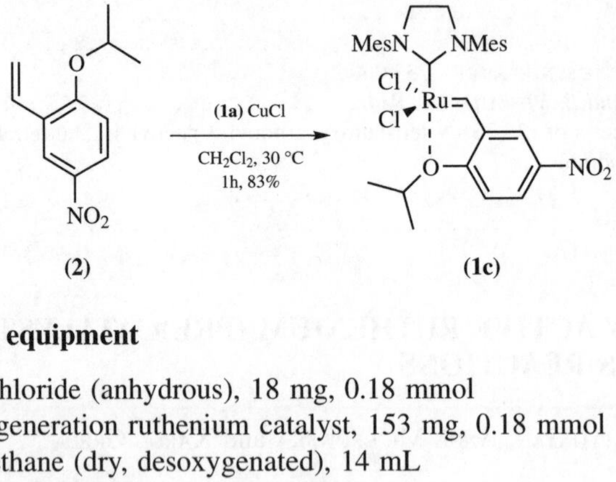

(2)   **(1c)**

Materials and equipment

- Copper(I) chloride (anhydrous), 18 mg, 0.18 mmol
- Grubbs' II-generation ruthenium catalyst, 153 mg, 0.18 mmol
- Dichloromethane (dry, desoxygenated), 14 mL
- 2-Isopropoxy-5-nitrostyrene, 38 mg, 0.18 mmol
- c-Hexane (HPLC grade), 300 mL
- Ethyl acetate (HPLC grade), 120 mL
- n-Pentane (dry, deoxygenated)

- Silica gel (Merk 60, 230-400 mesh), 15 g
- TLC plates, Polygram® SIL G/UV₂₅₄
- 30-ml Schlenk flask with magnetic stirrer bar
- Magnetic stirrer plate
- One glass column, diameter 3 cm
- Ten 50-mL Erlenmeyer flasks
- Rotary evaporator

Procedure

1. Grubbs' II-generation ruthenium catalyst (**1a**) (153 mg. 0.18 mmol), copper(I) chloride (18 mg, 0.18 mmol) and dichloromethane (10 mL) were placed in a Schlenk flask. A solution of 2-isopropoxy-5-nitrostyrene (**2**) (38 mg, 0.18 mmol) in dichloromethane (4 mL) was then added and the resulting solution was stirred under argon at 30 °C for 1 hour. From this point forth, all manipulations were carried out in air with reagent-grade solvents.

2. The reaction mixture was concentrated *in vacuo* and the resulting material was purified by column chromatography on silica. Elution with c-hexane:EtOAc (5:2) removes ruthenium complex containing a nitro group in *para*-position to the isopropoxy group as a green band ($Rf = 0.30$, c-hexane: EtOAc 8:2). Removal of solvent, washing with cold n-pentane and drying under vacuum afforded ruthenium (pre)catalyst (**1c**) as a green microcrystalline solid (100 mg, 83% yield).

^{1}H NMR (500 MHz, CD$_2$Cl$_2$) δ 16.42 (1H, s), 8.46 (1H, dd, J 9.1, 2.5 Hz), 7.80 (1H, d, J 2.5 Hz), 7.10 (4H, s), 6.94 (1H, d J 9.1 Hz,), 5.01 (1H, sept, J 6.1 Hz), 4.22 (4H, s), 2.47 (18H, 2s), 1.30 (6H, d, J 6.1 Hz).

^{13}C NMR (125 MHz, CD$_2$Cl$_2$) δ 289.1, 208.2, 156.8, 150.3, 145.0, 143.5, 139.6, 139.3, 129.8, 124.5, 117.2, 113.3, 78.2, 52.0, 21.3, 21.2, 19.4

IR (KBr) 2924, 2850, 1606, 1521, 1480, 1262, 1093,918, 745 cm^{-1}.

HRMS (IE) m/z calcd for C$_{31}$H$_{37}$N$_3$O$_3$^{35}Cl$_2$^{102}Ru: [M]$^{+\cdot}$. 671.1255, found 671.1229.

Elemental analysis calcd (%) for C$_{31}$H$_{37}$Cl$_2$N$_3$O$_3$Ru (671.63): C 55.44, H 5.55, N 6.26; found: C 55.35; H 5.70, N 6.09.

9.1.2 SYNTHESIS OF 1-[(4-METHYLPHENYL)SULFONYL]-2,3,6,7-TETRAHYDRO-1H-AZEPINE (4)

(3)   (4)

Materials and equipment

- *N,N*-Dibut-3-enylbenzenesulfonamide, 210 mg, 0.75 mmol
- Ruthenium (pre)catalysts (**1c**), 5 mg, 1 mol %
- Dichloromethane (dry, deoxygenated), 37 mL
- c-Hexane (HPLC grade), 240 mL
- Ethyl acetate (HPLC grade), 60 mL

- Silica gel (Merk 60, 230-400 mesh), 10 g
- TLC plates, SIL G-60 UV$_{254}$

- 30-ml Schlenk flask with magnetic stirrer bar
- Magnetic stirrer plate
- One glass column, diameter 4 cm
- Twenty 25-mL Erlenmeyer flasks
- Rotary evaporator

Procedure

1. To a solution of *N,N*-dibut-3-enylbenzenesulfonamide (**3**) (210 mg, 0.75 mmol) in dichloromethane (35 ml) a solution of ruthenium (**1c**) (5 mg, 1 mol %) in dichloromethane (2 ml) was added at 0°C. The resulting solution was stirred at the same temperature for 1 hour.
2. Solvent was then removed under vacuum and the residue was purified using silica gel column chromatography with c-hexane: EtOAc 8:2 to give 1-[(4-methylphenyl)sulfonyl]-2,3,6,7-tetrahydro-1H-azepine (186 mg, 99% yield) as colorless solid.

   ^{1}H NMR (200 MHz, CDCI$_3$) δ 2.28 (4H, m), 2.39 (3H, s), 3.25 (4H, m), 5.72 (2H, m), 7.25 (2H,.d, *J* 8.2 Hz), 7.64 (2H, d, *J* 8.2 Hz).

   ^{13}C NMR (50MHz, CDCI$_3$) δ 21.5, 29.948.2, 126.9, 129.5, 130.1, 136.2, 142.9.

   MS (EI) *m/z* 251 (5, [M]$^{+\cdot}$), 223 (2), 184 (6), 155 (4), 105 (2), 91 (19), 96 (16), 77 (1), 65 (13),42 (100).

   IR (KBr) 3030, 2942, 2899, 2855, 1657, 1596, 1450, 1332, 1286, 1162, 910, 816, 712 cm^{-1}.

   HRMS (EI) calculated for [M]$^{+\cdot}$ (C$_{13}$H$_{17}$O$_2$NS): 251.0980; found 2251.0979.

CONCLUSION

Although the second-generation ruthenium complex (**1a**) possesses in general very good application profile, the phosphine-free catalyst (**1b**), recently introduced by Hoveyda *et al.*,[3] displays even higher reactivity toward some electron-deficient substrates, such as acrylonitrile, fluorinated olefins and vinyl sulfones. Excellent air-stability, ease of storage and handling and possibility of a catalyst reuse render additional advantages of this system.

The Hoveyda-type catalysts has been shown that it can be significantly improved by changing electronic situation in the ruthenium-chelating isopropoxy-fragment. Namely, introduction of the strong electron-withdrawing group to the 2-isopropoxybenzylidene ligand leads to ruthenium complex (**1c**), which is similarly stable but dramatically more reactive than the Hoveyda-type catalyst. The active and easy to obtain (pre)catalyst (**1c**) is attractive from a practical point of view. This catalyst operates in very mild conditions (0 to 40 °C) and can be successfully applied in various types of metathesis (Table 9.1).[2]

Table 9.1 Metathesis reactions promoted by ruthenium complex (**1c**).

Substrate	Product	Yield [%] (1c, temp., time)
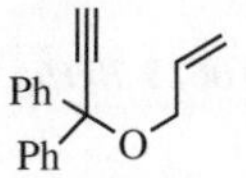 Ph, Ph—O (propargyl/allyl ether)	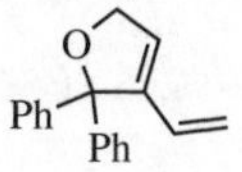 O, Ph, Ph (dihydrofuran)	98% (1 mol.%, 0 °C, 1 h)
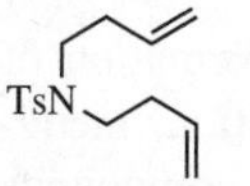 TsN (diallyl amine)	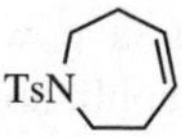 TsN (azepine)	99% (1 mol.%, 0 °C, 1 h)
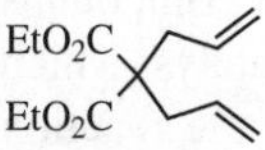 EtO₂C, EtO₂C (diene)	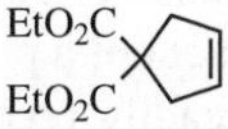 EtO₂C, EtO₂C (cyclopentene)	78% (2.5 mol.%, 0 °C, 8 h)
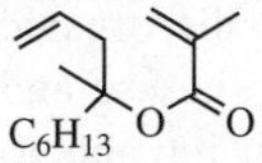 C₆H₁₃—O—O (acrylate ester)	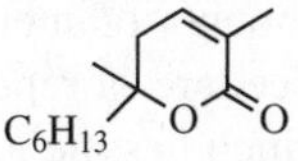 C₆H₁₃—O—O (dihydropyranone)	99% (2.5 mol.%, RT, 4 h)
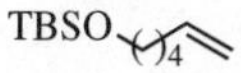 TBSO-( )₄ **5** + COCH₃ 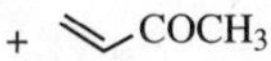	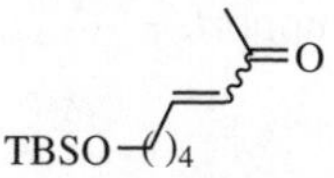 TBSO-( )₄ =O (E): (Z) = 99 : 1	82% (2.5 mol.%, RT, 30 min)
5 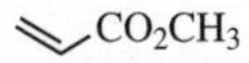 + CO₂CH₃	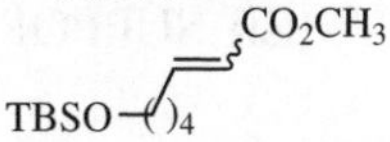 TBSO-( )₄ CO₂CH₃ (E): (Z) = 95 : 5	95% (1 mol.%, RT, 30 min)
5 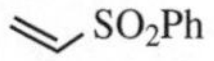 + SO₂Ph	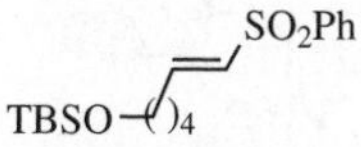 TBSO-( )₄ SO₂Ph (E):(Z) >99 : 1	90% (2.5 mol.%, RT, 16 h)
5 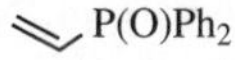 + P(O)Ph₂	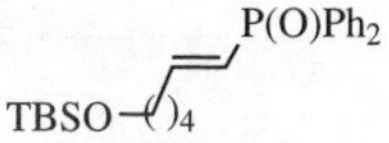 TBSO-( )₄ P(O)Ph₂ (E):(Z) >99 : 1	82% (5 mol.%, 40 °C, 16 h)
5 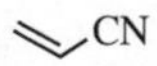 + CN	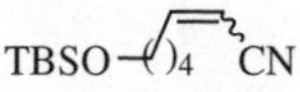 TBSO-( )₄ CN (E): (Z) = 1 : 2.7	83% (5 mol.%, 40 °C, 2 h)

REFERENCES

1. a) T. M. Trnka, and R. H. Grubbs, *Acc. Chem. Res.* **2001**, 34, 18; b) A. Fürstner, *Angew. Chem.* **2000**, 112, 3140; *Angew. Chem. Int. Ed.* **2000**, 39, 3012.
2. 2. a) K. Grela, S. Harutyunyan, and A. Michrowska, *Angew. Chem. Int. Ed. Engl.* **2002**, 41, 4038; b) For screening of the catalytic performance of Ru and Mo-based catalysts in cross-metathesis reactions, see: A. Michrowska, M. Bieniek, M. Kim, R. Klajn, and K. Grela *Tetrahedron,* **2003**, 59, 4525–31.
3. S. B. Garber, J. S. Kingsbury, B. L. Gray, and A. H. Hoveyda, *J. Am. Chem. Soc.* **2000**, 122, 8168.

9.2   A HIGHLY ACTIVE AND READILY RECYCLABLE OLEFIN METATHESIS CATALYST

STEPHEN J. CONNON, AIDEEN M. DUNNE and SIEGFRIED BLECHERT*

Institut für Chemie, Technische Universität Berlin, Strasse des 17. Juni 135, 10623 Berlin, Germany

Olefin metathesis has recently emerged as a powerful tool for the formation of carbon-carbon bonds. The widespread applicability of this transition metal-catalysed versatile reaction and the expense of ruthenium alkylidene compounds has highlighted the need for the development of a polymer supported catalyst which is easily assembled, highly active and readily recovered/reused after the reaction without leaving behind significant anounts of metal contaminants in the organic reaction product. Herein a facile procedure is reported for the synthesis, use and recovery of such a catalyst system, which has the added advantage of being able to generate its own soluble polymer support.

9.2.1   SYNTHESIS OF POLYMER SUPPORTED CATALYST (3)

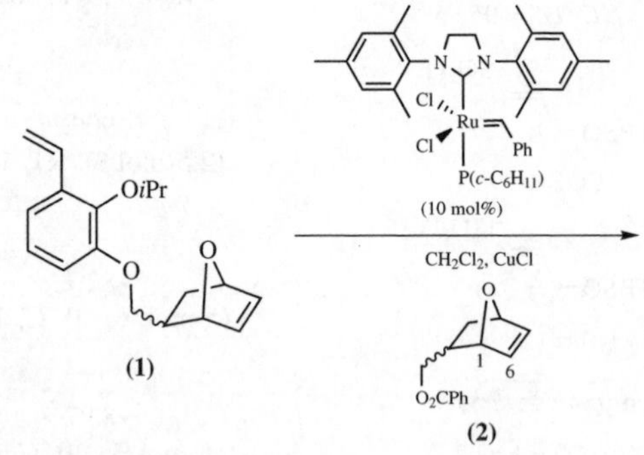

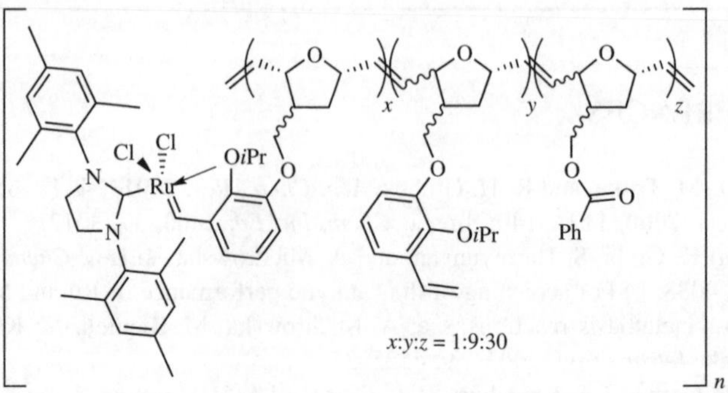

(3)

Materials and equipment

- Styrene derivative (1), 0.025 g, 0.087 mmol
- 2-Hydroxymethyl-(7-oxabicyclo[2.2.1]hept-5-enyl)benzoate (2), 0.060 g, 0.261 mmol
- Copper(I) chloride, 0.001g, 0.101 mmol
- 1,3-(Bis-(mesityl)-2-imidazolinylidene)dichloro(phenylmethylene)tricyclo-hexylphosphaneruthenium, 0.0074g, 0.0087 mmol
- Dry dichloromethane
- *n*-Hexane, dichloromethane, diethyl ether

- One two-necked 10-mL round-bottomed flask
- One 2-mL syringe
- One 25-mL round-bottomed flask
- Pre-washed cotton wool
- One Pasteur pipette
- Magnetic heater/stirrer plate, stirring bar and silicone oil bath
- One reflux condenser
- One rotary evaporator
- Nitrogen gas line

Procedure

1. To a solution of styrene (1)[1] (0.025 g, 0.087 mmol) and ester (2)[1] (0.060 g, 0.261 mmol) in dichloromethane (3 mL) under a nitrogen atmosphere in a 10-mL round-bottomed flask was added (2) (0.0074 g, 0.0087 mmol) in *dry* dichloromethane (2 mL) *via* syringe with vigorous stirring at room temperature. After 10 minutes ^{1}H NMR analysis of the red solution indicated complete consumption of both (1) and (2) as seen by the complete disappearance of the norbornene signals at δ 6.2–6.5 ppm. Copper(I) chloride (0.001 g, 0.101 mmol) was added in one portion, the flask was fitted with a condenser and the resulting solution heated under nitrogen at 45 °C for 1 hour, affording a light green solution.

2. After cooling, the reaction mixture was concentrated to dryness *in vacuo* (rotary evaporator) and taken up in (1:1) *n*-hexane:dichloromethane. The insoluble copper-phosphane salts were removed by passing the resulting suspension through a plug of pre-washed cotton wool in a Pasteur pipette into a 25-mL round-bottomed flask. The clear green solution was then evaporated to dryness *in vacuo* (rotary evaporator) and the solid washed successively with *n*-hexane (10 mL) and diethyl ether (10 mL) followed by drying under high vacuum to give catalyst (3) (0.0745 g, 93% yield) as a light green solid which readily adheres to surfaces such as glass or metal.

^{1}H NMR (500 MHz, CDCl$_3$): δ (ppm) 16.67 (1 H, bs, Ru = C$\underline{\text{H}}$), 7.99 (60 H, bs, *o*-Ar-ester), 7.50 (31 H, bs), 7.38 (62 H, bs), 7.04 (18 H, bs), 6.91 (9 H bs), 6.74 (9 H, bs), 5.7–5.6 (90 H, bs) 5.21 (10 H, bs), 4.7–3.7 (180 H, m), 2.78 (20 H, bs), 2.37 (61 H, bs), 2.01 (50 H, bs) 1.23 (60 H, bs).

The loading of the catalyst can be determined by comparison of the relative integration of the δ 16.67 (Ru alkylidene) and 7.99 (benzoate) ppm signals. A 1:60 ratio indicates quantitative loading of the catalyst. Typical catalyst loadings using this procedure are between 0.08 and 0.10 mmolg^{-1}.

Note: Due to overlapping and very broad signals some of the highfield signal integrals are approximate, but nevertheless consistent between polymer batches.

9.2.2 RING-CLOSING METATHESIS OF AN ACYCLIC DIENE AND SUBSEQUENT CATALYST RECOVERY/REUSE

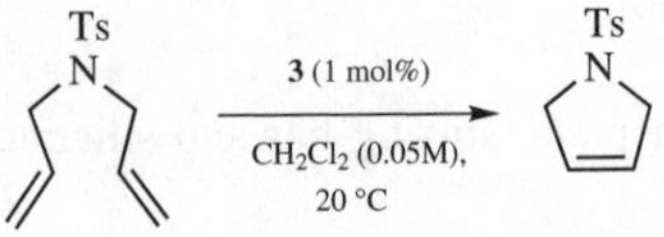

Materials and equipment

- N-Tosyldiallyamine, 0.0302 g, 0.12 mmol
- Catalyst (3), 0.0012 mmol (mass depends on catalyst loading determined by ^{1}H NMR)
- Dry dichloromethane
- n-Hexane, diethyl ether

- One 5-mL round-bottomed flask
- One 2-mL syringe
- Pre-washed cotton wool
- One Pasteur pipette
- Magnetic stirrer plate and stirring bar
  - One reflux condenser
  - One rotary evaporator
  - Nitrogen gas line

Procedure

1. To a solution of catalyst (3) (0.012 g, based on a loading of 0.1 mmolg^{-1}, 0.0012 mmol) in *dry* dichloromethane (1 mL) under nitrogen N-tosyldiallyla-mine[2,3] (0.0302 g, 0.12 mmol) in dichloromethane (1.4 mL) was added *via* syringe at room temperature. The resulting light-yellow/green solution was stirred under nitrogen until ^{1}H NMR or TLC analysis indicated quantitative conversion of the starting material (60 minutes). After reaction the catalyst was recovered by the addition of cold diethyl ether (10 mL), which precipitates the

catalyst as a green gum that adheres to the flask, allowing removal of the reaction mixture by Pasteur pipette. The flask was further rinsed with 1:1 diethyl ether:n-hexane $(2 \times 7\,mL)$ and the colourless organic extracts combined. Removal of the solvent *in vacuo* (rotary evaporator) furnished the pure white product in quantitative yield (26.3 mg).

2. Alternatively addition of cold n-hexane or n-hexane-diethyl ether (1:1) gives complete precipitation of the catalyst as a light green solid. The product can then be separated by a filtration/evaporation sequence.

3. Re-use of the catalyst: The flask containing the catalyst was subjected to high vacuum for 20 minutes, placed under a nitrogen atmosphere and a second charge of N-tosyldiallylamine (0.0302 g, 0.12 mmol) in dichloromethane (1.6 mL) was added *via* syringe at room temperature. After quantitative conversion was observed, the above procedure was repeated to isolate the product and catalyst respectively. Under these conditions it was possible to recycle the catalyst 7 times with quantitative conversion of the starting material in each cycle *(extended reaction times were required after cycle 5)*. TXRF analysis of the products from the first 4 cycles indicated a maximum of 0.004% ruthenium (wt%) was present in the ring-closed products.[4]

Note: While the catalyst itself is stable indefinitely in air, when wet/dissolved in solvent it decomposes slowly to give a brown solid/solution. Therefore, for best results every precaution should be taken to minimise the time taken to recover the catalyst from the reaction once it has been exposed to air.

CONCLUSION

The procedure reported above represents a facile and reproducible method for the execution of various olefin metathesis reactions using a highly active and recoverable catalyst. Table 9.2 gives a list of representative substrates compatible with the technology and the time taken for quantitative conversion of the substrate using the ring-closing metathesis procedure.

REFERENCES

1. Connon, S. J., Dunne, A. M., and Blechert, S. *Angew. Chem.* **2002**, *114*, 3989; *Angew. Chem. Int. Ed. Engl.* **2002**, *41*, 3835.
2. Gilbert, B. C., Kalz, W., Lindsay, C. I., McGrail, P. T., Parsons, A. F., and Whittaker, D. T. E. *J. Chem. Soc. Perkin Trans.1* **2000**, 1187.
3. The procedure can be used to perform ring-closing metathesis, ring-rearrangement metathesis and ring-opening cross-metathesis reactions involving a variety of substrates.
4. TXRF analysis performed by Prof. Dr. B. O. Kolbesen, Institut für Anorganische Chemie, Johann Wolfgang Goethe-Universität, Frankfurt am Main, Germany.

Table 9.2 Activity of (**3**) in various metathesis reactions using the above procedure (Reaction times (minutes) in parentheses).

Substrate	Product	Conversion (%)
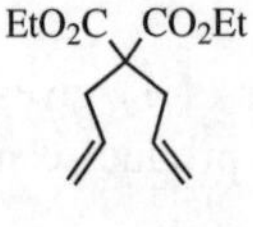	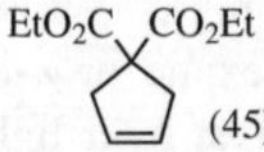	>98
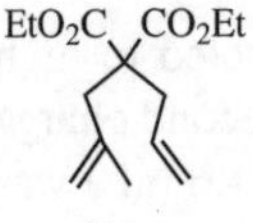	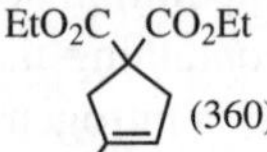	>98
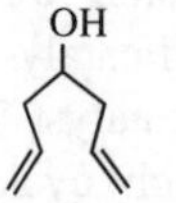	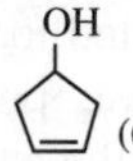	>98
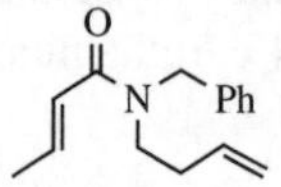	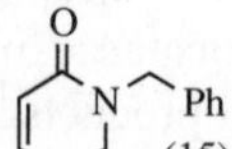	>98
	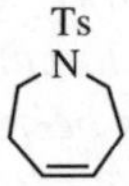	>98
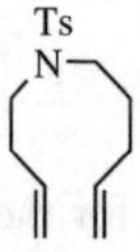	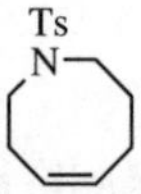	>98
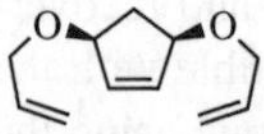	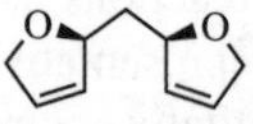	>98

9.3   STEREOSELECTIVE SYNTHESIS OF L-733,060

G. Bhaskar and B. Venkateswara Rao*

Organic Division III, Indian Institite of Chemical Technology, Hyderabad 500 007, India

The non-peptidic neurokinin NK1 receptor antagonist (**1**) (L-733,060) is known to have a variety of biological activities including neurogenic inflammation,[1] pain transmission and regulation of the immune response.[2] It has been implicated in a variety of disorders including migraine,[3] rheumatoid arthritis[4] and pain.[5] A stereoselective synthesis of L-733,060 using ring closing metathesis as key step, starting from L-phenylglycine has recently been reported.[6]

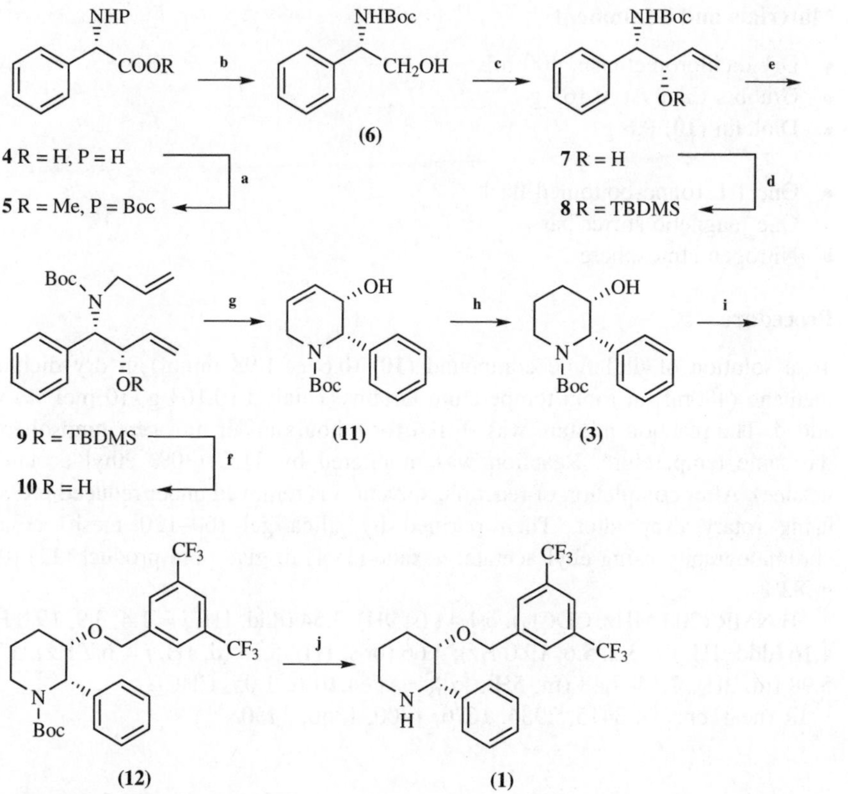

(6)

4 R = H, P = H

5 R = Me, P = Boc

a

7 R = H

8 R = TBDMS

d

9 R = TBDMS

10 R = H

f

(11)

(3)

(12)

(1)

a) AcCl, MeOH then (Boc)₂O, Et₃N, THF, 0 °C-rt, 8 h, 97%;
b) LiCl, NaBH₄, EtOH, THF, 0 °C-rt, 12 h, 87%;
c) DMSO, (COCl)₂, DCM, i-Pr₂NEt then CH₂=CHMgBr, THF, 2 h, rt, 61%;
d) TBDMS-Cl, imidazole, DCM, 0 °C-rt, 24 h, 90%;
e) CH₂=CHCH₂Br, NaH, DMF, 0 °C-rt, 24 h, 90%;
f) TBAF-AcOH, THF, 0 °C-rt, 24 h, 85%;
g) Grubbs' catalyst, DCM, rt, 6 h, 82%;
h) Pd/C, H₂, EtOH, 4 h, rt, 65%;
i) 3,5-bis(trifluoromethyl)benzyl bromide, NaH, DMF, 80 °C, 13 h, 80%;
j) trifluoroacetic acid, rt, 1 h, 79%.

9.3.1   SYNTHESIS OF (2S,3S)-N-tert-BUTOXYCARBONYL-2-PHENYL 1,2,3,6-TETRAHYDRO-3-PYRIDINOL

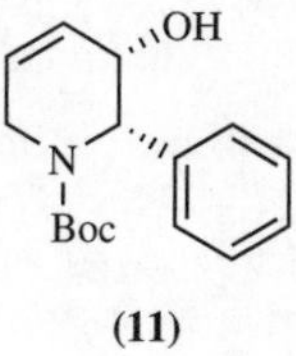

(11)

Materials and equipment

- Dry dichloromethane, 400 mL
- Grubb's Catalyst, 0.163 g
- Diolefin (10) 0.6 g

- One 1-L round-bottomed flask
- One magnetic stirrer bar
- Nitrogen atmosphere

Procedure

To a solution of diolefinic compound (10) (0.6 g, 1.98 mmol) in dry dichloromethane (400ml) at room temperature Grubbs' catalyst (0.163 g, 10 mol %) was added. The reaction mixture was stirred for 6 hours under nitrogen atmosphere at the same temperature. Reaction was monitered by TLC (30% ethyl acetate in hexane). After completion of reaction, solvent was removed under reduced pressure using rotary evaporator. Then purified by silica gel (60–120 mesh) column chromatography using ehyl acetate: hexane (1: 4) to give pure product (11) (0.45 g, 82%).

^{1}H NMR (200 MHz, CDCl$_3$): δ 1.44 (s, 9H), 3.54 (ddd, 1H, j = 1.4, 3.9, 19.0 Hz), 4.16 (ddd, 1H, j = 3.2, 5.6, 19.0 Hz), 4.66 (br s, 1H), 5.53 (d, 1H, j = 6.7 Hz), 5.70–5.98 (m, 2H), 7.14–7.48 (m, 5H). $[\alpha]_D^{25}$ = +64.10 (c 1.05, CHCl$_3$).

IR (neat, cm^{-1}): 3415, 2984, 1676, 1400, 1350, 1150.

CONCLUSION

The synthesis of L-733,060 (1) using ring-closing metathesis starting from L-phenylglycine has been accomplished.

REFERENCES

1. Lotz, M., Vaughan, J. H., and Carson, D. A. *Science* **1988**, *241*, 1218–1221.
2. Perianan, A., Snyderman, R., and Malfroy, B. *Biochem. Biophys. Res. Commun.* **1989**, *161*, 520–524.
3. Moskowitz, M. A. *Trends Pharmacol. Sci.* **1992**, *13*, 307–311.
4. Lotz, M, Vaughan, J. H., and Carson, D. A. *Science* **1987**, *235*, 893–895.
5. Otsuka, M. and Yanigasawa, M. *J. Physiol.* (*London*) **1988**, *395*, 255–270.
6. Bhaskar, G. and Venkateswara Rao, B. *Tetrahedron Lett.* **2003**, *44*, 915–917.

10 Cyclisation Reactions

CONTENTS

Catalysts for Fine Chemical Synthesis, Vol. 3, Metal Catalysed Carbon-Carbon Bond-Forming Reactions
Edited by S. M. Roberts, J. Xiao, J. Whittall, and T. Pickett
© 2004 John Wiley & Sons, Ltd ISBN: 0-470-86199-1

10.1 MOLECULAR SIEVES AS PROMOTERS FOR THE CATALYTIC PAUSON-KHAND REACTION

JAIME BLANCO-URGOITI, GEMA DOMÍNGUEZ and JAVIER PÉREZ-CASTELLS*

Department of Chemistry, Universidad San Pablo-CEU, Urb. Montepríncipe, 28668 Boadilla del Monte, Madrid, Spain

The Pauson-Khand reaction (PKR) is a powerful method for the construction of cyclopentenones.[1] Many efforts are currently being carried out to achieve an efficient catalytic procedure, in order to apply the reaction for industrial syntheses using small amounts of metal catalysts. Still, a carbon monoxide atmosphere is necessary. Recently, molecular sieves as promoters for the PKR have been introduced.[2] These zeolites can be used for a new efficient protocol of the catalytic version of the reaction. A procedure with no carbon monoxide atmosphere is also possible, although worse results are obtained.

10.1.1 SYNTHESIS OF 3AS*,5R*-5-HYDROXY-3,3A,4,5-TETRAHYDROCYCLOPENTA[A]NAPHTALEN-2-ONE

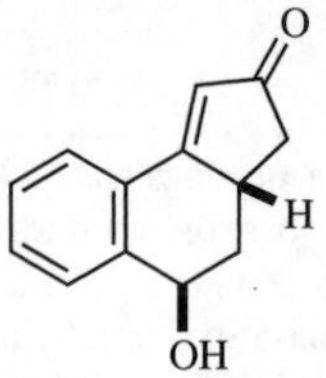

Materials and equipment

- 1-(2-Ethynylphenyl)-3-buten-1-ol (95%), 0.2 g, 1.16 mmol
- Dicobalt octacarbonyl (98%), 39.6 mg, 0.11 mmol
- Powdered molecular sieves 4Å, 0.4 g
- Carbon monoxide, 1 atmosphere
- Toluene, 15 mL

- One balloon
- Celite®, 10 g
- Silica gel (Matrex 60A, 37–70 μm), 20 g
- TLC plates, SIL G-60 UV$_{254}$
- One 25-mL two-necked round-bottomed flask with magnetic stirrer bar
- Condenser
- Magnetic stirrer plate
- One flask, 50 mL
- One glass sintered funnel, diameter 7 cm
- Water aspirator

- One glass column, diameter 1.5 cm
- Rotary evaporator

Procedure 1

1. 1-(2-Ethynylphenyl)-3-buten-1-ol (0.2 g, 1.16 mmol), powdered molecular sieves 4A (0.8 g), dicobalt octacarbonyl (39.6 mg, 0.11 mmol) and toluene (10 mL) were placed in a 25-mL two-necked round-bottomed flask equipped with a magnetic stirrer bar, a condenser and a septum. The mixture was submitted to three cycles of vacuum-argon and vacuum-carbon monoxide. The carbon monoxide atmosphere was kept with a balloon and the mixture was stirred at 70 °C for 18 hours.
2. The molecular sieves and the catalyst were then removed by filtration through Celite® into a glass sintered funnel with the aid of a water aspirator. The filtration was completed by rinsing the packing with toluene (5 mL).
3. Solvent evaporation in a rotary evaporator afforded the crude product, which was purified by column chromatography (silica gel, hexane/EtOAc: 1/1) (0.14 g, 70% yield).

^{1}H NMR (300 MHz, CDCl$_3$) δ 1.79 (td, 1H, $J_1 = 13.7$ Hz, $J_2 = 3.3$ Hz), 2.17 (dd, 1H, $J_1 = 18.1$ Hz, $J_2 = 3.3$ Hz), 2.38–2.44 (m, 1H), 2.73 (dd, 1H, $J_1 = 18.1$ Hz, $J_2 = 6.6$ Hz), 3.17 (bs, 1H), 3.52–3.58 (m, 1H), 4.92–4.94 (m, 1H), 6.36 (d, 1H, $J = 2.2$ Hz), 7.36–7.40 (m, 1H), 7.45 (d, 2H, $J = 3.8$ Hz), 7.64 (d, 1H, $J = 7.7$ Hz)

Materials and equipment

- 1-(2-Ethynylphenyl)-3-buten-1-ol (95%), 0.2 g, 1.16 mmol
- Dicobalt octacarbonyl (98%), 39.6 mg, 0.11 mmol
- Powdered molecular sieves 4Å, 0.4 g
- Carbon monoxide
- Toluene, 10 mL

- One balloon
- Oven
- Celite®, 10 g
- Silica gel (Matrex 60A, 37–70 μm), 20 g
- TLC plates, SIL G-60 UV$_{254}$
- One 25-mL two-necked round-bottomed flask with magnetic stirrer bar
- Condenser
- Magnetic stirrer plate
- Two flasks, 50 mL
- One glass sintered funnel, diameter 7 cm
- Water aspirator
- One glass column, diameter 1.5 cm
- Rotary evaporator

Procedure 2

1. A flask containing powdered 4Å molecular sieves was heated in an oven (200 °C, 3 hours) and cooled under a carbon monoxide atmosphere. These molecular sieves (0.4 g) were added to a 25-mL two-necked round-bottomed flask equipped with a magnetic stirrer bar and a condenser, containing a solution

Table 10.1 PKR with different substrates using molecular sieves as promoters.

Entry	Substrate	Product	Yield (%)[a]	
			Proc. 1	Proc. 2
1	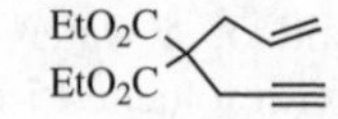	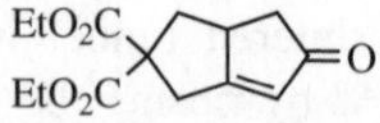	72	65
2	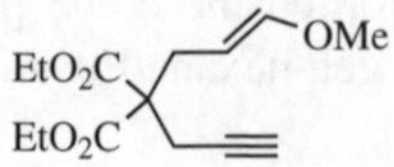	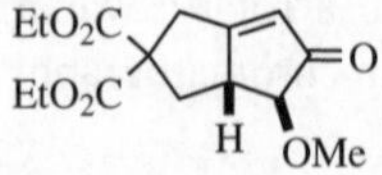	88	73
3	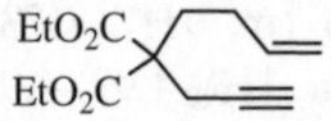	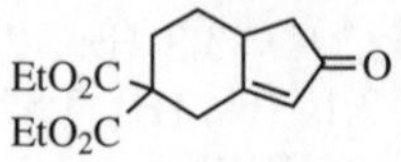	85	70
4	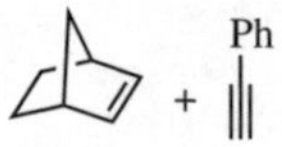	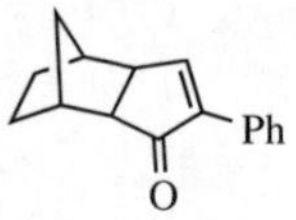	70	60
5	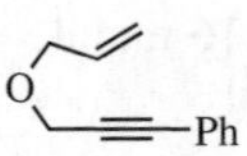	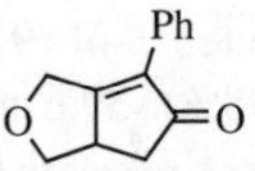	75	59
6	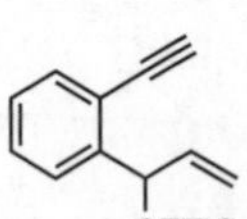	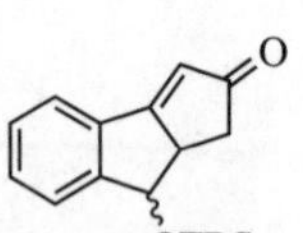	80[b]	70[b]
7[c]	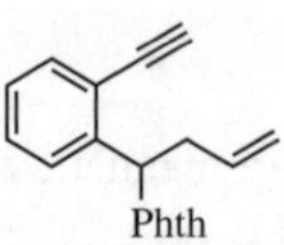	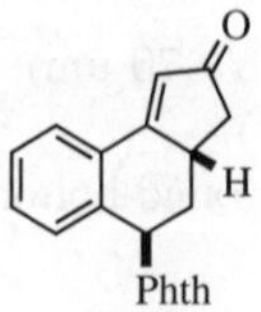	77	60
8	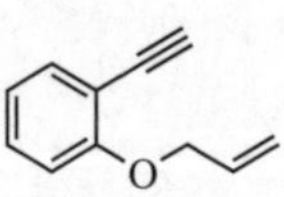	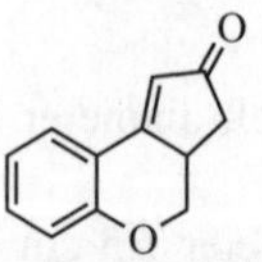	63	40

a) Pure product (^{1}H, ^{13}C RMN, IR). b) Pure mixture of diastereomers (1:1). c) Phth = phthalimido

of 1.00 mmol of 1-(2-ethynylphenyl)-3-buten-1-ol (0.2 g, 1.16 mmol) in toluene (10 mL) at room temperature. To this solution, 0.10 mmol of dicobalt octacarbonyl was added. The flask was stopped and the resulting mixture stirred for 18 hours at 70 °C.

2. The reaction was treated as in procedure 1, steps 2 and 3 (0.12 g, 58% yield).

CONCLUSION

The procedures are very easy to reproduce. They have been applied to a wide range of substrates with good results. Procedure 1 improves most literature protocols while procedure 2 usually gives slightly worse results but with the advantage of less use of carbon monoxide. Table 10.1 shows the results of the reaction of different substrates under both procedures.

REFERENCES

1. Recent reviews on the Pauson-Khand reaction: (a) Sugihara, T., Yamaguchi, M., and Nishizawa, M. *Chem. Eur. J.* **2001**, *7*, 1589–1595. (b) Brummond, K. M., and Kent, J. L. *Tetrahedron* **2000**, *56*, 3263–3282. (c) Keun Chung, K. *Coord. Chem. Rev.* **1999**, *188*, 297–341. (d) Geis, O. and Schmalz, H-G. *Angew. Chem., Int. Ed. Engl.* **1998**, *37*, 911–914.

2. Pérez-Serrano, L., Casarrubios, L., Domínguez, G. and Pérez-Castells, J. *Org. Lett.* **1999**, *1*, 1187–1188.

10.2 PALLADIUM(II)-CATALYSED CYCLIZATION OF ALKYNES WITH ALDEHYDES, KETONES OR NITRILES INITIATED BY ACETOXYPALLADATION OF ALKYNES

LIGANG ZHAO and XIYAN LU*

State Key Laboratory of Organometallic Chemistry, Shanghai Institute of Organic Chemistry, Chinese Academy of Sciences, 354 Fenglin Lu, Shanghai 200032, China

As a facile method of carbon-carbon bond formation, the insertion of carbon-carbon multiple bonds into carbon-transition-metal bonds is a very important fundamental reaction in organotransition-metal chemistry. However, in contrast to the tremendous number of reports about the insertion of carbon-carbon multiple bonds into carbon-transition-metal bonds, direct insertion of carbon-heteroatom multiple bonds, such as carbonyl and nitrile groups, without using stoichiometric organometallic reagents, has received scant attention.[1] A palladium(II)-catalysed cyclization reaction of alkynes with carbon-heteroatom multiple bonds under mild conditions has been developed, using insertion of carbon-heteroatom multiple bonds into the carbon-palladium bond as the key step.[2]

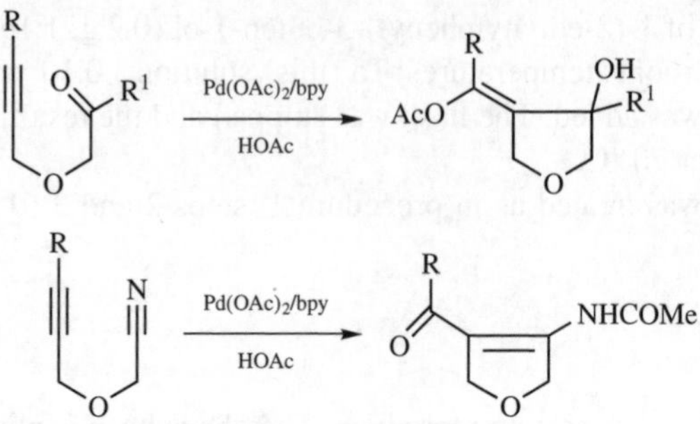

10.2.1   SYNTHESIS OF 3-PHENYL-3-HYDROXY-4-(1′-ACETOXYHEXYLIDENE)TETRAHYDROFURAN (2)

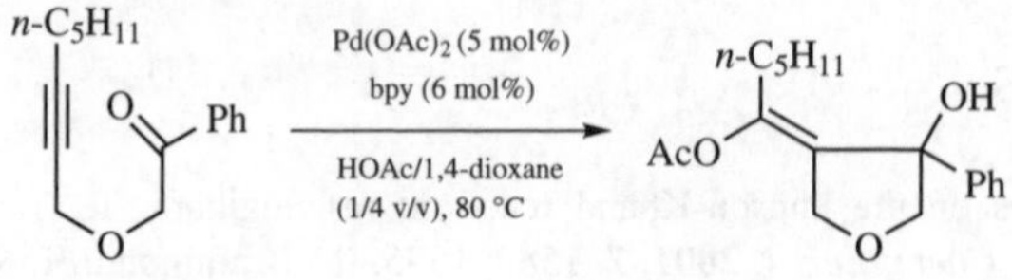

Materials and equipment

- Dry nitrogen gas
- 2-(2′-Octynyloxy)-1-phenylethanone, 136 mg, 0.5 mmol
- Glacial acetic acid, 1 mL
- 1,4-Dioxane, 4 mL
- Palladium acetate, 5.6 mg, 0.025 mmol
- 2,2′-Bipyridine, 4.5 mg, 0.03 mmol
- Saturated solution of sodium hydrogencarbonate
- Diethyl ether, 60 mL
- Saturated solution of sodium chloride
- Anhydrous sodium sulfate
- Ethyl acetate (dried)
- Petroleum ether (dried over sodium)

- Schlenk tube (25 mL) with magnetic stirrer bar
- Magnetic stirrer
- TLC plates, G-60
- Silica gel (60A), 250 g
- One 250-mL separating funnel
- Erlenmeyer flask (250 mL)
- One glass column, (diameter 2.5 cm)
- Rotary evaporator

Procedure

Under a nitrogen atmosphere, 2-(2′-octynyloxy)-1-phenylethanone (1) (136 mg, 0.5 mmol) was added to a mixture of Pd(OAc)$_2$ (5.6 mg, 0.025 mmol), 2,2′-bipyridine (4.5 mg, 0.030 mmol), HOAc (1 mL), and 1,4-dioxane (4 mL). The solution was stirred at 80 °C for 14 hours until the reaction was complete as monitored by TLC. On cooling, the reaction mixture was neutralised with saturated sodium hydrogencarbonate, and then extracted with diethyl ether (3 × 20 mL). The combined ether solution was washed with saturated sodium chloride, dried with sodium sulfate and concentrated. The residue was purified by flash chromatography on silica gel (EtOAc/petroleum ether: 1/4) to give the product 3-phenyl-3-hydroxy-4-(1′-acetoxyhexylidene)tetrahydrofuran (2) in 80% yield as a white solid; m.p. 79.5–80.5 °C.

^{1}H NMR (300 MHz, CDCl$_3$) δ 7.57 (d, $J = 8.7$ Hz, 2H), 7.40–7.35 (m, 2H), 7.31–7.28 (m, 1H), 4.65 (d, $J = 13.3$ Hz, 1H), 4.40 (d, $J = 13.3$ Hz 1H), 4.23 (d, $J = 9.7$ Hz, 1H), 3.82 (d, $J = 9.7$ Hz, 1H), 2.59 (br, 1H), 2.17 (s, 3H), 1.93 (t, $J = 7.5$ Hz, 2H), 1.07 – 0.89 (m, 6H), 0.75 (t, $J = 7.1$ Hz, 3H).

^{13}C NMR (75 MHz, CDCl$_3$)δ 168.3, 146.3, 142.1, 133.3, 128.1, 127.1, 125.6, 84.9, 80.8, 70.4, 31.2, 30.4, 25.2, 22.1, 13.8.

IR (KBr) v 3388, 3033, 2935, 2861, 1744, 1604, 1496, 1151, 697 cm^{-1}.

MS (EI) m/z: 287 [(M^++1)-H$_2$O], 246, 244, 232 (100), 77.

Anal. Calcd for C$_{18}$H$_{24}$O$_4$: C, 71.03; H, 8.95. Found: C, 71.25; H, 7.78.

10.2.2 SYNTHESIS OF DIMETHYL 3-ACETYLAMINO-4-BUTYRYLCYCLOPENT-3-ENE-1,1-DICARBOXYLATE (4)

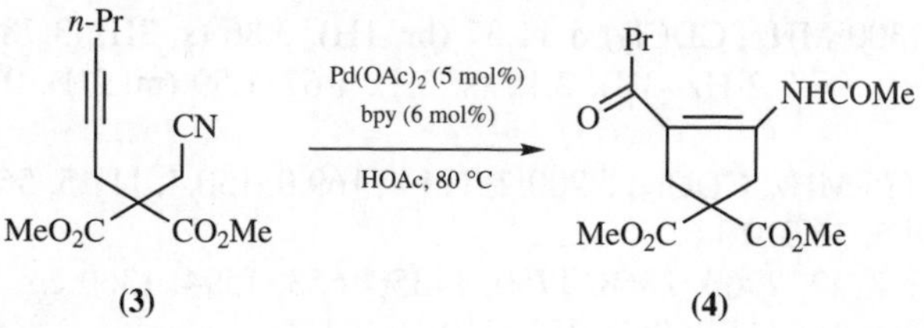

Materials and equipment

- Dry nitrogen gas
- Dimethyl 2-cyanomethyl-2-(hex-2′-ynyl)malonate, 126 mg, 0.5 mmol
- Glacial acetic acid, 4 mL
- Palladium acetate, 5.6 mg, 0.025 mmol
- 2,2′-Bipyridine, 4.5 mg, 0.03 mmol
- Saturated solution of sodium hydrogencarbonate
- Diethyl ether, 60 mL
- Saturated solution of sodium chloride
- Anhydrous sodium sulfate

- Ethyl acetate (dried)
- Petroleum ether (dried over sodium)

- Schlenk tube (25 mL) with magnetic stirrer bar
- Magnetic stirrer
- TLC plates, G-60
- Silica gel (60A), 250 g
- One 250-mL separating funnel
- Erlenmeyer flask (250 mL)
- One glass column, (diameter 2.5 cm)
- Rotary evaporator

Procedure

Under a nitrogen atmosphere, dimethyl 2-cyanomethyl-2-(hex-2′-yny)malonate (**3**) (136 mg, 0.5 mmol) was added to a mixture of Pd(OAc)$_2$ (5.6 mg, 0.025 mmol), 2,2′-bipyridine (4.5 mg, 0.03 mmol), HOAc (1 mL) and 1,4-dioxane (4 mL). The solution was stirred at 80 °C for 14 hours until the reaction was complete as monitored by TLC. On cooling, the reaction mixture was neutralised with saturated sodium hydrogencarbonate, and then extracted with diethyl ether (3 × 20 mL). The combined ether solution was washed with saturated sodium chloride, dried with sodium sulfate, and concentrated. The residue was purified by flash chromatography (EtOAc/petroleum ether: 1/4) to give the product dimethyl 3-acetylamino-4-butyrylcyclopent-3-ene-1,1-dicarboxylate (**4**) in 95% yield as white solid; m.p. 89.5–90.5 °C.

The present catalytic reactions can be performed even in an atmosphere of oxygen/moisture due to the stability of Pd(OAc)/bpy.

^{1}H NMR (300 MHz, CDCl$_3$) δ 11.37 (br, 1H), 3.86 (s, 2H), 3.78 (s, 6H), 3.23 (s, 2H), 2.42 (t, $J = 7.2$ Hz, 3H), 2.15 (s, 3H), 1.67–1.59 (m, 2H), 0.96 (t, $J = 7.2$ Hz, 3H).

^{13}C NMR (75 MHz, CDCl$_3$) δ 200.2, 171.4, 169.0, 150.4, 111.5, 56.3, 53.1, 43.4, 40.9, 36.5, 24.5, 16.7, 13.6.

IR (KBr) ν 3242, 2960, 2868, 1759, 1735, 1653, 1594, 1300 cm^{-1}.

MS (EI) *m/z*: 311 [M^+], 268, 252, 210 (100).

Anal. Calcd for C$_{15}$H$_{21}$NO$_6$: C, 57.87; H, 6.80; N, 4.50. Found: C, 57.94; H, 6.86; N, 4.26.

CONCLUSION

This is a reaction with high atom economy and without the use of organometallic reagents, additives, or redox systems. Such an acetoxypalladation-initiated carbon-heteroatom multiple bond insertion-protonolysis system may extend the scope of transition metal-catalysed reactions pertaining to the insertion of carbon-heteroatom multiple bonds into metal-carbon bonds, and provide a new methodology in organic synthesis. The generality of the present catalytic system is shown in Table 10.2.[3]

Table 10.2 Palladium(II)-catalysed cyclization of alkynes with aldehydes, ketones or nitriles.

Entry	Substrate	Condition (°C/h)	Product	Yield (%)
1	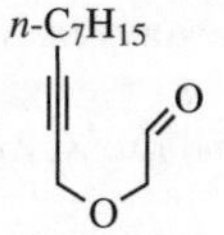	80/10	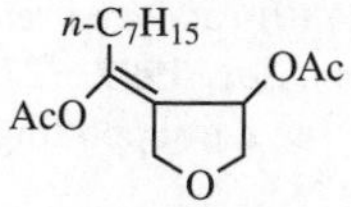	50
2	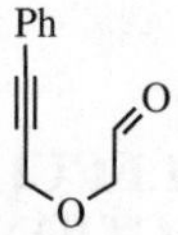	80/24	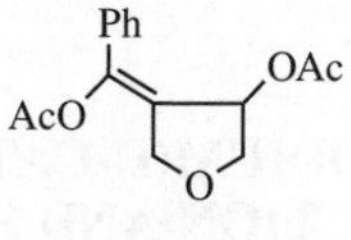	22
3	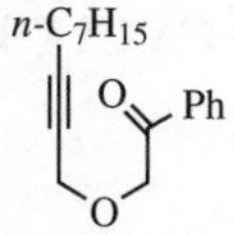	80/14	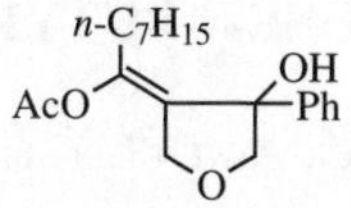	71
4	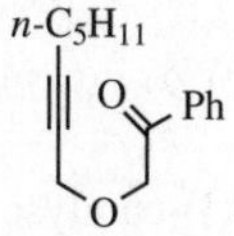	80/11	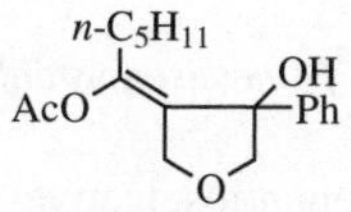	80
5	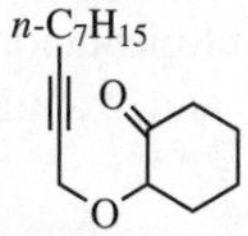	80/8	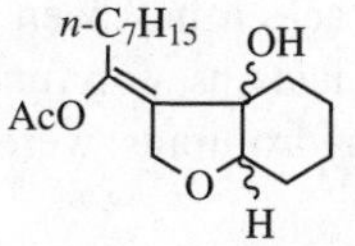 cis : trans = 2.2:1	65
6	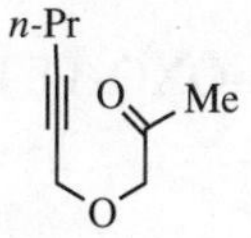	80/11	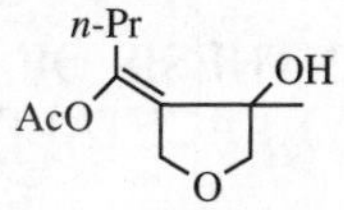	66
7	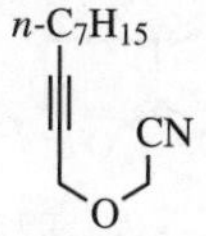	80/16	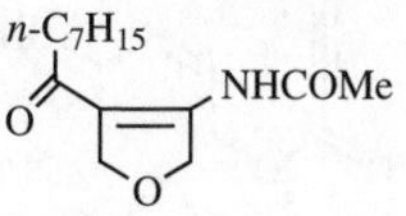	72
8	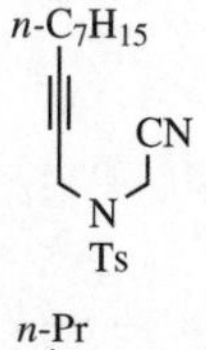	80/16	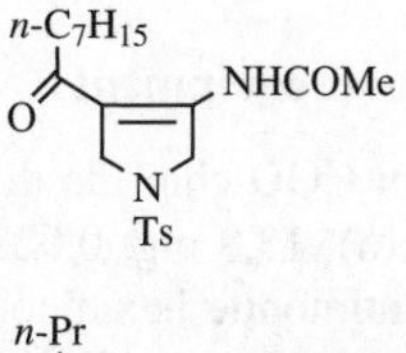	79
9	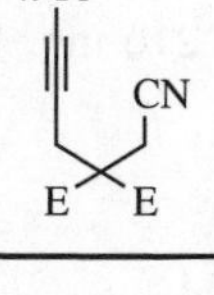	100/16	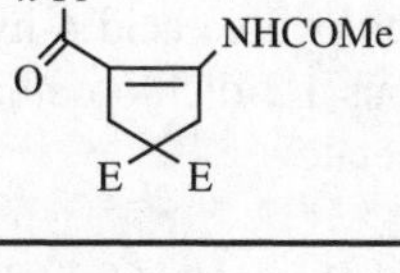	95

REFERENCES

1. Hegedus, L. S. In *Comprehensive Organic Synthesis*, Trost, B. M. and Fleming, L., Eds.; Pergamon Press: Oxford, **1990**; *4*, 571.
2. For some Pd(0)-catalyzed reactions, see: Quan, L. G., Gevorgyan, V. and Yamamoto, Y. *J. Am. Chem. Soc.* **1999**, *121*, 3545 and references therein.
3. For details, see the supporting information in: Zhao, L. and Lu, X. *Angew. Chem. Int. Ed.* **2002**, *41*, 4343.

10.3   RHODIUM(I)-CATALYSED INTRAMOLECULAR ALDER-ENE REACTION AND SYNTHESES OF FUNCTIONALISED α-METHYLENE-γ-BUTYROLACTONES AND CYCLOPENTANONES

MINSHENG HE[a], AIWEN LEI[b] and XUMU ZHANG[a*]

[a]*Department of Chemistry, 152 Davey Laboratory, Penn State University, University Park, PA 16802, USA*
[b]*Department of Chemistry, Stanford University, Stanford, CA 94305–5080, USA*

A highly enantioselective and general rhodium(I)-catalysed intramolecular Alder-ene reaction has been developed. α-Methylene-γ-butyrolactones, α-methylene-γ-butyrolactams, polyfunctional cyclopentanes, cyclopentanones, and functionalized tetrahydrofurans were formed in high yields and excellent enantiomeric excesses.[1–4]

10.3.1   SYNTHESIS OF (4-BENZYLIDENE-5-OXO-TETRAHYDRO-FURAN-3-YL)-ACETALDEHYDE

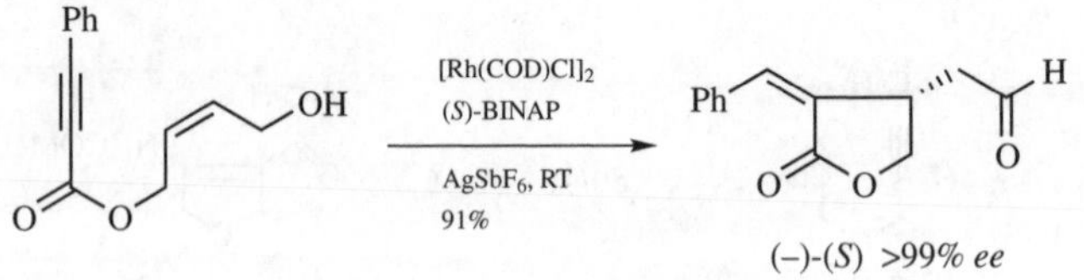

$(-)$-(S) >99% *ee*

Materials and equipment

- Rhodium COD chloride dimer, 4.9 mg, 0.01 mmol
- BINAP (S), 13.8 mg, 0.022 mmol
- Silver antimonite hexafluoride, 13.7 mg, 0.04 mmol
- Phenyl-propynoic acid 4-hydroxy-but-2-enyl ester, 216 mg, 1.0 mmol
- Anhydrous 1,2-dichloroethane, 11 mL
- Ethyl acetate
- Hexane
- Silica gel (Sorbent, 60A, 32–63 μm), 5g

- 25-mL Schenck tube
- Magnetic stirrer plate
- TLC plates (Sorbent, UV$_{254}$)

Procedure

In a dried 25-mL Schlenk tube under a nitrogen atmosphere, a mixture of [Rh(COD)Cl]$_2$ (4.9 mg, 0.01 mmol) and (S)-BINAP (13.8 mg, 0.022 mmol) was dissolved in 10 mL of 1,2-dichloroethane and stirred at room temperature for 1 minute. Then phenyl-propynoic acid 4-hydroxy-but-2-enyl ester (216 mg, 1.0 mmol) was added into the solution at room temperature under a nitrogen atmosphere. After stirring for 1 minute, a solution of silver antimonite hexafluoride (13.7 mg, 0.04 mmol) in 0.8 mL of 1,2-dichloroethane was added into the mixture in one portion *via* a syringe. The resulting turbid solution was stirred at room temperature for 10 minutes. TLC indicated that all the starting material was consumed and the reaction completed. The reaction mixture was directly subjected to column chromatography (silica gel, hexane/ethyl acetate: 80/20) to afford 196.7 mg (91% yield, 99% ee (S)) of pure white solid.

^{1}H NMR (360 MHz, CDCl$_3$) δ 9.80 (s, 1H), 7.80–7.77 (m, 2H), 7.38–7.32 (m, 3H), 6.91(d, $J = 2.1$, 1H), 4.57 (dd, $J = 7.9$, 9.3 Hz, 1H), 3.94 (dd, $J = 4.9$, 9.3 Hz, 1H), 3.60–3.58 (m, 1H), 2.96 (dd, $J = 5.0$, 18.9 Hz, 1H), 2.82 (dd, $J = 8.9$, 18.9 Hz, 1H).

^{13}C NMR (90 MHz, CDCl$_3$) δ 199.74, 168.77, 140.79, 133.60, 131.17, 130.24, 128.57, 127.35, 70.78, 49.11, 36.63.

MS (APCI) m/z: [M$^+$+1], 217.1.

HRMS (APCI), Cacld for C$_{13}$H$_{13}$O$_3$ [M$^+$+1]: 217.0865. Found: 217.0881.

The enantiomeric excess was determined by GC on a Chiral Select 1000 column at 170 °C, 1 mL/min; t_1 = 195.05 min, t_2 = 198.55 min, >99% ee; (R)-(+): $[\alpha]_D^{25} = +25.33$, (c = 1.0, CHCl$_3$), (S)-(−): $[\alpha]_D^{25} = -24.27$, (c = 2.0, CHCl$_3$).

10.3.2 SYNTHESIS OF (3-OXO-2-PENTYLIDENE-CYCLOPENTYL)-ACETALDEHYDE

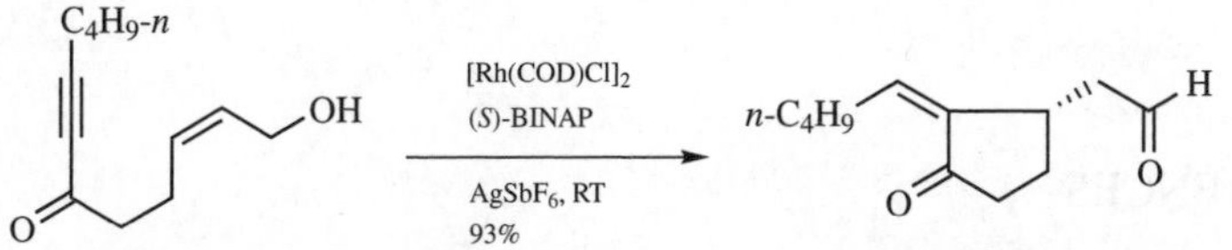

(+)-(S) >99% ee

Materials and equipment

- Rhodium (COD) chloride dimer, 25 mg, 0.05 mmol
- BINAP (S), 69 mg, 0.11 mmol

- Silver antimonite hexafluoride, 68 mg, 0.2 mmol
- 1-Hydroxy-dodec-*cis*-2-en-7-yn-6-one, 194 mg, 1.0 mmol
- Anhydrous 1,2-dichloroethane, 12 mL
- Ethyl acetate
- Hexane
- Silica gel (Sorbent, 60A, 32–63 µm), 5 g

- 25-mL Schenck tube
- Magnetic stirrer plate
- TLC plates (Sorbent, UV$_{254}$)

Procedure

In a dried 25-mL Schlenk tube under a nitrogen atmosphere, a mixture of [Rh(COD)Cl]$_2$ (25 mg, 0.05 mmol) and (*S*)-BINAP (69 mg, 0.11 mmol) was dissolved in 10 mL of 1,2-dichloroethane and stirred at room temperature for 1 minute. Then 1-hydroxy-dodec-*cis*-2-en-7-yn-6-one (194 mg, 1.0 mmol) was added into the solution at room temperature under a nitrogen atmosphere. After stirring for 1 minute, a solution of silver antimonite hexafluoride (68 mg, 0.2 mmol) in 2 mL of 1,2-dichloroethane was added into the mixture in one portion *via* a syringe. The resulting turbid solution was stirred at room temperature for 5 minutes. TLC indicated that all the starting material was consumed and the reaction completed. The reaction mixture was directly subjected to column chromatography (silica gel, hexane/ethyl acetate: 85/15) to afford 180 mg (93% yield, 99% ee (*S*)) of colorless oil.

^{1}H NMR (360 MHz, CDCl$_3$) δ 9.83 (s, 1H), 5.54 (dt, J = 2.3, 7.4 Hz, 1H), 3.22–3.20 (m, 1H), 2.70–2.49 (m, 4H), 2.34–2.16 (m, 3H), 1.57–1.47 (m, 1H), 1.39–1.20 (m, 4H), 0.87 (t, J = 7.1 Hz, 3H).

^{13}C NMR (90 MHz, CDCl$_3$) δ 207.61, 138.05, 49.28, 38.74, 36.85, 31.85, 27.83, 27.31, 22.76, 14.27.

MS m/z: 195.1 [M$^+$ + 1].

HRMS (APCI) Calcd. for C$_{12}$H$_{19}$O$_2$ [M$^+$ + 1]: 195.1385. Found: 195.1395.

The enantiomeric excess was determined by GC on a Gama 225 column at 170 °C, 1.5 mL/min; t$_1$ = 12.92 min, t$_2$ = 13.23 min, >99% ee; (S)-(+): [α]$_D^{25}$ = +4.20, (c = 1.0, CHCl$_3$).

REFERENCES

1. Lei, A., He, M. and Zhang, X. *J. Am. Chem. Soc. Comm.* **2002**, *124*, 8198.
2. Lei, A., He, M. Wu, S. and Zhang, X. *Angew. Chem. Int. Eng. Ed.* **2002**, *41*, 3607.
3. Lei, A., He, M. and Zhang, X. *J. Am. Chem. Soc. Comm.* **2003**, *125*, 11472.
4. Lei, A., Waldkirch, J. P., He, M. and Zhang, X. *Angew. Chem. Int. Eng. Ed.* **2002**, *41*, 4526.

10.4 RHODIUM-CATALYSED [2 + 2 + 2] CYCLOTRIMERIZATION IN AN AQUEOUS–ORGANIC BIPHASIC SYSTEM

HIROSHI SHINOKUBO* and KOICHIRO OSHIMA*

Department of Material Chemistry, Graduate School of Engineering, Kyoto University, Kyoto 615–8510, Japan

Transition metal catalysed [2 + 2 + 2] cyclotrimerization of alkynes is a powerful tool to construct substituted aromatic compounds.[1] Recently, cyclotrimerization of alkynes in a water–ether biphasic system with the use of a water soluble rhodium catalyst has been reported.[2,3] Distribution of a hydrophobic substrate between the two phases keeps the concentration of substrate in the aqueous phase low. The biphasic system enables medium and large-sized ring systems to be synthesised on the basis of the concentration control.

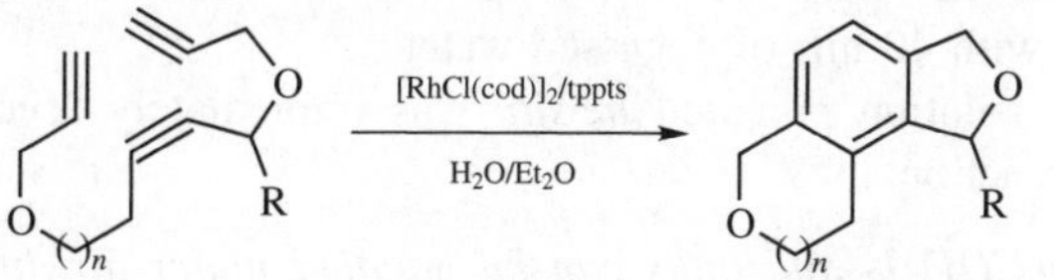

10.4.1 *IN SITU* PREPARATION OF A WATER-SOLUBLE RHODIUM CATALYST FROM [RhCl(COD)]₂ AND TRISODIUM SALT OF TRIS(m-SULFONATOPHENYL)PHOSPHINE (tppts)

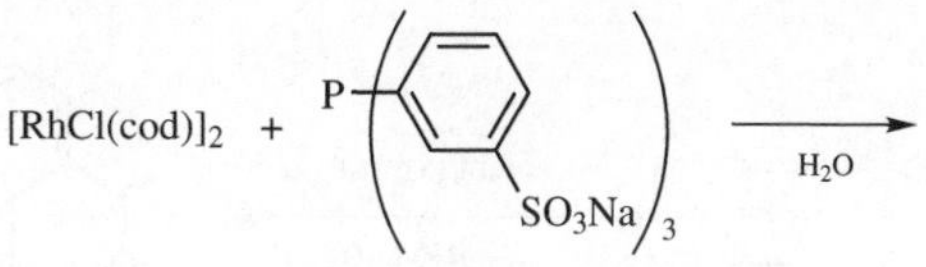

tppts

Materials and equipment

- [RhCl(COD)]₂, 6.2 mg, 0.0125 mmol
- Trisodium salt of tris(*m*-sulfonatophenyl)phosphine, 45.5 mg, 0.08 mmol
- Distilled water, 10 mL

- 100-mL round-bottomed flask
- 100-mL two-necked round-bottomed flask
- Magnetic stirrer bar
- A balloon filled with argon
- Three-way inlet
- Rubber septum
- Magnetic stirrer plate
- Oil bath
- Ultrasonic cleaner

- Vacuum pump
- Syringes
- Needles

Procedure

1. Water (approx. 60 mL) was introduced in a 100-mL flask and the flask was placed in an ultrasonic cleaner containing water in the tank. Nitrogen gas was bubbled into water through a needle with sonication for 30 minutes.
2. In a two-necked 100-mL flask equipped with a magnetic stirrer bar, a septum, and a three-way inlet attached to a balloon filled with argon, [RhCl(COD)]$_2$ (6.2 mg, 0.0125 mmol) and trisodium salt of tris(m-sulfonatophenyl)phosphine (tppts, 45.5 mg, 0.08 mmol) were placed. The flask was evacuated with argon three times. Degassed pure water (10 mL) was introduced *via* a syringe, and the mixture was stirred vigorously at 75 °C for about 1 hour until the solution was homogeneous. The clear yellow solution was cooled to room temperature and then diluted with 40 mL of degassed water.
3. The catalyst solution prepared *in situ* was immediately used for the macro-cyclization reaction.

Note: [RhCl(COD)]$_2$ and tppts can be handled under air. Inert gas line and Schlenk-type technique are not required.

10.4.2   SYNTHESIS OF 1,3,6,8,9,10,11,12,13-NONAHYDRO-2,7-DIOXACYCLODECA[E]INDENE

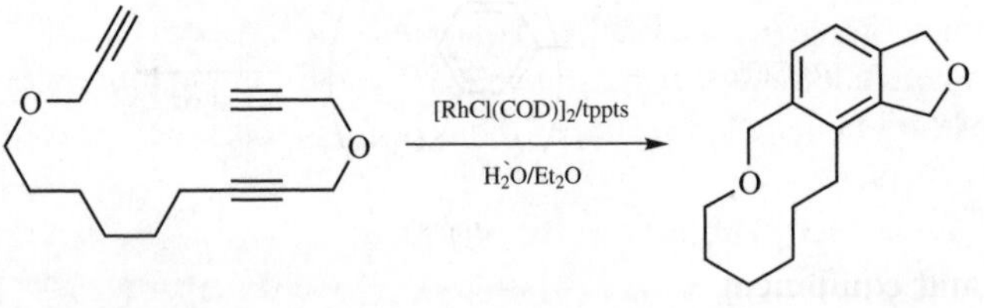

Materials and equipment

- 4,14-Dioxaheptadeca-1,6,16-triyne, 116.2 mg, 0.5 mmol
- Diethyl ether, 10 mL
- Hexane
- Ethyl acetate
- Silica gel (Wakogel 200 mesh)
- Sodium sulfate

- TLC plates, Merk Silica gel 60F$_{254}$
- Magnetic stirrer plate
- 100-mL Separating funnel
- Glass filter
- Rotary evaporator

- Syringes
- Needles

Procedure

1. Without stirring the catalyst solution prepared above, diethyl ether (5 mL) was introduced gently *via* a syringe. The system became biphasic. A solution of 4,14-dioxaheptadeca-1,6,16-triyne (116 mg, 0.5 mmol) in ether (5 mL) was syringed to the diethyl ether phase. The whole mixture was then vigorously stirred (2000 rpm) for 22 hours.
2. The reaction mixture was then transferred into a separating funnel. The mixture was extracted three times with hexane/ethyl acetate (1/1, 10 mL, each time). The organic extracts were dried through a short pad of silica gel on a sodium sulfate layer on a glass filter.
3. The organic extracts were concentrated in a rotary evaporator under reduced pressure. The residue was purified on silica gel column chromatography using hexane/ethyl acetate (5/1) as an eluent. Elution of the desired product was detected by TLC analysis. Concentration of the fraction containing the product provided 1,3,6,8,9,10,11,12,13-nonahydro-2,7-dioxacyclodeca[e]indene (103.4 mg, 0.445 mmol) in 89% yield as a white solid (m.p. 44 °C).
    ^{1}H NMR (CDCl$_3$) δ 1.35–1.43 (m, 2H), 1.52–1.78 (m, 6H), 2.70 (t, $J =$ 6.9 Hz, 2H), 3.64 (t, $J = 5.3$ Hz, 2H), 4.60 (s, 2H), 5.11 (s, 2H), 5.15 (s, 2H), 7.01 (d, $J = 7.5$ Hz, 1H), 7.12 (d, $J = 7.5$ Hz, 1H).

CONCLUSION

With this protocol, medium and large rings can be obtained in excellent yields with only trace amounts of polymeric compounds. In addition, no dimerization products were detected in the reaction mixture. Table 10.3 gives different substrates that can

Table 10.3 Macrocyclisation of triynes in a water-diethyl ether biphasic system.

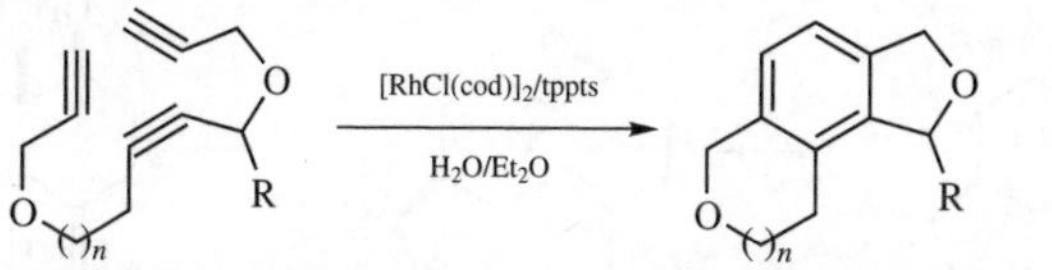

Entry	n	R	Time (h)	Yield (%)
1	1	H	10	93
2	2	H	19	84
3	3	H	19	88
4	4	H	19	91
5	5	H	22	89
6	2	Me	19	85

be cyclised with the catalyst consisting of [RhCl(COD)₂]/tppts in a water–ether biphasic system.

REFERENCES

1. Vollhardt, K. P. C. *Angew. Chem., Int. Ed. Engl.* **1984**, *23*, 539.
2. Kinoshita, H., Shinokubo, H. and Oshima, K. *J. Am. Chem. Soc.* **2003**, *125*, 7784.
3. For biphasic catalysis, see: (a) Sinou, D. In *Modern Solvent in Organic Synthesis*, Knochel, P. (Ed.) Springer-Verlag, Berlin, 1999; p. 41. (b) Cornils, B. and Herrmann, W. A. *Aqueous-Phase Organometallic Chemistry*, Wiley-VCH: Weinheim, 1998. (c) Li, C.-J. and Chan, T.-H. *Organic Reactions in Aqueous Media*, John Wiley & Sons, New York, 1997.

10.5 TITANOCENE-CATALYSED TRANSANNULAR CYCLISATION OF EPOXYGERMACROLIDES: ENANTIOSPECIFIC SYNTHESIS OF EUDESMANOLIDES

ANTONIO ROSALES, JUAN M. CUERVA, and J. ENRIQUE OLTRA*

Department of Organic Chemistry, Faculty of Sciences, University of Granada, Campus Fuentenueva s/n, E-18071 Granada, Spain

Eudesmanolides (such as (**3**)) constitute a large family of natural sesquiterpenoids, with more than 500 members described to date.[1] Their pharmacological properties include antifungal, anti-inflammatory, and antitumor activities among others,[2] but many of them are scarce in nature. To facilitate their chemical synthesis, a novel procedure has recently been developed that provides satisfactory overall yields.[3] Our method is based on the titanocene-catalyzed cyclization of epoxygermacrolides (such as (**2**)) easily prepared by selective epoxidation of accessible germacrolides such as (+)-11β,13-dihydrocostunolide (**1**)[4].

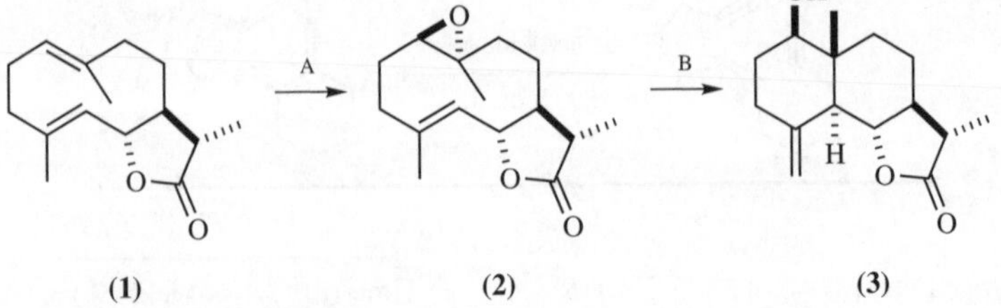

A: Selective epoxidation. B: Titanocene(III)-catalysed cyclization.

The cyclisation presumably proceeds *via* carbon-centered radicals. Nevertheless, the termination step of the process is subject to water-dependent control, which is unusual in free-radical chemistry. Thus the reaction can be controlled to afford either exocyclic alkenes or the corresponding reduction products, simply by excluding or adding water to the medium, and by choosing an adequate titanocene-regenerating agent i.e. collidine hydrochloride for the aqueous medium but Me_3SiCl/collidine for reactions under anhydrous conditions.[3]

10.5.1 PREPARATION AND TITANOCENE-CATALYSED CYCLIZATION OF EPOXYGERMACROLIDE (2): SYNTHESIS OF (+)-11β,13-DIHYDROREYNOSIN (3)

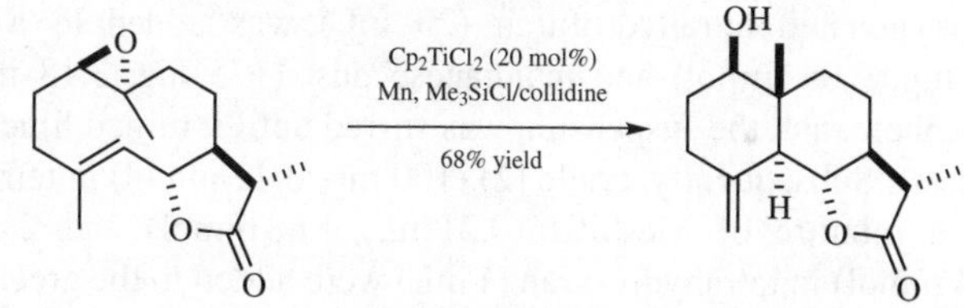

Cp_2TiCl_2 (20 mol%)
Mn, Me_3SiCl/collidine

68% yield

(2)   (3)

Materials and equipment

- *m*-Chloroperbenzoic acid (*m*-CPBA) (70%), 386 mg, 1.56 mmol
- (+)-11β,13-Dihydrocostunolide (**1**), 260 mg, 1.12 mmol, obtained from commercial Costus Resinoid[4]
- Pyridine, 0.2 mL
- Dichloromethane, 10 mL
- Cp_2TiCl_2 (bicyclopentadienyl titanium dichloride) (97%), 20 mg, 0.08 mmol
- Manganese dust (99.9%), 175 mg, 3.18 mmol
- Me_3SiCl (trimethylsilylchloride), 0.21 mL, 1.68 mmol
- 2,4,6-Collidine, 0.35 mL, 2.94 mmol
- Tetrahydrofuran, 36 mL
- *t*-BuOMe, 50 mL
- Hydrochloric acid (2 N), 5 mL
- Brine
- Anhydrous sodium sulfate
- Argon atmosphere

- Vacuum line system
- 50-mL Schlenk flask with a magnetic stirrer bar
- Magnetic stirrer plate
- 10-mL and 50-mL round-bottomed flasks
- One stainless steel cannula (1 mm diameter, 25 cm length)
- One 100-mL separating funnel
- Silica gel and flash chromatography equipment
- Twenty 100-mL Erlenmeyer flasks
- Filter paper

- One glass funnel
- Rotary evaporator

Procedure

Compound (**1**) (260 mg, 1.12 mmol), dissolved in dichloromethane (10 mL), was stirred with 70% *m*-CPBA (386 mg, 1.56 mmol) and pyridine (0.2 mL) for 3 hours. The solution was then washed with a saturated solution of Na_2SO_3 and brine. The solvent was removed and the residue (280 mg) analyzed by ^1H NMR, which revealed that it was made up of (**2**) and a trace of pyridine.

Strictly deoxygenated tetrahydrofuran (25 mL) was added to a mixture of Cp_2TiCl_2 (20 mg, 0.08 mmol) and manganese dust (175 mg, 3.18 mmol) under an argon atmosphere, and the suspension was stirred until it turned lime green (after about 15 minutes). Subsequently, crude (**2**) (100 mg, 0.42 mmol) in tetrahydrofuran (10 mL), and a mixture of Me_3SiCl (0.21 mL, 1.68 mmol) and 2,4,6-collidine (0.35 mL, 2.94 mmol) in tetrahydrofuran (1 mL) were added to the green suspension and the mixture was stirred at room temperature for 7 hours. The suspension was then filtered and the solvent was removed from the filtrate. *t*-BuOMe was then added to the residue and the ethereal solution was washed with 2N hydrochloric acid and brine. The organic layer was dried over anhydrous sodium sulfate and the solvent was removed. Flash chromatography (hexane/*t*-BuOMe: 2/3) of the residue afforded (**3**) (68 mg, 68% yield from (**1**)) as colorless needles: m.p. mp 120–122 °C; $[\alpha]^{25}_D + 98.7°$ (*c* 0.02, CHCl$_3$).

^{13}C NMR (CDCl$_3$, 100 MHz) δ 179.6 (C), 143.0 (C), 110.0 (CH$_2$), 79.4 (CH), 77.9 (CH), 52.3 (CH), 52.2 (CH), 42.8 (C), 41.1 (CH), 35.8 (CH$_2$), 33.5 (CH$_2$), 31.0 (CH$_2$), 22.9 (CH$_2$), 12.4 (CH$_3$), 11.5 (CH$_3$).

IR and ^1H NMR spectra matched those reported for natural 11β,13-dihydror-eynosin found in the plant *Michelia compresa*.[5]

10.5.2 TITANOCENE-CATALYSED CYCLIZATION OF EPOXYGERMACROLIDE (2) IN AQUEOUS MEDIUM

(2)   (4)

Materials and equipment

- 1,10-Epoxy-11β,13-dihydrocostunolide (**2**), 55 mg, 0.22 mmol
- Cp_2TiCl_2 (97%), 11 mg, 0.044 mmol
- Manganese dust (99.9%), 95 mg, 1.73 mmol

- Water, 20 µL, 1.1 mmol
- 2,4,6-Collidine (99%), 0.13 mL, 1.1 mmol
- 2,4,6-Collidine hydrochloride (>95%), 158 mg, 1.0 mmol
- Tetrahydrofuran, 20 mL
- t-BuOMe, 30 mL
- Hydrochloric acid (2 N), 5 mL
- Brine
- Anhydrous sodium sulfate
- Argon atmosphere

- Vacuum line system
- 50-mL Schlenk flask with a magnetic stirrer bar
- Magnetic stirrer plate
- 10-mL and 50-mL round-bottomed flasks
- One stainless still cannula (1 mm diameter, 25 cm length)
- One 100-mL separating funnel
- Silica gel and flash chromatography equipment
- Twenty 100-mL Erlenmeyer flasks
- Filter paper
- One glass funnel
- Rotary evaporator

Procedure

Strictly deoxygenated tetrahydrofuran (15 mL) was added to a mixture of commercially available Cp_2TiCl_2 (11 mg, 0.044 mmol) and manganese dust (95 mg) in an Schlenk-type flask under an argon atmosphere, and the suspension was stirred at 25 °C until it turned lime green (after about 15 minutes). Then a mixture of (2) (55 mg, 0.22 mmol), water (20 µL, 1.1 mmol) and 2,4,6-collidine (0.13 mL, 1.1 mmol) in tetrahydrofuran (5 mL) and, subsequently, 2,4,6-collidine hydrochloride (158 mg, 1.0 mmol) were added to the green suspension, giving a deep-blue mixture which was stirred at 25 °C for 7 hours. The solvent was then removed; t-BuOMe (30 mL) was added to the residue and the ethereal solution was washed with 2N hydrochloric acid and brine. The organic layer was dried over anhydrous sodium sulfate and the ether was removed. Flash chromatography (hexane/t-BuOMe: 1/9) of the residue afforded (+)-4α,11β,13,15-tetrahydroreynosin (4) (40 mg, 72% yield). Spectroscopic data for (4) are found in reference 4.

REFERENCES

1. a) Connolly, J. D. and Hill, R. A. *Dictionary of Terpenoids, Vol. 1*; Chapman and Hall, London, **1991**; pp. 340–371; b) Fraga, B. M. *Nat. Prod. Rep.* **2002**, *19*, 650–672, and previous issues in this series.
2. a) Hehner, S. P., Heinrich, M., Bork, P. M., Vogt, M., Ratter, F., Lehmann, Schulze-Osthoff, V., K., Dröge, W. and Schmitz, M. L. *J. Biol. Chem.* **1998**, *273*, 1288–1297;

b) Dirsch, V. M., Stuppner, H., Ellmerer-Müller, E. P. and Vollmar, A. M. *Bioorg. Med. Chem.* **2000**, *8*, 2747–2753; c) Skaltsa, H., Lazari, D., Panagouleas, C., Georgiadou, E., García, B. and Sokovic, M. *Phytochemistry* **2000**, *55*, 903–908.

3. Barrero, A. F., Rosales, A., Cuerva, J. M. and Oltra, J. E. *Organic Lett.* **2003**, *5*, 1935–1938.

4. Homochiral (+)-11β,13-dihydrocostunolide **1** can be obtained in multi-gram quantities from commercially available Costus Resinoid, see: Barrero, A. F., Oltra, J. E., Cuerva, J. M. and Rosales, A. *J. Org. Chem.* **2002**, *67*, 2566–2571.

5. Ogura, M., Cordell, G. A. and Farnsworth, N. R. *Phytochemistry*, **1978**, *17*, 957–961.

11 Asymmetric Aldol and Michael Reactions

CONTENTS

Catalysts for Fine Chemical Synthesis, Vol. 3, Metal Catalysed Carbon-Carbon Bond-Forming Reactions
Edited by S. M. Roberts, J. Xiao, J. Whittall, and T. Pickett
© 2004 John Wiley & Sons, Ltd   ISBN: 0-470-86199-1

11.1 DIRECT CATALYTIC ASYMMETRIC ALDOL REACTION OF A α-HYDROXYKETONE PROMOTED BY A Et₂Zn/LINKED-BINOL COMPLEX

MASAKATSU SHIBASAKI,* SHIGEKI MATSUNAGA and NAOYA KUMAGAI

Graduate School of Pharmaceutical Sciences, The University of Tokyo, Hongo, Bunkyo-ku, Tokyo 113-0033, Japan

The catalytic asymmetric aldol reaction is one of the most powerful and efficient asymmetric carbon-carbon bond-forming reactions. The *direct* catalytic asymmetric aldol reaction, in which *unmodified ketones* are used as substrates, is the ideal form of this reaction in terms of atom economy.[1] A first generation zinc catalyst, prepared from diethyl zinc/linked-BINOL^2 = 2/1 (Figure 11.1), efficiently

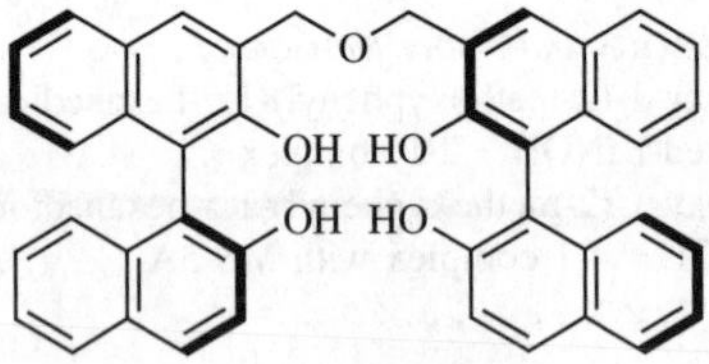

(S,S)-linked-BINOL

Figure 11.1 The structure of (S,S)-linked-BINOL.

promoted the direct catalytic asymmetric aldol reaction of α-hydroxyketones to provide *syn*-1,2-dihydroxyketones in high yields (up to 95%), good diastereomeric ratios (up to 97/3), and excellent enantiomeric excesses (up to 99% ee).[3] The second generation catalyst system, prepared from diethyl zinc/linked-BINOL = 4/1 with activated molecular sieves 3A, has much higher reactivity and as little as 0.1 mol% of catalyst is enough to complete the reaction.[4]

11.1.1 SYNTHESIS OF (2R,3S)-2,3-DIHYDROXY-1-(2-METHOXY-PHENYL)-5-PHENYL-1-PENTANONE BY THE FIRST GENERATION Et₂Zn/LINKED-BINOL = 2/1 COMPLEX

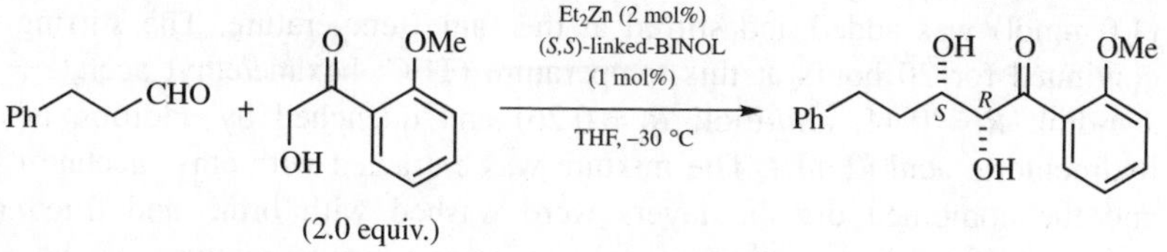

Materials and equipment

- (S,S)-Linked-BINOL (95.9 wt%, diethyl ether and hexane included), 6.41 mg, 0.01 mmol
- Diethyl zinc (Et₂Zn) (1.0 M solution in hexanes), 20 μL, 0.02 mmol
- Hydrocinnamaldehyde (99%, distilled), 132 μL, 1 mmol
- 2-Hydroxy-2'-methoxyacetophenone (>99%, recrystallised), 322.3 mg, 2 mmol
- 2,2-Dimethoxypropane, 4.5 mL
- p-Toluenesulfonic acid monohydrate, 30 mg
- Dry tetrahydrofuran, 5 mL
- Dry dimethyl formamide, 4.5 mL
- Ethyl acetate, diethyl ether, acetone, hexane
- 1 M aqueous hydrochloric acid
- Saturated aqueous solution of sodium hydrogen carbonate
- Saturated aqueous solution of sodium chloride
- Anhydrous sodium sulfate
- Silica gel (Merck 60 (230–400 mesh ASTM)), 250 g
- TLC plates, (Merck Silica Gel 60 F₂₅₄)

- 20-mL test tube with a magnetic stirrer bar
- 30-mL round-bottomed flask with magnetic stirrer bar
- Magnetic stirrer plate
- Heat gun
- One 3-way tap
- One 50-mL Erlenmeyer flask
- Glass funnels, diameter 3 cm
- One 50-mL separating funnel
- One glass column, diameter 2 cm
- Low temperature controller or dry ice with appropriate solvent
- Rotary evaporator

Procedure

1. A 20-mL, flame-dried test tube equipped with a magnetic stirrer bar and a 3-way tap with an argon balloon was charged with (S,S)-linked-BINOL (6.41 mg, 4.1 wt% diethyl ether and hexane included, 0.01 mmol) and dry tetrahydrofuran (0.3 mL) under argon. The mixture was cooled to −78 °C and diethyl zinc

(Et$_2$Zn) (20 μL, 0.02 mmol, 1.0 M in hexanes) was added at the same temperature. After stirring for 30 minutes at $-20\,°C$, a solution of 2-hydroxy-2′-
methoxyacetophenone (322.3 mg, 2.0 mmol) in tetrahydrofuran (4.7 mL) was
added. The resulting mixture was cooled to $-30\,°C$ and hydrocinnamaldehyde
(1.0 mmol) was added and stirred at the same temperature. The stirring was
continued for 20 hours at this temperature (TLC: hexane/ethyl acetate = 2/1,
syn-diol: $R_f = 0.44$, *anti*-diol: $R_f = 0.26$) and quenched by addition of 1 M
hydrochloric acid (2 mL). The mixture was extracted with ethyl acetate (× 3)
and the combined organic layers were washed with brine and dried over
sodium sulfate. Evaporation of solvent gave a crude mixture of the aldol
products.

Notes: *The purity of Et$_2$Zn solution and 2-hydroxy-2′-methoxyacetophenone was
critical for catalytic efficiency.*

*Et$_2$Zn solution should be released inside the tetrahydrofuran solution of
linked-BINOL via a dry syringe needle.*

Et$_2$Zn solution can be added and stirred for 30 minutes at $-30\,°C$.

*Longer reaction times (especially with higher catalyst loading) causes a
decrease in dr and ee.*

2. The crude residue in a 30-mL round-bottomed flask was treated with *p*-
toluenesulfonic acid monohydrate (30 mg) in dry dimethyl formamide/2,2-
dimethoxypropane (4.5 mL/4.5 mL) at room temperature for 2 hours (TLC:
hexane/acetone = 2/1, acetal derived from ketone: $R_f = 0.82$, *syn*-acetonide:
$R_f = 0.76$, *anti*-acetonide: $R_f = 0.56$). Saturated aqueous sodium hydrogen
carbonate (12 mL), water and ether were added to the mixture and the aqueous
layer was separated and extracted with diethyl ether (x 2). The combined
organic layers were washed with water and with brine (x 2) and dried over
sodium sulfate. The solvent was removed under reduced pressure and the
resulting residue was analysed to determine the diastereomeric ratio of the
aldol products by ^{1}H NMR (in CDCl$_3$, $\underline{H}$ at C-2, *syn*-acetonide: δ 4.95,
anti-acetonide: δ 5.48). The crude residue was purified by flash silica gel
column chromatography (hexane/ether/acetone 30/1/1) to afford the acetonides.
The diastereomers were separated by this procedure. The enantiomeric excesses
of the acetonides were determined by HPLC after cleavage of the
acetonides (DAICEL CHIRALCEL OD, 2-propanol/hexane 20/80, flow
1.0 mL/min, detection at 254 nm, t_R 13.6 min (minor(2*S*,3*R*)) and 15.9 min
(major(2*R*,3*S*)). The results obtained from various aldehydes are summarized in
Table 11.1.

Note: *Diastereomers can be isolated before conversion to the acetonides.*

^{1}H NMR (*syn*-acetonide) (500 MHz, CDCl$_3$) δ 1.34 (s, 3H), 1.49 (s, 3H),
1.87–1.99 (m, 2H), 2.65 (ddd, $J = 6.9, 9.8, 14.1$ Hz, 1H), 2.80 (ddd, $J = 5.5,
10.0, 14.1$ Hz, 1H), 3.82 (s, 3H), 4.21–4.23 (m, 1H), 4.95 (d, $J = 6.7$ Hz, 1H),
6.92 (brd, $J = 8.3$ Hz, 1H), 6.99 (ddd, $J = 0.9, 7.5, 7.5$ Hz, 1H), 7.10–7.12 (m,
2H), 7.14–7.17 (m, 1H), 7.22–7.25 (m, 2H), 7.42–7.48 (m, 2H).

Table 11.1 Direct catalytic asymmetric aldol reaction with the first generation $Et_2Zn/$ linked-BINOL = 2/1 system.

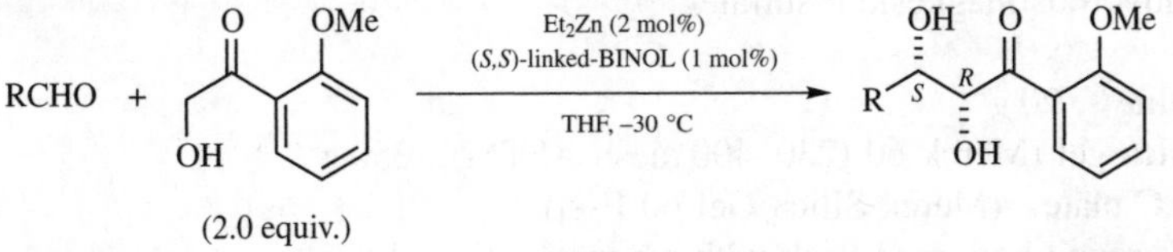

$$RCHO + \text{(2.0 equiv.)} \xrightarrow[\text{THF, }-30\text{ °C}]{\substack{Et_2Zn\ (2\ mol\%) \\ (S,S)\text{-linked-BINOL}\ (1\ mol\%)}}$$

Entry	R	Time (h)	Yield (%)	dr (*syn/anti*)	ee (*syn/anti*)
1	$Ph(CH_2)_2$	20	94	89/11	92/89
2	$CH_3(CH_2)_4$	18	88	88/12	95/91
3	$CH_3(CH_2)_5$	18	84	87/13	96/87
4	$(CH_3)_2CHCH_2$	18	84	84/16	93/87
5	$CH_3C(O)CH_2CH_2$	12	91	93/7	95/–
6	$(E)\text{-}CH_3(CH_2)_4CH{=}CH(CH_2)_2$	24	94	86/14	87/92
7	$BnO(CH_2)_2$	18	81	86/14	95/90
8	$BnOCH_2$	16	84	72/28	96/93
9	$BOMO(CH_2)_2$	14	93	84/16	90/84
10	$(CH_3)_2CH$	24	83	97/3	98/–
11	$(CH_3CH_2)_2CH$	16	92	96/4	99/–
12	*c*-Hex	18	95	97/3	98/–

11.1.2   SYNTHESIS OF (2*R*,3*S*)-3-CYCLOHEXYL-2,3-DIHYDROXY-1-(2-METHOXYPHENYL)-1-PROPANONE BY THE SECOND GENERATION $Et_2Zn/$LINKED-BINOL = 4/1 COMPLEX WITH MS3A

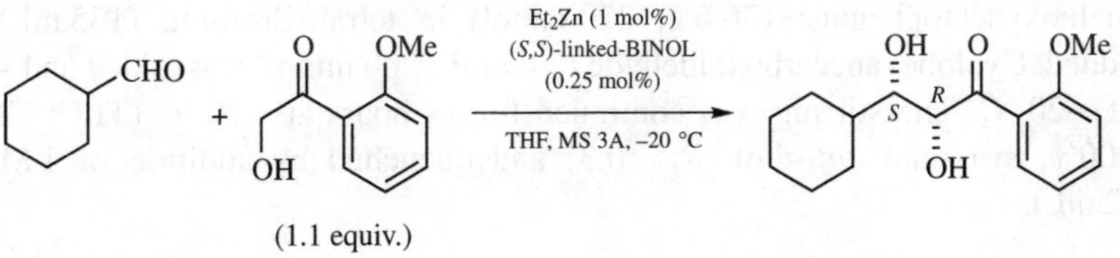

$$\xrightarrow[\text{THF, MS 3A, }-20\text{ °C}]{\substack{Et_2Zn\ (1\ mol\%) \\ (S,S)\text{-linked-BINOL} \\ (0.25\ mol\%)}}$$

(1.1 equiv.)

Materials and equipment

- (*S*,*S*)-Linked-BINOL (90w/w%, diethyl ether and hexane included), 341.5 mg, 0.5 mmol
- Molecular sieves 3A (MS 3A) (Fluka, UOP type, powder), 40 g
- Diethyl zinc (1.0 M solution in hexanes), 2.0 mL, 2.0 mmol
- Cyclohexanecarboxaldehyde (99%, distilled), 24.2 mL, 200 mmol
- 2-Hydroxy-2′-methoxyacetophenone (>99%), 36.6 g, 220 mmol
- Dry tetrahydrofuran, 335 mL
- Ethyl acetate, dichloromethane
- 1 M aqueous hydrochloric acid

- Saturated aqueous solution of sodium hydrogencarbonate
- Saturated aqueous solution of sodium chloride
- Anhydrous magnesium sulfate

- Celite®, 20 g
- Silica gel (Merck 60 (230–400 mesh ASTM)), 250 g
- TLC plates, (Merck Silica Gel 60 F_{254})
- 1-L round-bottomed flask with magnetic stirrer bar
- Magnetic stirrer plate
- Heat gun
- One 3-way tap
- One 1-eL Erlenmeyer flask
- Glass funnel, diameter 8 cm,
- One 1-L separating funnel
- One glass column, diameter 7 cm
- Low temperature controller or dry ice with appropriate solvent
- Rotary evaporator

Procedure

1. A 1-L, flame-dried round-bottomed flask equipped with a large magnetic stirrer bar and a 3-way tap was charged with a molecular sieves 3A (40 g) under argon. The molecular sieve 3A was activated under reduced pressure (ca. 0.7 kPa) at 160 °C for 3 hours. After cooling, a solution of (S,S)-linked-BINOL (341.5 mg, 10w/w% diethyl ether and hexane included, 0.5 mmol) in tetrahydrofuran (200 mL) was added under argon. The mixture was cooled to −20 °C. To the mixture was added diethyl zinc (Et$_2$Zn) (2.0 mL, 2.0 mmol, 1.0 M in hexanes) at −20 °C. After stirring for 10 minutes at −20 °C, a solution of 2-hydroxy-2'-methoxyacetophenone (36.6 g, 220 mmol) in tetrahydrofuran (135 mL) was added. Cyclohexanecarboxaldehyde (24.2 mL, 200 mmol) was added and stirred at −20 °C. The stirring was continued for 18 hours at −20 °C (TLC: CH$_2$Cl$_2$ 100%, syn- and anti-diol: $R_f = 0.3$) and quenched by addition of 1 M HCl (2 mL).

   Notes: The activation of the molecular sieve 3A is essential.
   The same precautions for Et$_2$Zn described in Section 11.1.1 should be noted.

2. The mixture was passed through a pad of Celite® and extracted with ethyl acetate and the combined organic layers were washed with saturated aqueous sodium hydrogencarbonate, brine and dried over magnesium sulfate. Evaporation of solvent gave a crude mixture of the aldol products. The diastereomeric ratio of the aldol products was determined by ¹H NMR of the crude product. After purification by silica gel flash column chromatography (hexane/CH$_2$Cl$_2$ 3/1–ethyl acetate 100%), a diastereomixture (syn/anti = 98/2) of the title compound was obtained (53.7 g, 0.898 mmol, yield 96%) as a colourless oil.

   Note: The diastereomers are inseparable at this stage.

3. The diastereomers were separated after conversion to the corresponding acetonides in the same procedure as described in Section 11.1.1 (TLC: hexane/ether/
CH_2Cl_2 4/1/1, acetal derived from ketone: $R_f = 0.80$, syn-acetonide: $R_f = 0.75$,
anti-acetonide: $R_f = 0.64$.; flash silica gel column chromatography: hexane/
ether/CH_2Cl_2 20/1/1). The enantiomeric excess of the syn-diastereomer was
determined by HPLC after cleavage of the acetonide (DAICEL CHIRALCEL
OD, 2-propanol/hexane 20/80, flow 1.0 mL/min, detection at 254 nm, t_R 6.5 min
(minor($2S,3R$)) and 12.9 min (major($2R,3S$)). The results obtained from other
aldehydes are summarized in Table 11.2.

Table 11.2 Direct catalytic asymmetric aldol reaction with the second generation Et_2Zn/
linked-BINOL $= 4/1$ with MS3A system.

Formulae for Table 2

Entry	R	Et_2Zn (x mol%)	Ligand (y mol%)	Time (h)	Yield (%)	dr (syn/anti)	ee (syn)
1	Ph(CH$_2$)$_2$	1	0.25	18	90	89/11	96
2	Ph(CH$_2$)$_2$	0.4	0.1	36	84	89/11	92
3	c-Hex	1	0.25	17	92	98/2	96

^{1}H NMR (syn-acetonide) (500 MHz, CDCl$_3$) δ 0.99–1.25 (m, 5H), 1.22 (s,
3H), 1.38 (s, 3H), 1.40–1.75 (m, 5H), 1.81–1.88 (m, 1H), 3.83 (s, 3H), 4.13 (dd,
$J = 6.4$, 6.4 Hz, 1H), 4.94 (d, $J = 6.4$ Hz, 1H), 6.91 (brd, $J = 8.5$ Hz, 1H), 6.95
(ddd, $J = 0.6$, 7.6, 7.6 Hz, 1H), 7.40 (ddd, $J = 1.5$, 7.6, 8.5 Hz, 1H), 7.45 (dd,
$J = 1.5$, 7.6 Hz, 1H).

CONCLUSION

The procedure is very easy to reproduce and applicable to various aldehydes
including readily enolizable aliphatic aldehydes. With the second generation
catalyst, the substrate/ligand ratio reached 1000, to afford the aldol product in
high chemical yield, diastereomeric ratio, and ee. The ketone moiety in the aldol
product was successfully transformed into ester and amide units via regioselective
rearrangement.[3,4] Therefore, the procedure is one of the most efficient methods to
give syn-1,2-dihydroxycarboxylic acid derivatives in a highly enantioselective
manner.

REFERENCES

1. For recent reviews on the direct catalytic asymmetric aldol reactions: (a) Alcaide, B. and Almendros, P. *Eur. J. Org. Chem.* **2002**, 1595. (b) List, B. *Tetrahedron* **2002**, *58*, 5573.
2. Linked-BINOL is commercially available from Wako Pure Chemical Industries, Ltd. Cat. No. 152-02431 for (*S,S*)-ligand, No. 155-02421 for (*R,R*)-ligand. Fax +1-804-271-7791 (USA), +81-6-6201-5964 (Japan), +81-3-5201-6590 (Japan).
3. Kumagai, N., Matsunaga, S., Yoshikawa, N., Ohshima, T. and Shibasaki, M. *Org. Lett.* **2001**, *3*, 1539.
4. Kumagai, N., Matsunaga, S., Kinoshita, T., Harada, S., Okada, S., Sakamoto, S., Yamaguchi, K. and Shibasaki, M. *J. Am. Chem. Soc.* **2003**, *125*, 2169.

11.2   HIGHLY ENANTIOSELECTIVE DIRECT ALDOL REACTION CATALYSED BY A NOVEL SMALL ORGANIC MOLECULE

ZHUO TANG, LIU-ZHU GONG*, AI-QIAO MI and YAO-ZHONG JIANG

Key Laboratory for Asymmetric Synthesis and Chirotechnology of Sichuan Province, Chengdu Institute of Organic Chemistry, Chinese Academy of Sciences, Chengdu 610041, China

Novel organic molecules derived from L-proline and amines or amino alcohols, were found to catalyse the asymmetric direct aldol reaction with high efficiency. Notably those containing L-proline amide moiety and terminal hydroxyl group could catalyse direct asymmetric aldol reactions of aldehydes in neat acetone with excellent results[1]. Catalyst (**1**), prepared from L-proline and (*1S, 2S*)-diphenyl-2-aminoethanol, exhibits high enantioselectivities of up to 93% ee for aromatic aldehydes and up to >99% ee for aliphatic aldehydes.

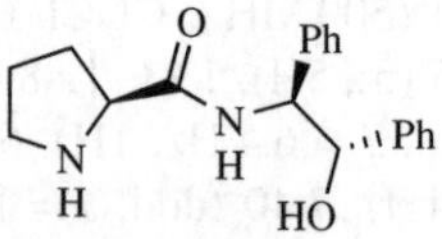

(**1**)

11.2.1   SYNTHESIS OF (*S,S,S*)-PYROLIDINE-2-CARBOXYLIC ACID (2′-HYDROXYL-1′,2′-DIPHENYL-ETHYL)-AMINE (1)

Materials and equipment

- *N*-Carbobenzyloxy-L-proline, 2.0 g, 8.0 mmol
- TEA (triethylamine), 0.81 g, 8.0 mmol
- Tetrahydrofuran, 30 mL
- Ethyl chloroformate, 0.88 g, 8.0 mmol
- (*1S,2R*)-2-amino-1,2-diphenyl-ethanol, 1.7 g, 8.0 mmol
- Ethanol, 20 ml

- 5% Pd/C, 0.1 g
- Methanol, 30 mL
- Anhydrous magnesium sulfate, 50 g

- 100-mL round-bottomed flasks
- 100-mL two-necked round-bottomed flask
- Magnetic stirring plate and magnetic stirring bars
- TLC plates, UV_{254}
- Rotary evaporator

Procedure

1. *N*-Carbobenzyloxy-L-proline (2.0 g, 8.0 mmol) and TEA (0.81 g, 8.0 mmol) were dissolved in tetrahydrofuran (30 mL). The solution was cooled to 0 °C and ethyl chloroformate was added (0.88 g, 8.0 mmol) dropwise over 15 minutes. After stirring for 30 minutes, (1*S*,2*R*)-2-amino-1,2-diphenyl-ethanol (1.7 g, 8.0 mmol) was added for 15 minutes. The resulting solution was stirred at 0 °C for 1 hour, at room temperature for 16 hours, and then refluxed for 3 hours. After cooling to room temperature, the solution was washed with ethyl acetate. The solvent was evaporated and the residue was recrystallised from ethanol to give pure (2*S*,1'*S*,2'*S*)-*N*-carbobenzyloxy -pyrrolidine-2-carboxylic acid (2-hydroxy-1,2-diphenyl-ethyl)-amide.

2. (2*S*,1'*S*,2'*S*) -*N*-Carbobenzyloxy -pyrrolidine-2-carboxylic acid (2-hydroxy-1,2-diphenyl-ethyl) -amide (1.0 g), 5% Pd/C (0.1 g) and methanol (30 mL) were introduced into a 100-mL two-necked flask. The solution was stirred under hydrogen (1 atmosphere) at 50 °C for 2 hours. After filtration, the solvent was evaporated to provide pure **(1)**. Yield: 74%, m.p. 141–144 °C; $[\alpha]^{22}_D = -23.8$ (c = 0.52, CH_3CH_2OH).

   ^{1}HNMR (300 MHz, CD_3OD) δ (ppm) 1.37 (m, 1H), 1.56 (m, 1H), 1.66 (m, 1H), 1.93 (m, 1H), 2.93 (m, 1H), 3.64 (m, 1H), 4.94 (d, *J* = 6.9 Hz, 1H), 5.16 (d, *J* = 6.9 Hz, 1H), 7.26 (m, 10H).

   ^{13}CNMR (75 MHz, CD_3OD) δ (ppm) 25.9, 31.2, 47.3, 59.3, 60.9, 76.8, 127.6, 127.8, 128.1, 128.44, 128.47, 128.5, 140.0, 142.2, 174.2.

   HRMS (FT-ICRMS) exact mass calcd for $(C_{19}H_{22}N_2O_2 + H)^+$ requires m/z 311.1760, found m/z 311.1751 (MH^+)

11.2.2 DIRECT ALDOL REACTION

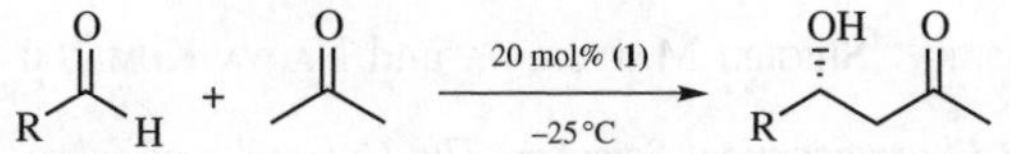

Materials and equipment

- Anhydrous acetone, 1 mL
- Aldehyde, 0.5 mmol

- Anhydrous magnesium sulfate, 50 g
- Hexane, 500 mL
- Ethyl acetate, 500 mL
- Silica gel 60 μ, 5 g

- One glass column, diameter 1.5 cm
- Cooling vessel
- Rotary evaporator

Procedure

To anhydrous acetone (1 mL) the corresponding aldehyde (0.5 mmol) and catalyst (1) (20–30 mol%) were added and the resulting mixture was stirred at −25 °C for 12–48 hours. The reaction mixture was treated with saturated ammonium chloride solution, the layers were separated, and the aqueous layer was extracted with ethyl acetate. The combined organic layer was washed with brine and dried over anhydrous magnesium sulfate. After removal of the solvent, the residue was purified by flash column chromatography over silica gel (eluent: hexane: ethyl acetate = 1 : 3) to give the pure adducts.

Entry	R	Yield(%)	ee (%) (Absolute Configuration)
1	4-NO$_2$Ph	66	93(S)
2	4-ClPh	75	93(S)
3	c-C$_6$H$_{11}$	85	97(S)
4	i-Pr	43	98(S)
5	t-Bu	57	>99(S)

REFERENCE

1. Tang, Z., Jiang, F., Yu, L.-T., Cui, X., Gong, L.-Z., Mi, A.-Q., Jiang, Y.-Z. and Wu, Y.-D. *J. Am. Chem. Soc.* **2003**, *125*, 5262–5263.

11.3   DIRECT CATALYTIC ASYMMETRIC MICHAEL REACTION OF α-HYDROXYKETONE PROMOTED BY Et$_2$Zn/LINKED-BINOL COMPLEX

MASAKATSU SHIBASAKI,* SHIGEKI MATSUNAGA and NAOYA KUMAGAI

Graduate School of Pharmaceutical Sciences, The University of Tokyo, Hongo, Bunkyo-ku, Tokyo 113-0033, Japan

The catalytic asymmetric Michael reaction is one of the most widely used synthetic methods for the asymmetric carbon-carbon bond formation.[1] Among various approaches, the *direct* use of *unmodified ketone* as a nucleophile provides an efficient, atom-economical process to afford 1,5-dicarbonyl chiral building blocks. A first generation zinc catalyst, prepared from diethyl zinc/linked-BINOL[2] = 2/1 (Figure 11.2), efficiently promoted the direct catalytic asymmetric Michael reaction of α-hydroxyketone to provide 2-hydroxy-1,5-diketones in high yields (up to 90%)

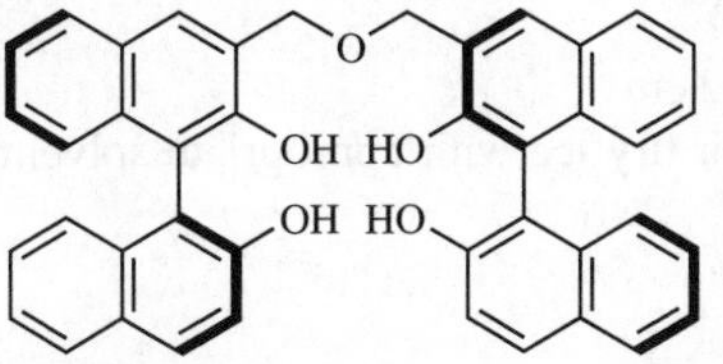

(S,S)-linked-BINOL

Figure 11.2 The structure of (S,S)-linked-BINOL.

and excellent enantiomeric excesses (up to 99% ee).[3] The second generation catalyst system, prepared from diethyl zinc/linked-BINOL = 4/1 with activated molecular sieves 3A, has a much higher reactivity and as little as 0.01 mol% of catalyst can promote the reaction.[4]

11.3.1 SYNTHESIS OF (2R)-2-HYDROXY-1-(2-METHOXY-PHENYL)-1,5-HEXANEDIONE BY THE FIRST GENERATION Et$_2$Zn/LINKED-BINOL = 2/1 COMPLEX

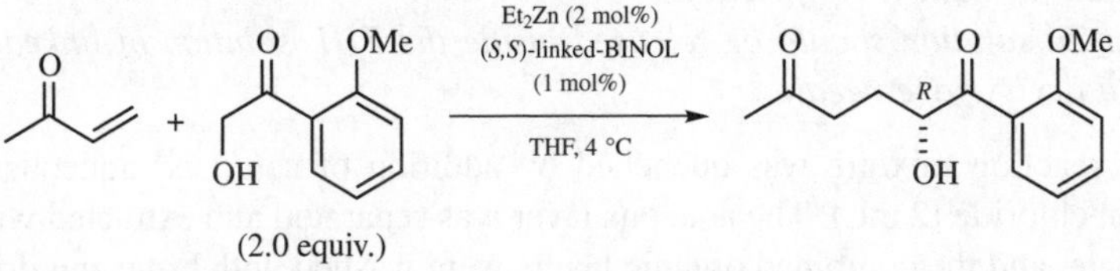

(2.0 equiv.)

Materials and equipment

- (S,S)-Linked-BINOL (95.9 w/w%, diethyl ether and hexane included), 6.41 mg, 0.01 mmol
- Diethyl zinc (1.0 M solution in hexanes), 20 μL, 0.02 mmol
- Methyl vinyl ketone (99%, distilled), 82 μL, 1 mmol
- 2-Hydroxy-2'-methoxyacetophenone (>99%), 322.3 mg, 2 mmol
- Dry tetrahydrofuran, 5 mL
- Ethyl acetate, hexane
- Saturated aqueous ammonium chloride
- Saturated aqueous solution of sodium chloride
- Anhydrous sodium sulfate
- Silica gel (Merck 60 (230–400 mesh ASTM)), 15 g
- TLC plates, (Merck Silica Gel 60 F$_{254}$)

- 20-mL test tube with a magnetic stirrer bar
- Magnetic stirrer plate
- Heat gun
- One 3-way tap
- One 50-mL Erlenmeyer flask
- Glass funnels, diameter 3 cm
- One 50-mL separating funnel
- One glass column, diameter 2 cm
- Low temperature controller or dry ice with appropriate solvent
- Rotary evaporator

Procedure

1. A 20-mL, flame-dried test tube equipped with a magnetic stirrer bar and a 3-way tap with an argon balloon was charged with (S,S)-linked-BINOL (6.41 mg, 4.1 wt% diethyl ether and hexane included, 0.01 mmol) and dry tetrahydrofuran (0.3 mL) under argon. The mixture was cooled to $-78\,°C$ and diethyl zinc (Et_2Zn) (20 µL, 0.02 mmol, 1.0 M in hexanes) was added at the same temperature. After stirring for 30 minutes at $-20\,°C$, a solution of 2-hydroxy-2'-methoxyacetophenone (322.3 mg, 2.0 mmol) in tetrahydrofuran (4.7 mL) was added. The resulting mixture was warmed to $4\,°C$, and methyl vinyl ketone (82 µL, 1.0 mmol) was added. The stirring was continued for 4 hours at this temperature (TLC: hexane/ethyl acetate = 2/1, Michael product $R_f = 0.30$).

   Notes: The purity of Et_2Zn solution and 2-hydroxy-2'-methoxyacetophenone was critical for catalytic efficiency.

   Et_2Zn solution should be released inside the THF solution of linked-BINOL via a dry syringe needle.

2. The reaction mixture was quenched by addition of saturated aqueous ammonium chloride (2 mL). The aqueous layer was separated and extracted with ethyl acetate, and the combined organic layers were washed with brine and dried over sodium sulfate. Evaporation of oraganic solvent gave a crude mixture of the Michael product.

3. The crude residue was purified by flash silica gel column chromatography (hexane/acetone 10/1–5/1) to give the title compound (203.6 mg, 0.86 mmol, 86% yield, 93% ee) as a colourless oil. The enantiomeric excess was determined by HPLC (DAICEL CHIRALCEL OD column, 2-propanol/hexane 20/80 as eluent, flow 1.0 mL/min, detection at 254 nm, t_R 8.0 min (minor) and 10.1 min (major)). The results obtained from various α,β-unsaturated ketones are summarised in Tables 11.3 and 11.4.

   ^{1}H NMR (500 MHz, $CDCl_3$) δ 1.55–1.63 (m, 1H), 2.07–2.17 (m, 1H), 2.11 (s, 3H), 2.48 (ddd, $J = 5.2, 9.5, 17.7$ Hz, 1H), 2.64 (ddd, $J = 6.4, 9.2, 17.7$ Hz, 1H), 3.84 (d, $J = 5.8$ Hz, 1H), 3.90 (s, 3H), 5.09 (ddd, $J = 3.7, 5.8, 8.3$ Hz, 1H), 6.96 (brd, $J = 8.2$ Hz, 1H), 7.03 (ddd, $J = 0.9, 7.3, 7.6$ Hz, 1H), 7.51 (ddd, $J = 1.8, 7.3, 8.2$ Hz, 1H), 7.79 (dd, $J = 1.8, 7.6$ Hz, 1H).

Table 11.3 Direct catalytic asymmetric Michael reaction with the first generation Et_2Zn/linked-BINOL = 2/1 system (1).

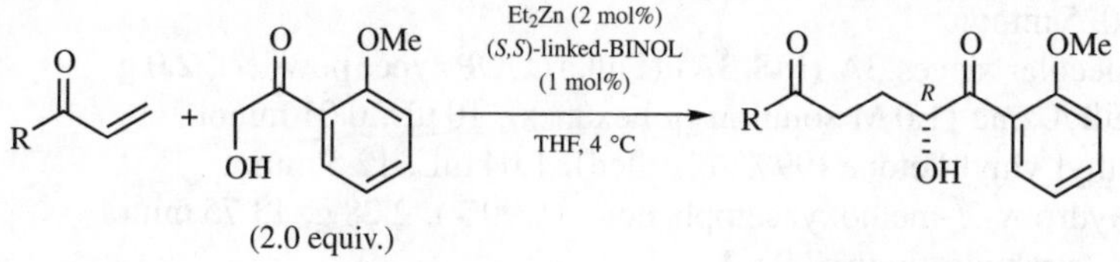

(2.0 equiv.)

Entry	R	Time (h)	Yield (%)	ee
1	p-MeOC$_6$H$_4$	8	83	95
2	C$_6$H$_5$	4	86	93
3	o-MeOC$_6$H$_4$	12	90	94
4	p-ClC$_6$H$_4$	12	84	92
5	CH$_3$	4	86	93
6	CH$_3$CH$_2$	4	82	91

Table 11.4 Direct catalytic asymmetric Michael reaction with the first generation Et_2Zn/linked-BINOL = 2/1 system (2).

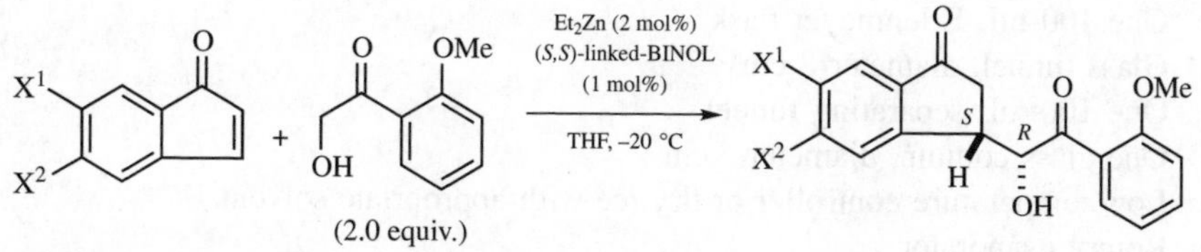

(2.0 equiv.)

Entry	X^1	X^2	Time (h)	Yield (%)	dr (*syn/anti*)	ee (*syn/anti*)
1	H	H	4	74	98/2	99
2	Br	H	4	74	98/2	99
3	H	OMe	4	65	97/3	97

11.3.2 SYNTHESIS OF (2R)-2-HYDROXY-1-(2-METHOXYPHENYL)-1,5-HEXANEDIONE BY THE SECOND Et$_2$Zn/LINKED-BINOL = 4/1 COMPLEX WITH MS 3A

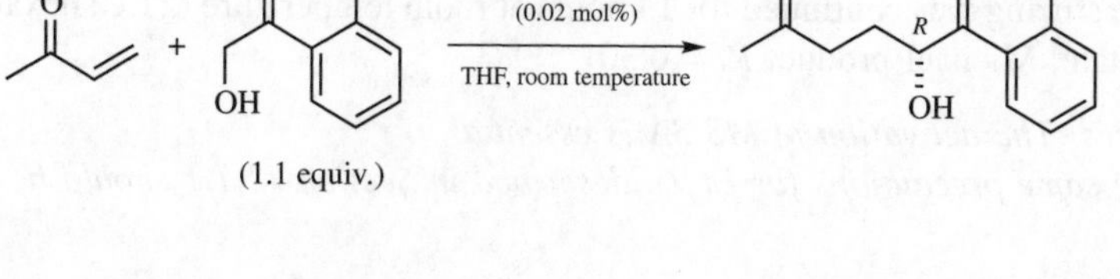

(1.1 equiv.)

Materials and equipment

- (S,S)-Linked-BINOL (90w/w%, diethyl ether and hexane included), 1.5 mg, 0.0025 mmol
- Molecular sieves 3A (MS 3A) (Fluka, UOP type, powder), 2.0 g
- Diethyl zinc (1.0 M solution in hexanes), 10 µL, 0.01 mmol
- Methyl vinyl ketone (99%, distilled), 1.04 mL, 12.5 mmol
- 2-Hydroxy-2'-methoxyacetophenone (>99%), 2.28 g, 13.75 mmol
- Dry tetrahydrofuran, 7.9 mL
- Ethyl acetate, hexane, acetone
- 1 M aqueous hydrochloric acid
- Saturated aqueous solution of sodium hydrogencarbonate
- Saturated aqueous solution of sodium chloride
- Anhydrous magnesium sulfate
- Celite®, 10 g
- Silica gel (Merck 60 (230–400 mesh ASTM)), 120 g
- TLC plates, (Merck Silica Gel 60 F_{254})

- 50-mL round-bottomed flask with magnetic stirrer bar
- Magnetic stirrer plate
- Heat gun
- One 3-way tap
- One 100-mL Erlenmeyer flask
- Glass funnel, diameter 8 cm
- One 100-mL separating funnel
- One glass column, diameter 5 cm
- Low temperature controller or dry ice with appropriate solvent
- Rotary evaporator

Procedure

1. A 50-mL, flame-dried round-bottomed flask equipped with a magnetic stirrer bar and a 3-way tap was charged with MS 3A (2.0 g) under argon. MS 3A was activated under reduced pressure (ca. 0.7 kPa) at 160 °C for 3 hours. After cooling, a solution of (S,S)-linked-BINOL (1.5 mg, 10 wt% diethyl ether and hexane included, 0.0025 mmol) in tetrahydrofuran (1.0 mL) was added under argon. The mixture was cooled to −20 °C. To the mixture was added diethyl zinc (10 µL, 0.01 mmol, 1.0 M in hexanes) at −20 °C. After stirring for 10 minutes at −20 °C, a solution of 2-hydroxy-2'-methoxyacetophenone (2.28 g, 13.75 mmol) in THF (6.9 mL) was added. The mixture was warmed to room temperature and methyl vinyl ketone (1.04 mL, 12.5 mmol) was added. The stirring was continued for 13 hours at room temperature (TLC: hexane/ethyl acetate, Michael product $R_f = 0.30$).

Notes: The activation of MS 3A is essential.
The same precautions for Et_2Zn described in Section 11.1.1 should be noted.

2. The reaction mixture was quenched with 1 M hydrochloric acid. The mixture was passed through a pad of Celite® and extracted with ethyl acetate and the combined organic layers were washed with saturated aqueous sodium hydrogen carbonate, brine and dried over magnesium sulfate. Evaporation of organic solvent gave a crude mixture of the Michael product.
3. The crude residue was purified by flash silica gel column chromatography (hexane/acetone 10/1–5/1) to afford the title compound (2.66 g, 11.3 mmol, 90% yield, 91% ee) as a colourless oil. The enantiomeric excess was determined by HPLC (HPLC conditions were same as Section 11.1.1). The results obtained from various α,β-unsaturated ketones are summarised in Tables 11.5 and 11.6.

Table 11.5 Direct catalytic asymmetric Michael reaction with the second generation $Et_2Zn/$ linked-BINOL = 4/1 system with MS 3A (1).

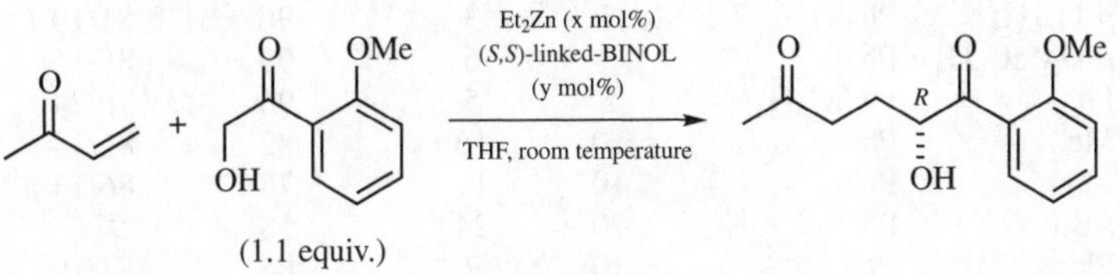

(1.1 equiv.)

Entry	Et_2Zn (x mol%)	Ligand (y mol%)	Substrate/ligand	Time (h)	Yield (%)	ee
1	0.08	0.02	5000	13	90	91
2	0.04	0.01	10000	28	78	89

CONCLUSION

The procedure is very easy to reproduce and applicable to various α,β-unsaturated ketones including β-substituted substrates. With the second generation catalyst, the substrate/ligand ratio reached 10 000 to afford the Michael product in good chemical yield and high ee. The aromatic ketone moiety in the Michael product was successfully transformed into the corresponding ester and amide units *via* regioselective rearrangement.[3,4] The procedure provides a practical method for the synthesis of 2-hydroxy-1,5-dicarbonyl chiral building blocks.

REFERENCES

1. For recent reviews on the catalytic asymmetric Michael reactions: (a) Krause, N. and Hoffmann-Röder, A. *Synthesis* **2001**, 171. (b) *Comprehensive Asymmetric Catalysis*; Jacobsen, E. N., Pfaltz, A. and Yamamoto, H., (Eds.), Springer, Berlin, **1999**, Chapter 31.

Table 11.6 Direct catalytic asymmetric Michael reaction with the second generation Et_2Zn/ linked-BINOL = 4/1 system with MS 3A(2).

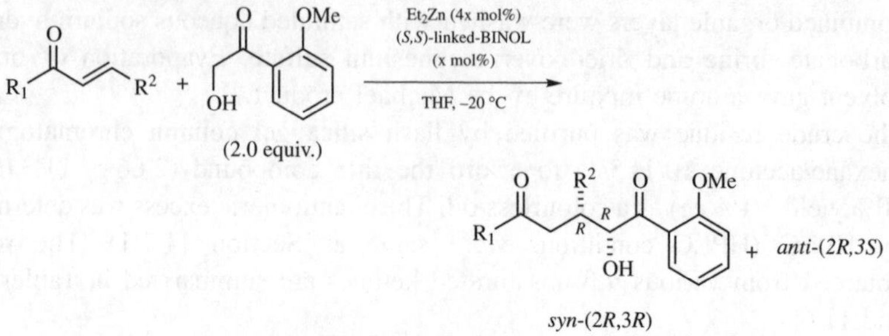

Entry	R^1	R^2	Cat. (mol%)	Time (h)	Yield (%)	dr (syn/anti)	ee (syn/anti)
1	Ph	Ph	5	3	93	78/22	95/93
2	p-ClC$_6$H$_4$	Ph	5	3	95	79/21	97/83
3	p-FC$_6$H$_4$	Ph	5	3	96	81/19	97/80
4	p-OMeC$_6$H$_4$	Ph	5	6	99	85/15	97/52
5	Ph	p-ClC$_6$H$_4$	5	5	96	76/24	95/71
6	Me	Ph	10	16	82	86/14	99/7
7	i-Pr	Ph	10	17	73	86/14	87/–
8	t-Bu	Ph	10	24	58	93/7	74/–
9	Ph	Et	10	7	85	81/19	97/79
10	Ph	i-Bu	5	7	95	69/31	97/65
11	Ph	CH$_2$OTBDPS	10	24	93	61/39	81/52
12	Ph	(CH$_2$)$_2$BOM	10	24	72	77/23	80/–
13	Me	(CH$_2$)$_2$CH$_3$	10	24	39	68/32	93/86

In entry 9, 12, and 13, major isomer had syn-(2R, 3S) configuration.

2. Linked-BINOL is commercially available from Wako Pure Chemical Industries Ltd., Cat. No. 152-02431 for (S,S)-ligand, No. 155-02421 for (R,R)-ligand.
3. Kumagai, N., Matsunaga, S. and Shibasaki, M. *Org. Lett.* **2001**, *3*, 4251.
4. Harada, S., Kumagai, N., Kinoshita, T., Matsunaga, S. and Shibasaki, M. *J. Am. Chem. Soc.* **2003**, *125*, 2582.

11.4   CATALYTIC ENANTIOSELECTIVE MICHAEL REACTION CATALYSED BY WELL-DEFINED CHIRAL RUTHENIUM-AMIDO COMPLEXES

MASAHITO WATANABE, KUNIHIKO MURATA and TAKAO IKARIYA*

Department of Applied Chemistry, Graduate School of Science and Engineering, Tokyo Institute of Technology and Frontier Collaborative Research Center, Ookayama, Meguro-ku, Tokyo 152-8552, Japan

Catalytic asymmetric Michael reactions represent one of the most important organic synthetic procedures for stereoselective carbon-carbon bond forming reaction, partly due to their high atom economy. There are many reports on enantioselective Michael-type reactions catalysed by chiral metal catalyst systems including copper, nickel, cobalt, rhodium, palladium and heterobimetallic complexes as well as with chiral organic compounds as catalysts.[1–7] It has now been found that the chiral ruthenium-amido complexes (**1**)[8b] (Figure 11.3, the ruthenium-amido complex, Ru[(*R*,*R*)-Tsdpen](η^6-arene), (*R*,*R*)-TsDPEN = (1*R*,2*R*)-*N*-(*p*-toluenesulfonyl)-1,2-diphenylethylenediamine), efficiently initiate catalytic enantioselective Michael addition of malonates to cyclic enones to provide the corresponding Michael adducts in high yields and with excellent ee's (Figure 11.4).[9]

chiral Ru cat:

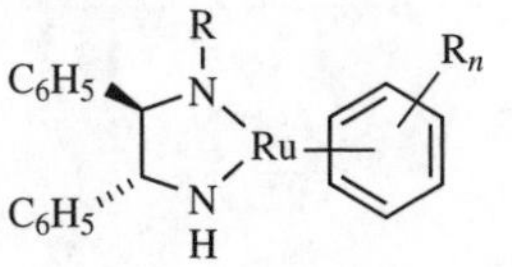

1a: R =	Ts	R_n = 1-CH_3-4-$CH(CH_3)_2$: *p*-cymene
1b:	Ts	1,3,5-$(CH_3)_3$: mesitylene
1c:	Ts	1,2,4,5-$(CH_3)_4$: durene
1d:	Ts	1,2,3,4,5-$(CH_3)_5$: pmb
1e:	Ts	1,2,3,4,5,6-$(CH_3)_6$: hmb
1f:	Ms	1,2,3,4,5,6-$(CH_3)_6$: hmb

Figure 11.3 The structure of ruthenium-amido complexes.

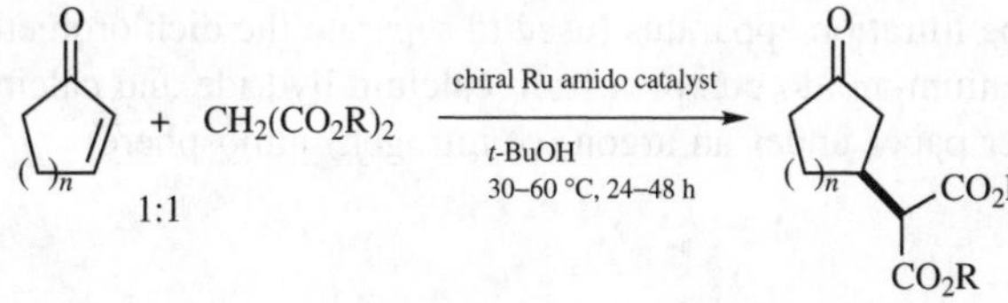

S/C = 50–100

99% yield
up to 99% ee

acceptor: n = 1, **2a**
n = 2, **2b**
n = 3, **2c**
n = 1 (4,4 dimethyl), **2d**

donor: $CH_2(CO_2CH_3)_2$, **3a**
$CH_2(CO_2C_2H_5)_2$, **3b**
$CH(CH_3)(CO_2CH_3)_2$, **3c**

Figure 11.4 Enantioselective Michael addition of malonates to cyclic enones.

11.4.1 SYNTHESIS OF Ru[(R,R)-TsDPEN](η^6-ARENE); Ru[(*R*,*R*)-TsDPEN](p-CYMENE), ((*R*,*R*)-TSDPEN = (1*R*,2*R*)-*N*-(*P*-TOLUENESULFONYL)-1,2-DIPHENYLETHYLENEDIAMINE), (*P*-CYMENE = η^6-1-CH_3-4-$CH(CH_3)_2C_6H_4$)[8B], Ru[(*R*,*R*)-TsDPEN](HMB), AND Ru[(*R*,*R*)-MsDPEN](HMB)

Materials and equipment

- $[RuCl_2(p\text{-cymene})]_2$[10], 306.2 mg, 0.50 mmol
- (*R*,*R*)-TsDPEN[11], 366.4 mg, 1.0 mmol

- Potassium hydroxide (86%), 462 mg, 7.1 mmol
- Dichloromethane, 7 mL
- Water, degassed, *ca* 30 mL
- Sodium sulfate, *ca* 1 g
- Calcium hydride, *ca* 200 mg
- $[RuCl_2(hmb)]_2^{[12]}$, (134 mg, 0.20 mmol) × 2
- (*R,R*)-TsDPEN, (134 mg, 0.40 mmol)
- (*R,R*)-MsDPEN, (116 mg, 0.40 mmol)
- Potassium hydroxide (86%), (183 mg, 2.80 mmol) × 2
- Dichloromethane, 4.0 mL × 2
- Water, degassed, *ca* 27 mL × 2
- Sodium sulfate, *ca* 1 g × 2
- Calcium hydride, *ca* 200 mg × 2

- Vacuum line
- Magnetic stirrer plate
- Schlenk flask, 20 mL
- Syringes, 5 ml and 10 mL
- Schlenk type filtration apparatus (used to separate the dichloromethane solution of the ruthenium-amido complex from calcium hydride and calcium hydroxide using a filter paper under an argon (or nitrogen) atmosphere)
- Filter paper

Procedure for synthesis of the ruthenium-amido complexes; Ru[(*R,R*)-Tsdpen](*p*-cymene), Ru[(*R,R*)-Tsdpen](hmb), and Ru[(*R,R*)-Msdpen](hmb), (hmb = η^6-hexamethylbenzene)

Ru[(*R,R*)-Tsdpen](*p*-cymene) was prepared according to a published procedure with a slight modification.[8b] A mixture of $[RuCl_2(p\text{-cymene})]_2$ (306.2 mg, 0.5 mmol), (*R,R*)-TsDPEN, (366.4 mg, 1.0 mmol), and potassium hydroxide (462 mg, 7.1 mmol) in dichloromethane was stirred at room temperature for 5 minutes. On addition of water to the reaction mixture, the colour changed from orange to deep purple. The purple organic layer was washed 8 times with degassed water (4 mL × 8). The washing of the organic layer by water should be repeated to get the almost neutral water layer because the residual contaminants such as potassium hydroxide or potassium chloride in the catalyst may promote a non-stereoselective Michael reaction. After drying over sodium sulfate, the organic layer was dried over calcium hydride (*ca* 200 mg, 4.8 mmol), filtered though a paper filter using a Schlenk type filteration apparatus under an argon (or nitrogen) atmosphere. The filtrate was concentrated to dryness to afford deep purple Ru[(*R,R*)-Tsdpen](*p*-cymene) (516 mg, 86% yield).

^{1}H NMR (400 MHz, toluene-d_8): δ 1.20, 1.25 (d × 2, $^3J = 7$ Hz, 3H × 2, CH(C*H*$_3$)$_2$), 2.05 (s, 3H, C*H*$_3$ of *p*-cymene), 2.22 (s, 3H, C*H*$_3$ of Ts), 2.53 (m, 1H, C*H*(CH$_3$)$_2$), 4.08 (d, $^3J = 4.4$ Hz, 1H, C*H*NH), 4.89 (s, 1H, C*H*NHTs), 5.11, 5.27, 5.28, 5.39 (d × 4, $^3J = 6$ Hz, 1H × 4, aromatic ring protons of *p*-cymene),

6.64 (br d, 1H, NH), 6.87, 7.67 (d × 2, $^3J = 8$ Hz, 2H × 2, aromatic ring protons of Ts), 7.2–7.7 (m, 10H, aromatic ring protons).

Anal. calcd. for $C_{31}H_{34}N_2O_2RuS$: C 62.09, H 5.71, N 4.67, Ru 16.85; found: C 62.06, H 5.77, N 4.66, Ru 16.47.

The molecular structure of Ru[(S,S)-Tsdpen](p-cymene) was determined by the X-ray single crystal structural analysis.[8b]

- [RuCl$_2$(hmb)]$_2$ was prepared from the reaction of commercially available [RuCl$_2$(p-cymene)]$_2$ and hexamethylbenzene prepared according to the published procedure.[12]
- (R,R)-MsDPEN was prepared from the reaction of (R,R)-DPEN and MsCl in the presence of triethylamine in dichloromethane at 0 °C.[11]
- Ru[(R,R)-Tsdpen](hmb) was obtained as a deep purple complex in a similar manner to that described above (213 mg 85% yield).
- Analytical data, ^{1}H NMR (300 MHz, CD$_2$Cl$_2$): δ 2.14 (s, 3H, CH_3 of Ts), 2.25 (s, 18H, CH_3 of hmb), 3.84 (d, $^3J = 4.2$ Hz, 1H, CHNH), 4.20 (s, 1H, CHNHTs), 5.84 (br d, 1H, NH), 6.70–7.50 (m, 14H, aromatic ring protons).
- In a similar procedure, Ru[(R,R)-Msdpen](hmb) was obtained in 90% yield (199 mg).
- Analytical data, ^{1}H NMR (300 MHz, CD$_2$Cl$_2$): δ, 2.21 (s, 3H, CH_3 of Ms), 2.30 (s, 18H, CH_3 of hmb), 3.87 (d, $^3J = 4.2$ Hz, 1H, CHNH), 4.22 (s, 1H, CHNHMs), 5.97 (br d, 1H, NH), 7.05–7.50 (m, 10H, aromatic ring protons).
- This procedure has been scaled up to 200 g scale.

11.4.2 SYNTHESIS OF (S)-3-DI(METHOXYCARBONYL)METHYL-1-CYCLOPENTANONE FROM THE MICHAEL REACTION OF DIMETHYL MALONATE AND 2-CYCLOPENTEN-1-ONE CATALYSED BY Ru[(R,R)-TsDPEN](HMB)

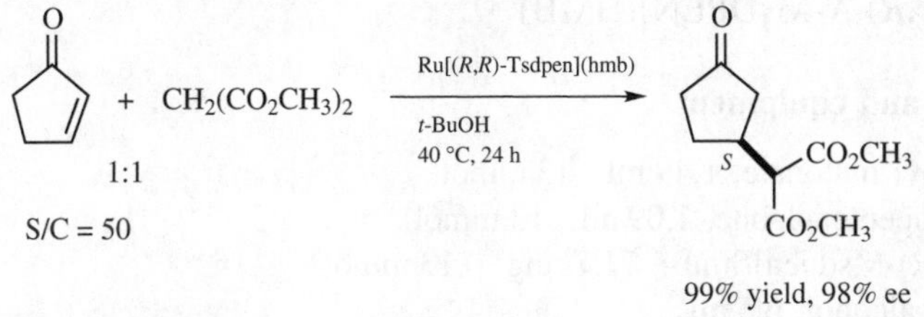

99% yield, 98% ee

Materials and equipment

- Dimethyl malonate, 114 μL, 1.0 mmol
- 2-Cyclopenten-1-one, 84 μL, 1.0 mmol
- Ru[(R,R)-Tsdpen](p-cymene) (12.6 mg, 0.02 mmol)
- t-Butyl alcohol, 1.0 mL
- n-Hexane
- Ethyl acetate

- Vacuum line
- Magnetic stirrer plate
- Schlenk flask, 20 mL
- Syringes 100 μL, 250 μL, 1 mL
- Water bath
- Siliga gel 60 (Merck 9385)
- Rotary evaporator

Procedure

Dimethyl malonate (114 μL, 1.0 mmol), 2-cyclopenten-1-one (84 μL, 1.0 mmol) and *t*-butyl alcohol (1.0 mL) were added to Ru[(R,R)-Tsdpen](hmb) (12.6 mg, 0.02 mmol) and the mixture was degassed by freeze-thaw cycles. The mixture was stirred at 40 °C for 24 hours, then evaporated with a vacuum pump and purified by column chromatography (silica gel, eluent: *n*-hexane:acetone = 9:1) to give (*S*)-3-di(methoxycarbonyl)methyl-1-cyclopentanone in 98% isolated yield (98% ee).

^{1}H NMR (300 MHz, CDCl$_3$): δ 1.60–1.75 (m, 1H, cyclopentanone ring proton), 2.01 (ddd, $J = 1.2, 11.2, 18.3$ Hz, 1H, cyclopentanone ring proton), 2.12–2.44 (m, 3H, cyclopentanone ring protons), 2.52 (dd, $J = 7.8, 18.6$ Hz, 1H, cyclopentanone ring proton), 2.78–2.97 (m, 1H, cyclopentanone ring proton), 3.39 (d, $J = 9.3$ Hz, 1H, C*H*(COOCH$_3$)$_2$), 3.75, 3.78 (s, 3H × 2, C*H$_3$* × 2).

HPLC separation conditions (column: CHIRALPAK AS (4.6 mm i.d. × 250 mm), eluent: Hexane/2-propanol = 80/20, flow rate: 1.0 mL/min, temp: 30 °C, detection UV 210 nm); retention time, (*R*), 25.1 min, (*S*), 33.0 min. $[\alpha]_D^{21}$ −96.3 (*c* 0.54 CHCl$_3$) (lit. $[\alpha]_D^{24}$ +98.8 (*c* 0.54 CHCl$_3$), >99% ee (*R*)[13]).

11.4.3   SYNTHESIS OF (*S*)-3-DI(METHOXYCARBONYL)METHYL-1-CYCLOPENTANONE FROM THE MICHAEL REACTION OF DIMETHYL MALONATE AND CYCLOPENTENONE CATALYZED BY Ru[(*R,R*)-*N*-MsDPEN](HMB)

Materials and equipment

- Dimethyl malonate, 1.49 mL, 13 mmol
- 2-Cyclopenten-1-one, 1.09 mL, 13 mmol
- Ru[(*R,R*)-Msdpen](hmb), 71.7 mg, 0.13 mmol
- *t*-Butyl alcohol, 6.5 mL
- *n*-Hexane
- Ethyl acetate

- Vacuum line
- Magnetic stirrer plate
- Schlenk flask, 20 mL
- Syringes, 2 ml, 10 mL
- Water bath

- Siliga gel 60 (Merck 9385)
- Rotary evaporator

Procedure

Dimethyl malonate (1.49 mL, 13 mmol), 2-cyclopenten-1-one (1.09 mL, 13 mmol) and *t*-butyl alcohol (6.5 mL) were added to Ru[(*R,R*)-Msdpen](hmb) (71.7 mg, 0.13 mmol) and the mixture was degassed by freeze-thaw cycles. The mixture was stirred at 30 °C for 50 hours, then evaporated with a vacuum pump and purified by column chromatography (silica gel, eluent: *n*-hexane:acetone = 9:1) to give (*S*)-3-di(methoxycarbonyl)methyl-1-cyclopentanone in 99% isolated yield (98% ee).

The reaction under similar conditions except for heating the mixture at 60 °C for 24 hours gave (*S*)-3-di(methoxycarbonyl)methyl-1-cyclopentanone in 99% isolated yield (97% ee).

11.4.4 SYNTHESIS (*S*)-3-DI(METHOXYCARBONYL)METHYL-1-CYCLOHEXANONE FROM THE MICHAEL REACTION OF DIMETHYL MALONATE AND CYCLOHEXENONE CATALYSED BY Ru[(*R,R*)-MsDPEN](HMB)

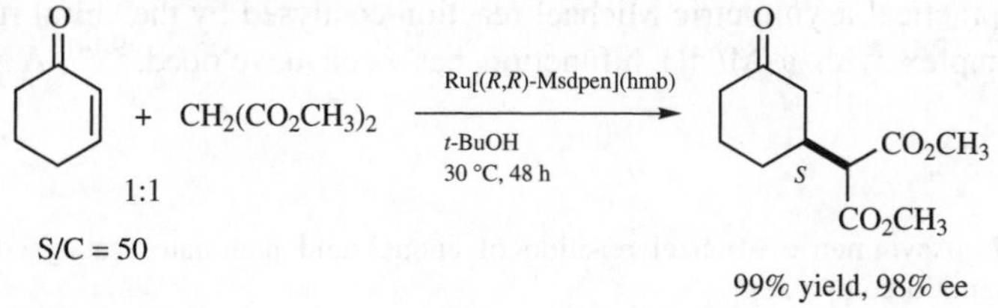

Materials and equipment

- Dimethyl malonate, 114 μL, 1.0 mmol
- 2-Cyclohexen-1-one, 97 μL, 1.0 mmol
- Ru[(*R,R*)-Msdpen](hmb), 11.0 mg, 0.02 mmol
- *t*-Butyl alcohol, 1.0 mL
- *n*-Hexane
- Ethyl acetate

- Vacuum line
- Magnetic stirrer plate
- Schlenk flask, 20 mL
- Syringes, 100 μL, 250 μL, 1 mL
- Water bath
- Siliga gel 60 (Merck 9385)
- Rotary evaporator

Procedure

Dimethyl malonate (114 µL, 1.0 mmol), 2-cyclohexen-1-one (97 µL, 1.0 mmol) and *t*-butyl alcohol (1.0 mL) were added to Ru[(*R*,*R*)-Msdpen](hmb) (11.0 mg, 0.02 mmol) and the mixture was degassed by freeze-thaw cycles. The mixture was stirred at 30 °C for 48 hours, then evaporated with a vacuum pump and purified by column chromatography (silica gel, eluent: *n*-hexane:acetone = 9:1) to give (*S*)-3-di(methoxycarbonyl)methyl-1-cyclohexanone in 99% isolated yield (98% ee).

^{1}H NMR (300 MHz, CDCl$_3$): δ, ppm 1.40–1.60 (m, 1H, cyclohexanone ring proton), 1.60–1.80 (m, 1H, cyclohexanone ring proton), 1.89–2.01 (m, 1H, cyclohexanone ring proton), 2.02–2.15 (m, 1H, cyclohexanone ring proton), 2.19–2.35 (m, 2H, cyclohexanone ring protons), 2.35–2.65 (m, 3H, cyclohexanone ring protons), 3.35 (d, *J* = 7.8 Hz, 1H, CH(COOCH$_3$)$_2$), 3.75, 3.76 (s, 3H × 2, CH$_3$ × 2).

HPLC separation conditions (column: CHIRALPAK AS (4.6 mm i.d. × 250 mm), eluent: Hexane/2-propanol = 85/15, flow rate: 0.5 ml/min, temp: 30 °C, detection UV 210 nm); retention time, (*R*), 32.9 min, (*S*), 48.8 min.

$[\alpha]_D^{23}$ −3.45 (c 2.44 CHCl$_3$) (lit. $[\alpha]_D^{24}$ + 3.99 (c 2.10 CHCl$_3$), >99% ee (*R*).[13]

CONCLUSION

A highly practical asymmetric Michael reaction catalysed by the chiral ruthenium-amido complex with a M/NH bifunction has been developed.[9,14] A variety of

Table 11.7 Asymmetric Michael reaction of enones and malonates catalysed by chiral ruthenium-amido complexes^a

run	enone	malonate	catalyst	solvent	Temp (°C)	Time (h)	Yield (%)^b	ee (%)^c	config
1	2a	3a	1a	(CH$_3$)$_3$COH	30	48	87	85	S
2	2a	3a	1b	(CH$_3$)$_3$COH	40	24	99	89	S
3	2a	3a	1c	(CH$_3$)$_3$COH	40	24	99	95	S
4	2a	3a	1d	(CH$_3$)$_3$COH	40	24	99	97	S
5	2a	3a	1e	(CH$_3$)$_3$COH	40	24	98	98	S
6	2a	3a	1f	(CH$_3$)$_3$COH	40	24	99	98	S
7	2a	3a	1f^d	(CH$_3$)$_3$COH	60	24	99	97	S
8	2a	3b	1e	(CH$_3$)$_3$COH	40	24	96	96	S
9	2a	3c	1f	toluene	30	48	51	97	−^{e}
10	2b	3a	1e	(CH$_3$)$_3$COH	30	48	93	96	S
11	2c	3a	1e	(CH$_3$)$_3$COH	30	48	53	>99^f	S
12	2c	3a	1f	(CH$_3$)$_3$COH	30	72	75	>99^f	S
13	2d	3a	1f	(CH$_3$)$_3$COH	30	72	83	>99	−^{e}

^{a}Unless otherwise noted, the reaction was carried out using 1.0 mmol of Michael acceptors and donors (1 : 1) in 1.0 mL of solvent. The molar ratio of acceptor:donor:ruthenium is 50 : 50 : 1 (S/C = 50). ^{b}Isolated yield after flash chromatography on the silica gel. ^{c}Determined by HPLC analysis. ^{d}S/C = 100. ^{e}Not determined. ^{f}Determined by ^{13}C NMR of the ketals derived from the products and (2*R*,3*R*)-butanediol.

cyclic enones **(2b–d)** and malonates **(3a-c)** (Section 11.4.2) can be successfully transformed with the amido catalysts to corresponding optically active Michael adducts with high ee's (Table 11.7). The outcome of the reaction was found to be delicately influenced by the structures of the ligands in the ruthenium-amido complexes (Table 11.7).

REFERENCES

1. [Cu] (a) Kanai, M., Nakagawa, Y. and Tomioka, K. *Tetrahedron* **1999**, 3843. (b) Feringa, B. L., Pineschi, M., Arnold, L. A., Imbos, R. and de Vries, A. H. M. *Angew. Chem., Int. Ed. Engl.* **1997**, *36*, 2620. (c) Hu, H., Chen, H. and Zhang, X. *Angew. Chem., Int. Ed. Engl.* **1999**, *38*, 3518. (d) Alexakis, A. and March, S. *J. Org. Chem.* **2002**, *67*, 8753.

2. [Ni] (a) Soai, K., Hayasaka, T. and Ugajin, S. *J. Chem. Soc., Chem. Commun.* **1989**, 516. (b) de Vries A, H. M., Imobos, R. and Feringa, B. L. *Tetrahedron: Asymmetry* **1997**, *8*, 1467.

3. [Co] (a) Reddy, C. K. and Knochel, P. *Angew. Chem., Int. Ed. Engl.* **1996**, *35*, 1700. (b) de Vries A, H. M., Imobos, R. and Feringa, B. L. *Tetrahedron: Asymmetry* **1997**, *8*, 1467.

4. [Rh] (a) Takaya, Y., Ogasawara, M. and Hayashi, T. *J. Am. Chem. Soc.* **1998**, *120*, 5579. (b) Hayashi, T. *Synlett* **2001**, 879. (c) Sakura, S., Sasai, M., Itooka, R. and Miyaura, N. *J. Org. Chem.* **2000**, *65*, 5951. (d) Kuriyama, M. and Tomioka, K. *Tetrahedron Lett.* **2001**, *42*, 921. (e) Hayashi, T., Takahashi, M., Takaya, Y. and Ogasawara, M. *J. Am. Chem. Soc.* **2002**, *124*, 5052. (f) Suzuki, T. and Torii, T. *Tetrahedron Asymmetry*, **2001**, *12*, 1077. (g) Hayashi, T., Ueyama, K., Tokunaga, N. and Yoshida, K. *J. Am. Chem. Soc.* **2003**, *125*, 11508. (h) Hayashi, T. and Yamasaki, K. *Chem. Rev.* **2003**, *103*, 2829.

5. [Pd] (a) Sodeoka, M., Ohrai, K. and Shibasaki, M. *J. Org. Chem.* **1995**, *60*, 2468. (b) Hamashima, Y., Hotta, D. and Sodeoka, M. *J. Am. Chem. Soc.* **2002**, *124*, 11240.

6. [Ln] (a) Arai, T., Sasai, H., Aoe, K., Okamura, K., Date, T. and Shibasaki, M. *Angew. Chem. Int. Ed. Engl.* **1996**, *35*, 104. (b) Arai, T., Yamada, Y., Yamamoto, N., Sasai, H. and Shibasaki, M. *Chem. Eur. J.* **1996**, *2*, 1368. (c) Shibasaki, M. and Yoshikawa, N. *Chem. Rev.* **2002**, *102*, 2187.

7. [Organic compounds] Review: (a) List, B. *Tetrahedron* **2002**, *58*, 5573. (b) Halland, N., Aburel, P. S. and Jørgensen. *Angew. Chem., Int. Ed. Engl.* **2003**, *42*, 661.

8. (a) Hashiguchi, S., Fujii, A., Takehara, J., Ikariya, T. and Noyori, R. *J. Am. Chem. Soc.* **1995**, *117*, 7562. (b) Haack, K.–J., Hashiguchi, S., Fujii, A., Ikariya, T. and Noyori, R. *Angew. Chem., Int. Ed. Engl.* **1997**, *36*, 285. (c) Noyori, R. and Hashiguchi, S. *Acc. Chem. Res.* **1997**, *30*, 97. (d) Noyori, R., Yamakawa, M. and Hashiguchi, S. *J. Org. Chem.* **2001**, *66*, 7931.

9. (a) Watanabe, M., Murata, K. and Ikariya, T. *J. Am. Chem. Soc.* **2003**, *125*, 7508. (b) Ikariya, T., Wang, H., Watanabe, M., Murata, K. *J. Organometal. Chem.* **2004**, *689*, 1377.

10. Bennett, M. A., Huang, T.–N., Matheson, T. W. and Smith, A. K. *Inorg. Synth.* **1982**, *21*, 74.

11. Oda, T., Irie, R., Katsuki, T. and Okawa, H. *Synlett.* **1992**, 641.

12. Bennett, M. A., Matheson, T. W., Robertson, G. B., Smith, A. K. and Trucker, P. A. *Inorg. Chem.* **1980**, *19*, 1014.

13. Kim, S. Y., Matsunaga, S., Das, J., Sekine, A., Ohshima, T. and Shibasaki, M. *J. Am. Chem. Soc.* **2000**, *122*, 6506.

14. Murata, K., Konishi, H., Ito, M. and Ikariya, T. *Organometallics* **2002**, *21*, 253.

12 Stereoselective Hydroformylation, Carbonylation and Carboxylation Reactions

CONTENTS

Catalysts for Fine Chemical Synthesis, Vol. 3, Metal Catalysed Carbon-Carbon Bond-Forming Reactions
Edited by S. M. Roberts, J. Xiao, J. Whittall, and T. Pickett
© 2004 John Wiley & Sons, Ltd   ISBN: 0-470-86199-1

12.1   ORTHO-DIPHENYLPHOSPHANYLBENZOYL-(o-DPPB) DIRECTED DIASTEREOSELECTIVE HYDROFORMYLATION OF ALLYLIC ALCOHOLS

BERNHARD BREIT

Institut für Organische Chemie und Biochemie, Albert-Ludwigs-Universität, Albertstr. 21, 79104 Freiburg i. Br., Germany

Hydroformylation of alkenes is one of those industrially important reactions which rely on homogeneous catalysis.[1] Additionally, it is a potentially useful synthetic method meeting the criteria of atom economy. However, the simultaneous control of chemo-, regio- and stereoselectivity is a challenge.[2] One solution to this problem employs a substrate-bound catalyst-directing group such as *ortho*-diphenylphosphanylbenzoate (o-DPPB). Attachment of the o-DPPB group through an ester linkage to an allylic alcohol system allows chemo-, regio- and stereoselective hydroformylation of allylic alcohols.[3]

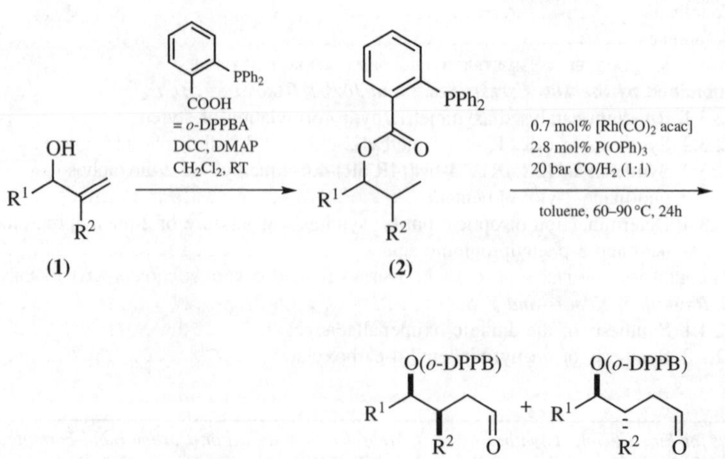

12.1.1 SYNTHESIS OF 1RS-(±)-[(1-ISOPROPYL-2-METHYL)PROP-2-ENYL] (2-DIPHENYLPHOSPHANYL)BENZOATE (2b)

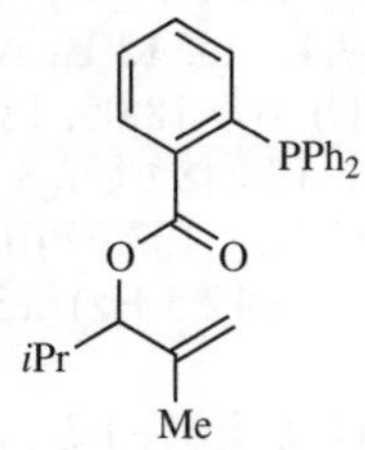

((±)-2b)

Materials and equipment

- 2,4-Dimethyl-pent-1-en-3-ol (>98%), 1.14 g, 10 mmol[4]
- *ortho*-Diphenylphosphanylbenzoic acid (*o*-DPPBA) (>98%), 3.06 g, 10 mmol[5]
- Dicyclohexylcarbodiimide (DCC), (98%), 2.27 g, 11 mmol
- *N,N*-Dimethylaminopyridine (DMAP), (99%), 0.12 g, 1 mmol
- Dichloromethane (dry), 20 mL
- Petroleum ether (40–60), (98%), 1800 mL
- *tert*-Butylmethyl ether, (98%), 200 mL

- Celite®, 10 g
- Silica gel Si 60, E. Merck AG, Darmstadt, 40–63 μm, 150 g
- TLC plates, Merck (silica gel 60) UV$_{254}$
- 50-mL round-bottomed flask with magnetic stirrer bar
- Magnetic stirrer plate
- Glass filter
- 250-mL round-bottomed flask
- Flash column, diameter 5 cm
- Rotary evaporator

General Procedure for Synthesis of o-DPPB Esters (2)

To a solution of allylic alcohol (1 equiv.) in dichloromethane (0.5 M) *o*-DPPBA, (1 equiv.), DMAP (0.1 equiv.) and DCC (1.1 equiv.) were successively added and the resulting mixture was stirred at room temperature until TLC analysis indicated complete consumption of the starting material. Subsequently, the reaction mixture was filtered through a plug of dichloromethane-wetted Celite® and washed with additional dichloromethane. An appropriate amount of silica gel was added to the filtrate, which was then concentrated to dryness. Flash chromatography with petroleum ether/*tert*-butyl methyl ether (9:1) provided the *o*-DPPB esters (2), usually as slightly yellow to colourless, highly viscous oils.

RS-(±)-*[(1-isopropyl-2-methyl)prop-2-enyl]* *(2-diphenylphosphanyl)benzoate* (**2b**): From 1.14 g (10 mmol) 2,4-dimethyl-pent-1-en-3-ol[4], 3.02 g (75%) of *o*-DPPB ester was obtained as colourless crystals, m.p. 98–99 °C.

¹H NMR (300 MHz, CDCl₃): δ = 0.9 (d, J = 6.8 Hz, 3 H, CH(CH₃)₂), 0.94 (d, J = 6.6 Hz, 3 H, CH(CH₃)₂), 1.74 (s, 3 H, CH₃), 2.0 (pseudo sext., J = 6.8 Hz, CH(CH₃)₂), 4.92 (s, 1 H, =CH₂), 4.99 (s, 1 H, =CH₂), 5.18 (d, J = 7.5 Hz, 1 H, OCH), 7.0 (m, 1 H, ArH), 7.32–7.42 (m, 12 H, ArH), 8.2 (m, 1 H, ArH).

¹³C NMR (75.469 MHz, CDCl₃): δ = 18.05, 18.52, 19.03, 29.87, 83.06, 114.08, 128.17, 128.39 (d, $J_{C,P}$ = 7.1 Hz), 128.48 (d, $J_{C,P}$ = 1.5 Hz), 130.43 (d, $J_{C,P}$ = 2.3 Hz), 131.77, 133.84 (d, $J_{C,P}$ = 20.7 Hz), 133.97 (d, $J_{C,P}$ = 20.8 Hz), 134.34, 134.73 (d, $J_{C,P}$ = 18.9 Hz), 138.14 (d, $J_{C,P}$ = 12.5 Hz), 138.26 (d, $J_{C,P}$ = 11.9 Hz), 140.75 (d, $J_{C,P}$ = 27.9 Hz), 141.9, 165.78.

³¹P NMR (161.978 MHz, CDCl₃): δ = −4.3.

$C_{26}H_{27}O_2P$ (402.5): calcd. C 77.59, H 6.76; found C 77.32, H 6.80.

12.1.2   SYNTHESIS OF (1R*,2R*)-(±)-[(1-ISOPROPYL-4-OXO-2-METHYL)BUTYL] (2-DIPHENYLPHOSPHANYL)BENZOATE (3b)

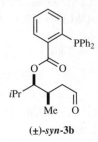

(±)-*syn*-3b

Materials and equipment

- 1RS-(±)-[(1-isopropyl-2-methyl)prop-2-enyl]   (2-diphenylphosphanyl)benzoate (2b) (>98%), 201 mg, 0.5 mmol
- [Rh(CO)₂acac], 0.9 mg, 3.5 · 10⁻³ mmol
- Triphenylphosphite (freshly distilled, stored under Argon), 4.5 mg, 1.4 · 10⁻² mmol
- Toluene (dry, degassed), 5 mL
- Petroleum ether (40–60), (98%), 180 mL
- *tert*-Butylmethyl ether, (98%), 50 mL
- Synthesis gas (premixed 1:1 mixture of carbon monoxide 2.0 and hydrogen 3.0, Messer Griesheim)

- Stainless steel autoclave (100 mL), equipped with a magnetic stir bar
- Schlenk flask (20 mL) for preparation of catalyst solution
- Syringe (5 mL)
- Silica gel (Merck 60, 40–63 µm), 50 g
- TLC plates, Merck (silica gel 60) UV₂₅₄
- Magnetic stirrer plate
- 250-mL round-bottomed flask

- One glass column, diameter 1.5 cm
- Rotary evaporator

General Procedure for Hydroformylation of allylic o-DPPB Esters (2)

To a solution of [Rh(CO)$_2$acac] (0.9 mg, 3.5 · 10^{-3} mmol) in toluene (3 ml) at room temperature (exclusion of air and moisture), triphenylphosphite (4.5 mg, 1.4 · 10^{-2} mmol) was added and stirred for 15 minutes. Subsequently, the corresponding allylic o-DPPB ester (0.5 mmol) was added and the resulting solution was cannulated *via* syringe into a stainless-steel autoclave followed by rinsing with additional toluene (2 ml). The autoclave was heated to 60 °C and then pressurised with 20 bar of a 1:1 mixture of hydrogen and carbon monoxide. The reaction was monitored by TLC and stopped after complete consumption of starting material. The autoclave was cooled to room temperature and the reaction mixture was filtered through a small plug of silica with *tert*-butyl methyl ether (30 ml). After evaporation of the solvent *in vacuo* the crude product was analyzed by NMR to determine conversion and diastereomer ratio. Subsequent flash chromatography with petroleum ether/*tert*-butyl methyl ether (9:1) provided the corresponding aldehydes (3) as highly viscous oils which in some cases could be crystallised from diethylether.

Attention: Working with carbon monoxide requires the use of a well ventilated fume hood due to its toxicity.

(1R*,2R*)-(±)-[(1-Isopropyl-4-oxo-2-methyl)butyl] *(2-diphenylphosphanyl)- benzoate (syn-3b)*: From 201 mg (0.5 mmol) **2b**, 210 mg (97%) of *syn-3b* was obtained. Diastereomer ratio 96: 4 (*syn/anti*).

^{1}H NMR (300 MHz, CDCl$_3$): δ = 0.9 (d, *J* = 6.59 Hz, 3 H, CH$_3$), 0.97 (d, *J* = 6.84 Hz, 3 H, CH(CH$_3$)$_2$), 0.98 (d, *J* = 6.77 Hz, 3 H, CH(CH$_3$)$_2$), 1.94 (m, 1 H, CH(CH$_3$)$_2$), 2.11 (ddd, *J* = 17.62, 8.26, 1.76 Hz, 1 H, CH$_2$CHO), 2.30 (dd, *J* = 17.62, 5.33 Hz, 1 H, CH$_2$CHO), 2.52 (m, 1 H, H at C2), 4.87 (dd, *J* = 8.89, 3.25 Hz, 1 H, OCH), 7.02 (m, 1 H, ArH), 7.22–7.47 (m, 12 H, ArH), 8.2 (m, 1 H, ArH), 9.57 (d, *J* = 0.73 Hz, 1 H, CHO-*syn*), [9.68 (CHO-*anti*)].

^{13}C NMR (75.469 MHz, CDCl$_3$): δ = 13.69, 19.15, 29.09, 29.83, 48.33, 82.07, 125.47, 128.62 (d, $J_{C,P}$ = 2.1 Hz), 128.81 (d, $J_{C,P}$ = 7.3 Hz), 130.72 (d, $J_{C,P}$ = 1.7 Hz), 132.23, 133.85 (d, $J_{C,P}$ = 18 Hz), 134.08 (d, $J_{C,P}$ = 20.8 Hz), 134.28, (d, $J_{C,P}$ = 21.1 Hz), 134.41, 138.04 (d, $J_{C,P}$ = 13.6 Hz), 138.48 (d, $J_{C,P}$ = 11.3 Hz), 141.54 (d, $J_{C,P}$ = 28.2 Hz), 166.62, 201.47 (CHO-*syn*), [202.03 (CHO-*anti*)].

^{31}P NMR (161.978 MHz, CDCl$_3$): δ = −3.1.

C$_{27}$H$_{29}$O$_3$P (432.5): calcd. C 74.98, H 6.76; found C 74.91, H 6.90.

CONCLUSION

The general procedure is easy to reproduce and allows stereoselective hydroformylation to be applied to a wide variety of 1,1 disubstituted allylic alcohols. Table 12.1 summarizes the result of o-DPPB-directed stereoselective hydroformylation of several different allylic alcohols, with high levels of stereoinduction.

Table 12.1 Results of o-DPPB-directed diastereoselective hydroformylation of 1,1-disubstituted allylic alcohols.

Entry	Product	R¹	R²	Yield (%)	dr (syn:anti)
1	3a	Ph	Me	98	90:10
2	3b	iPr	Me	97	96:4
3	3c	Cy	Me	81	95:5
4	3d	Et	iPr	81	84:16
5	3e	Bn	iPr	96	92:8
6	3f	Ph	iPr	97	99:1
7	3g	Et	tBu	71	94:6
8	3h	Bn	tBu	95	99:1

REFERENCES

1. C. D. Frohning and C. W. Kohlpaintner, in *Applied Homogeneous Catalysis with Organometallic Compounds* (Eds. B. Cornils and W. A. Herrmann) Weinheim, **2000**, pp 29–104.
2. B. Breit and W. Seiche, *Synthesis* **2001**, 1–36.
3. B. Breit, *Angew. Chem. Int. Ed. Engl.* **1996**, *35*, 2835–2837. B. Breit, *Liebigs Ann.* **1997**, 1841–1851. B. Breit, M. Dauber, and K. Harms, *Chem. Eur. J.* **1999**, *5*, 1819–2827. B. Breit and S. K. Zahn, *J. Org. Chem.* **2001**, *66*, 4870–4877. B. Breit, G. Heckmann, and S. K. Zahn, *Chem. Eur. J.* **2003**, *9*, 425. B. Breit, *Acc. Chem. Res.* **2003**, *36*, 264–275.
4. D. J. Faulkner and M. R. Petersen, *J. Am. Chem. Soc.* **1973**, *95*, 553.
5. J. E. Hootes, T. B. Rauchfuss, D. A. Wrobleski, and H. C. Knachel, *Inorg. Synth.* **1982**, *21*, 175–179.

12.2 THE SYNTHESIS AND APPLICATION OF ESPHOS: A NEW DIPHOSPHORUS LIGAND FOR THE HYDROFORMYLATION OF VINYL ACETATE

MARTIN WILLS and SIMON W. BREEDEN

Department of Chemistry, The University of Warwick, Coventry, CV4 7AL, UK

The asymmetric hydroformylation of alkenes is an exceptionally atom-efficient method for the synthesis of enantiomerically-pure carbonyl-containing compounds.[1] The hydroformylation of vinylacetate, in particular, represents an excellent method for the preparation of α-alkoxy aldehydes and, through their reduction, homochiral 1,2-diols. The use of the novel chiral ligand, ESPHOS (**1**),[2] in a rhodium(I) complex, results in hydroformylation of vinyl acetate in high branched:linear selectivity and exceptional ee (Figure 12.1).[3]

ESPHOS (**1**) may be prepared in a 5-step procedure (Figure 12.2) from 2-bromoaniline (**2**), through diazotization to give (**3**), followed by decomposition and

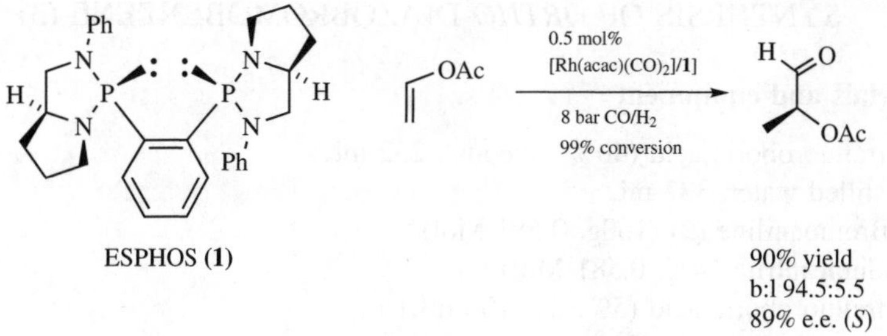

Figure 12.1

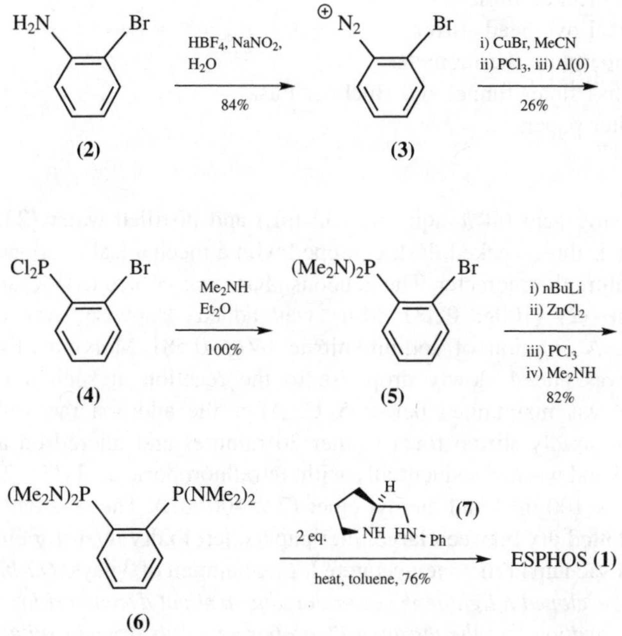

Figure 12.2 Synthesis of ESPHOS.

trapping with phosphorus trichloride to give (4), then reaction with excess dimethylamine to give (5). In the last two steps, (5) is converted to (6), which is then used in an exchange reaction with diamine (7) to furnish the ligand. In this summary the procedures for the conversion of (2) through to (1) will be described, as will the hydroformylation reaction. The method employed is a modification of that reported by Drewelies and Latscha[4] and later improved by Thomaier and Grutzmacher.[5] The synthesis of diamine (7), and of bis(dimethylamino)-chloro-phosphine has already been reported and will not be repeated here.[6] All the reactions described below must be carried out under an inert atmosphere (argon or nitrogen).

12.2.1 SYNTHESIS OF *ORTHO*-DIAZOBROMOBENZENE (3)

Materials and equipment

- Tetrafluoroboric acid (40% aqueous), 232 mL
- Distilled water, 332 mL
- 2-Bromoaniline (**2**) (100g, 0.581 Mol)
- Sodium nitrite (40g, 0.581 Mol)
- Tetrafluoroboric acid (5%, 2 × 100 mL)
- Methanol (2 × 100 mL)
- Diethyl ether (2 × 400 mL)

- 1-L three-necked flask
- Mechanical overhead stirrer
- Low temperature thermometer
- Large glass sinter funnel and Buchner flask
- Large filter papers

Procedure

Tetrafluoroboric acid (40% aqueous 232 mL) and distilled water (232 mL) were added to a 1-L three-necked flask equipped with a mechanical overhead stirrer and low temperature thermometer, The aqueous phase was cooled to 0 °C and molten 2-bromoaniline (**2**) (100g, 0.581 Mol) was added dropwise over a period of 10 minutes. A solution of sodium nitrite (40g, 0.581 Mols) in distilled water (100 mL) was added slowly dropwise to the reaction at such a rate that the temperature was maintained below 5 °C. After the addition the yellow viscous mixture was rapidly stirred for a further 30 minutes and filtered on a large glass sinter funnel and washed sequentially with: tetrafluoroboric acid (5%, 2 × 100 mL), methanol (2 × 100 mL) and diethyl ether (2 × 400 mL). The resultant pale yellow solid was blotted dry between large filter papers, left to dry in air for 1 day and then dried at high vacuum at room temperature for a minimum of 3 days. *During this period the product developed a light pink coloration due to slight decomposition which serves as a good indication that the diazotetrafluoroborate salt is dry and suitable for use in subsequent reactions.* After drying the product (**3**) was obtained (132g, 84% yield).

v_{max} solid/cm^{-1} 3089, 2287, 1552, 1462, 1299, 1267, 1022, 758; δ_H (300 MHz, CDCl$_3$, Me$_4$Si) 8.84 (1H, d, *J* 8.3, Ar-H); 8.31 (1H, d, *J* 8.1, Ar-H); 8.17 (1H, t, *J* 7.7, Ar-H); 7.99 (1H, t, *J*, 8.5, Ar-H); δ_C (75 MHz, CDCl$_3$, Me$_4$Si) 142.20, 135.62, 135.39, 130.75, 124.79 (*ipso* C); 119.04 (*ipso* C).

12.2.2 SYNTHESIS OF *ORTHO*-(DICHLOROPHOSPHINE) BROMOBENZENE (4)

Materials and equipment

- *ortho*-Diazobromobenzene (**3**), 60g, 0.219 Mol
- Anhydrous acetonitrile, 300 mL

- Copper (1) bromide, 1.88g, 0.0131 Mol
- Phosphorus trichloride, 33.5 mL 0.372 Mol
- Aluminium powder, 20 micron, 5.91g, 0.219 Mol

- Three-necked round-bottomed flask
- Large rugby ball shaped magnetic stirrer
- Reflux condenser
- Nitrogen bubbler
- Filter funnel
- Cotton wool
- Distillation apparatus

Procedure

An oven dried 1-L three-necked round-bottomed flask equipped with a large rugby ball shaped magnetic stirrer and a reflux condenser fitted with a nitrogen bubbler was charged with *ortho*-diazobromobenzene (**3**) (60g, 0.219 Mol) and placed under an atmosphere of dry nitrogen before dissolution of the salt with rapid and efficient stirring in anhydrous acetonitrile (300 mL). Under a positive flow of dry nitrogen copper (1) bromide (1.88g, 0.0131 Mol) was added in small spatula tip portions effecting decomposition of the diazotetrafluoroborate salt at a controlled rate. During the decomposition the solution becomes black as nitrogen continued to evolve. Phosphorus trichloride (33.5 mL 0.372 Mol) was added in one portion and the evolution of nitrogen continued. The solution was warmed slowly to 30 °C and maintained at that temperature for 30 minutes. **Careful control of the temperature is required at this stage to ensure a controlled release of nitrogen.** The reaction was then heated at reflux for 30 minutes and then allowed to cool to room temperature. Aluminium powder (20 micron, 5.91g, 0.219Mol) was added and the mixture stirred overnight under a positive pressure of nitrogen. After this period the grey green mixture was heated at reflux for one hour and then allowed to cool to ambient temperature. Filtration of the mixture through a plug of cotton wool allowed the separation of particulate aluminium and the filtrate was concentrated *in-vacuo* to give a brown semi solid. Distillation under reduced pressure provided several lower boiling fractions with the product (**4**) obtained as a clear oil (101 °C 1mmHg) (14.7g, 26% yield).

v_{max} solid/cm^{-1} 3058, 2353, 1563, 1450, 1419, 1330, 1248, 1123, 1018, 921, 870, 742; δ_H (300 MHz, CDCl$_3$, Me$_4$Si); 8.11–8.08 (1H, m, Ar-H); 7.61–7.51 (1H, m, Ar-H); 7.52 (1H, t, *J* 6.8, Ar-H); 7.39–7.34 (1H, m, Ar-H); δ_C (75 MHz, CDCl$_3$, Me$_4$Si); 133.5, 132.7, 131.6, 128.2, 126.5, 125.9 δ_P (121 MHz, CDCl$_3$) 155.09; *m/z* (EI) 257 (M$^+$); 224, 222, 220, 178, 176, 140, 107, 75, 50; found (EI) 255.8606 (M$^+$) C$_6$H$_4$BrCl$_2$P requires 255.8606 (^{79}Br, ^{35}Cl).

12.2.3 SYNTHESIS OF *ORTHO*-BIS(DIMETHYLAMINO)PHOSPHINO BROMOBENZENE (5)

Materials and equipment

- Technical grade diethyl ether (150 mL)

- *Ortho*(dichlorophosphine)bromobenzene **4** (5g, 0.01937 Mol)
- Anhydrous dimethylamine gas

- 1-L three-neck round-bottomed flask
- Low temperature thermometer
- Large magnetic stirrer bar
- Nitrogen bubbler
- Celite

Procedure

An oven-dried 1-L three-neck round-bottomed flask equipped with a large magnetic stirrer bar, low temperature thermometer and nitrogen bubbler was charged with fresh technical grade diethyl ether (150 mL) and sparged with dry nitrogen for 1 hour. *Ortho*(dichlorophosphine)bromobenzene (**4**) (5g, 0.01937 Mol) in ether (5 mL) was added *via* nitrogen purged syringe and the solution cooled to 0 °C. Whilst being maintained at 0 °C, excess anhydrous dimethylamine gas was passed over the surface of the etherial phosphine solution causing the precipitation of a white solid (dimethylamine hydrochloride). *The end point of the reaction may be monitored by taking phosphorus NMR spectra every 15 minutes.* After complete conversion the inorganic salt was removed by vacuum filtration through celite and the filtrate concentrated *in-vacuo* to give the product (**5**) as a clear oil. *This transformation generally gives 100% conversion and usually gives a synthetically pure product after simple removal of solvent. For absolute purity the product may be distilled under vacuum (106 °C, 0.2 mmHg).*

v_{max} solid/ cm^{-1} 3050, 2968, 2879, 2859, 2832, 2785, 1548, 1439, 1415, 1265, 1014, 976, 948, 750, 668, 637; δ_H (300 MHz, CDCl$_3$, Me$_4$Si) 7.53 (1H, ddd, *J* 1.1, 3.6, 4.7, Ar-H); 7.45 (1H, dt, *J* 2.1, 7.7, Ar-H); 7.33–7.26 (1H, m, Ar-H); 7.14–7.08 (1H, m, Ar-H); 2.70 (24H, d, J_{H-P} 9.0, Me); δ_C (75 MHz, CDCl$_3$, Me$_4$Si)141.08 (*ipso* C); 134.02, 133.10, 129.65, 127.27, 126.94 (*ipso* C); δ_P (121 MHz, CDCl$_3$) 101.68; *m/z* (EI) 274 (M$^+$) 232, 230, 189, 186, 119, 107, 76, 60; found (EI) 274.0228 (M$^+$) C$_{10}$H$_{16}$BrN$_2$P requires 274.0229 (^{79}Br,).

12.2.4   SYNTHESIS OF 1,2-BIS(DIMETHYLAMINOPHOSPHANYL) BENZENE (6)

Materials and equipment

- *ortho*-Bis(dimethylamino)phosphino-bromobenzene (**5**), 10.16g, 36.7 mMol
- Freshly distilled and anhydrous ether (200 mL)
- nBuLi (14.7mL of a 2.5M solution, 36.7 mMol)
- Anhydrous zinc chloride (5g, 36.7 mMol)
- Phosphorus trichloride (3.2 mL, 5.038g, 36.7 mMol)
- Dimethylamine gas
- Celite$^{®}$

- Three-neck 1-L round-bottomed flask
- Large magnetic stirrer
- Reflux condenser
- Nitrogen bubbler
- Filter funnel
- Cotton wool
- Distillation apparatus

Procedure

To a solution of *ortho*-bis(dimethylamino)phosphino-bromobenzene (**5**) (10.16g, 36.7 mMol) in ether (200 mL; *note this must be freshly distilled and anhydrous*) at −15 °C was added, dropwise *via* syringe, nBuLi (14.7mL of a 2.5M solution, 36.7 mMol). The temperature was maintained at −15 °C for 30 minutes. Anhydrous zinc chloride (5g, 36.7 mMol) was then added and the solution was allowed to warm to 0 °C, at which temperature it was maintained for 30 minutes. The reaction mixture was then cooled to −70 °C and phosphorus trichloride (3.2 mL, 5.038g, 36.7 mMol) was added *rapidly* in one portion. The reaction was allowed to warm up to room temperature over a period of 12 hours. *Note: a sticky solid may be formed and the stirrer bar may require freeing*. The reaction is then cooled to 0 °C and dimethylamine gas was bubbled through the solution for 20 minutes. A white precipitate was formed. This was filtered through Celite® and washed with ether. *It is very important to exclude air during this process*. The solvent was removed under vacuum to yield the product (**6**) as a clear oil (9.46g, 30.13 mMol, 82%).

δ_H (300 MHz, CDCl$_3$, Me$_4$Si) 7.45–7.55 (2H, m, ArH), 7.20–7.30 (2H, m, ArH), 2.66 (24H, dd, J 8.0, 4.0, Me); δ_C (75 MHz, CDCl$_3$, Me$_4$Si) 131.63 (t, J_{P-C} 5.8), 126.93 (t, J_{P-C} 1.2), 41.80 (dd, J_{P-C} 10.0, 10.2); δ_P (121 MHz, CDCl$_3$) 102.16.

12.2.5 SYNTHESIS OF ESPHOS (1)[2]

Materials and equipment

- (*S*)-2-(Phenylaminomethyl)pyrrolidine (3.99g, 22.7 mMol)
- Anhydrous, degassed toluene (30 mL)
- 1,2-Bis(dimethylaminophosphanyl)benzene **6** (3.57g, 11.3 mMol)
- 250-mL Two-necked round-bottomed flask
- Magnetic stirrer
- Reflux condenser
- Nitrogen bubbler

Procedure

To a stirred solution of (*S*)-2-(phenylaminomethyl)pyrrolidine[6] (3.99g, 22.7 mMol) in toluene (10 mL) a solution of 1,2-bis(dimethylaminophosphanyl)benzene (**6**)

(3.57g, 11.3 mMol) in toluene (20 mL) was added. The resulting mixture was refluxed for 72 hours under a positive stream of nitrogen. The solvent was then removed *in vacuo* to yield a pale yellow solid which was recrystallised from toluene (20 mL) to yield the product (1) as colourless crystals (4.15g, 76%); m.p. 172–174 °C.

[α]$_D$ −674.2 (c 0.5, CHCl$_3$); δ_H (400 MHz,C$_6$D$_6$, Me$_4$Si) 7.39 (2H, m, ArH), 7.23 (4H, m, ArH), 7.08 (4H, m, ArH), 6.90 (2H, dt, J 7.2 and 2.0, ArH), 6.81 (2H, t, J 7.2, ArH), 3.71 (2H, ddd, J 2.0, 7.9 and 14.8, CH),), 3.42–3.55 (2H, m, CH$_2$), 1.27–1.38 (2H, m, CH$_2$), 3.05–3.19 (4H, m, CH$_2$), 2.74 (2H, br.t, J 8.6, CH$_2$), 1.63–1.78 (2H, m, CH$_2$), 1.41–1.62 (4H, m, CH$_2$); δ_P (c (75 MHz, C$_6$D$_6$, Me$_4$Si) 101.9; δ_C (75 MHz, C$_6$D$_6$, Me$_4$Si) 148.3 (d, J_{C-P} 7.8), 147.6 (d, J_{C-P} 9.2), 129.8 (t, J_{C-P} 6.0), 129.3, 126.9, 118.0, 115.3 (d, J_{C-P} 6.9), 64.3 (d, J_{C-P} 4.6), 54.6 (d, J_{C-P} 2.0), 52.5 (d, J_{C-P} 14.1), 31.2, 25.9 (d, J_{C-P} 2.3); m/z (EI) 487 (M+H$^+$, 38%), 486 (M$^+$, 100).

Anal. Calcd for C$_{28}$H$_{32}$N$_4$P$_2$ (+0.5 C$_7$H$_8$,1/2 toluene of crystallisation); C, 71.04; H, 6.81; N, 10.52. Found; C, 71.00; H,6.79; N, 10.54.

12.2.6   HYDROFORMYLATION OF VINYL ACETATE[3]

Materials and equipment

- Mini-autoclave fitted with a system for substrate or catalyst injection
- [Rh(2,4-pentanedionato)(CO)$_2$] (5 × 10^{-5} Mol)
- ESPHOS, **ca** 10 mg
- Toluene, 2 cm^3

Procedure

Hydroformylation of vinyl acetate[3]

The reactions were carried out at the CATS catalyst evaluation service in St Andrews University using a specially designed mini-autoclave fitted with a system for substrate or catalyst injection and for measuring kinetics at constant pressure. The ligand, [Rh(CO)$_2$ (acac)] and toluene solvent were placed in the autoclave and flushed with carbon monoxide/hydrogen before pressurising to several bars below the operating pressure. The mixture was heated with stirring to the desired reaction temperature. The substrate was introduced *via* the substrate injection facility and the pressure raised to the desired reaction pressure. It took *ca* 90 seconds to stabilise the reaction conditions after the substrate injection. The pressure in the ballast vessel attached to the reaction vessel *via* a mass flow controller, which metered the gas to keep the constant pressure in the reactor, was monitored electronically with time. At the end of the reaction, the stirrer was stopped and the reactor cooled by plunging it into cold water. The liquid products were analysed by GC-FID (quantitative) and GC-MS (qualitative) using a chiral capillary column.

The catalyst was prepared *in situ* from [Rh(2,4-pentanedionato)(CO)$_2$] (5×10^{-5} Mol) and the ESPHOS (1.5 eq.) in toluene (2 cm^3) in a mini-autoclave fitted with a system for injection of substrate and for measuring kinetics at constant pressure. The autoclave was then flushed with carbon monoxide/hydrogen before pressurising to several bars below the operating pressure. The temperature was raised with stirring to the desired operating temperature and vinyl acetate (1.0g, 1.04×10^{-2} mMol) in toluene (2 cm^3) was injected and the pressure was raised to the desired reaction pressure. The pressure was maintained through the operation of a mass flow controller. The autoclave stirring speed was 500 r.p.m. At the end of the reaction, the stirrer was stopped and the reactor cooled rapidly. The liquid products were analysed by GC-FID (quantitative) and GC-MS (qualitative). The enantioselectivity of the reaction was determined by comparison with authentic product samples.

CONCLUSION

Using the combination of ESPHOS with rhodium(I) a catalytic hydroformylation of vinyl acetate may be achieved in high regio- and enantioselectivity using as little as 0.5 Mol% of catalyst. This ligand benefits from a selectivity similar to that exhibited by the very best ligands for this application.[1]

REFERENCES

1. I. Ojima, C.-Y. Tsai, M. Tzamarioudaki and D. Bonafoux, *Organic Reactions* **56**, pp 1–354, ed. L. Overman *et al.*, *John* Wiley and Sons, 2000.
2. M. Wills and S. W. Breeden, *J. Org. Chem.*, 1999, **64**, 9735–9738.
3. S. Breeden, D. J. Cole-Hamilton, D. F. Foster, G. J. Schwarz and M. Wills, *Angew. Chem., Int. Edn.*, 2000, **39**, 4106–9.
4. K. Drewelies and H. P. Latscha, *Angew. Chem., Int. Edn.*, 1982, **21**, 638–639; *Angew. Chem.*, 1982, **94**, 642–3.
5. J. Thomaier and H. Gruetzmacher, *Synthetic Methods of Organometallic and Inorganic Chemistry*, 1996, **3**, 73–75.
6. C. W. Edwards, M. R. Shipton, N. W. Alcock, H. Clase and M. Wills, *Tetrahedron*, 2003, **59**, 6473–6480.

Acknowledgement: We thank, D. J. Cole-Hamilton, D. F. Foster and G. J. Schwarz of the St Andrews catalyst evaluation service (CATS) for the hydroformylation tests.

12.3   PLATINUM-CATALYSED ASYMMETRIC HYDROFORMYLATION OF STYRENE

Submitted by Stefánia Cserépi-Szűcs[a], and József Bakos[*b]

[a]Research Group for Petrochemistry, Hungarian Academy of Science, H-8201 Veszprém, PO Box 158, Hungary. [b]Department of Organic Chemistry, University of Veszprém, H-8201 Veszprém, PO Box 158, Hungary

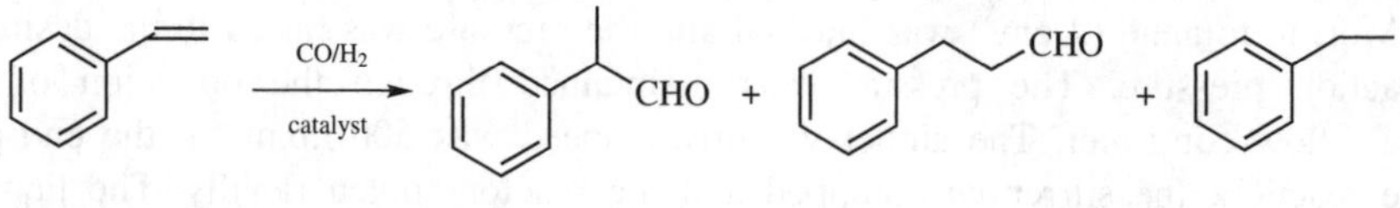

Materials and equipment

- 100-mL Stainless-steel autoclave
- Anhydrous toluene, 10 mL
- (2R,4R)-2,4-Bis[(4R,6R)-4,6-dimethyl-1,3,2-dioxaphosphorinane-2-yloxy]-pentane, 18.6 mg
- Anhydrous tin(II) chloride, 9.5 mg
- $Pt(PhCN)_2Cl_2$, 23.6 mg
- Styrene, 11.5 mL
- Decane, 1 mL (used as internal standard for GC analysis)

- Schlenk tube with magnetic stirrer bars
- Magnetic stirrer plate
- Distillation equipment

Procedure

1. (2R,4R)-2,4-Bis[(4R,6R)-4,6-dimethyl-1,3,2-dioxaphosphorinane-2-yloxy]-pent ane (18,6 mg), anhydrous tin(II) chloride (9.5 mg) and $Pt(PhCN)_2Cl_2$ (23.6 mg) were placed in a Schlenk tube equipped with magnetic stirrer bar, under argon. Toluene (10 mL) and decane (1 mL) were added to the mixture with stirring. The yellow solution was stirred for 30 minutes at room temperature.

2. The autoclave was purged five times with argon and then filled with the olefin (11.46 mL) and catalyst mixture. It was then purged with syngas (carbon monoxide:hydrogen = 1:1) and pressurised to 100 atmosphere initial pressure. The autoclave was heated in a thermostatically-controlled electric oven to the temperature indicated in Table 12.2, and agitated by an arm-shaker. The pressure was monitored throughout the reaction.

3. At the end of the reaction the autoclave was cooled and depressurised. Conversion to aldehydes and the regioselectivity of the reaction (2-phenylpro-panal/3-phenylpropanal) were determined without evaporation of the solvent. The reaction mixture was fractionally distilled under reduced pressure to give a mixture of regioisomers of aldehydes. The enantiomeric excess was determined

by measuring the optical rotation of the aldehyde and/or by chiral GC analysis of the corresponding acid obtained by oxidation.

(S)-2-Phenylpropanal $[\alpha]_D^{20} + 238°$ (neat)

Gas chromatographic analysis were run on a Hewlett Packard 5830A equipped with a flame ionisation detector (SPB-1 30 m column, film thickness 0.1 µm, carrier gas 2 mL/min).

12.3.1 RHODIUM-CATALYSED ASYMMETRIC HYDROFORMYLATION OF STYRENE

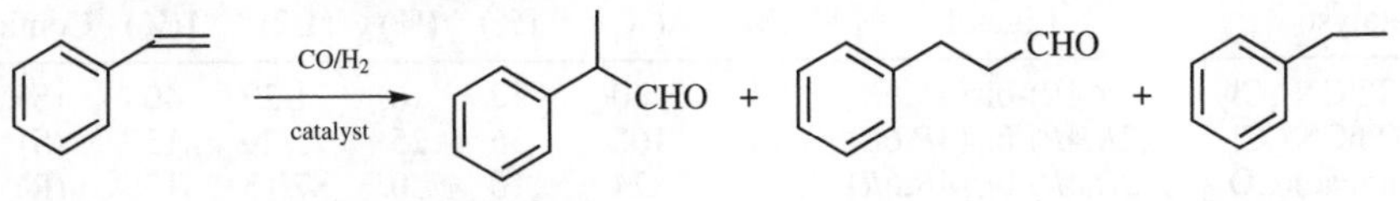

Materials and equipment

- 20-mL Stainless-steel autoclave
- Anhydrous benzene, 2 mL
- (2R,4R)-2,4-bis[(4R,6R)-4,6-dimethyl-1,3,2-dioxaphosphorinane-2-yloxy]-pentane, 24.6 mg
- Rh(acac)(CO)$_2$, 8.0 mg
- Styrene, 3 mL

- Schlenk tube with magnetic stirrer bars
- Magnetic stirrer plate
- Decane, 1 mL (used as internal standard for gc analysis)
- Distillation equipment

Procedure

1. (2R,4R)-2,4-bis[(4R,6R)-4,6-Dimethyl-1,3,2-dioxaphosphorinane-2-yloxy]-pentane (24.6 mg) and Rh(acac)(CO)$_2$ (8.0 mg) were placed in a Schlenk tube equipped with magnetic stirrer bar, under argon. Benzene (2 mL) and decane (1 mL) were added to the mixture with stirring. The yellow solution was stirred for 30 minutes at room temperature.
2. The autoclave was purged five times with argon and then filled with the olefin (3 mL) and catalyst mixture. It was then purged with syngas (CO:H$_2$=1:1) and pressurised to 100 atmosphere initial pressure. The autoclave was heated in an electric oven to the temperature indicated in the Table 12.2, and agitated by an arm-shaker. The pressure was monitored throughout the reaction.
3. The work-up of the reaction mixture and the analysis were conducted in the same way as in the platinum-catalysed reaction.

    (S)-2-Phenylpropanal $[\alpha]_D^{20} + 238°$ (neat)

    Gas chromatographic analysis were run on a Hewlett Packard 5830A gas chromatograph (SPB-1 30 m column, film thickness 0.1 µm, carrier gas 2 mL/min).

Table 12.2 Asymmetric hydroformylation of styrene with platinum and rhodium catalysts modified by (2R,4R)-2,4-bis[(4R,6R)-4,6-dimethyl-1,3,2-dioxaphosphorinane-2-yloxy]-pentane.

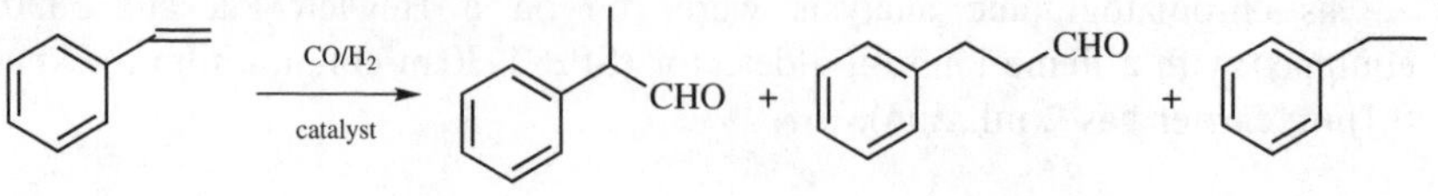

| | | | | | | | | |
| | | (1) | | | (2) | | (3) | |

Catalyst	Ligand	Time (h)	Temp. (°C)	Conv. (%)	**3** (%)	**(1/2)**	ee of **1**(%)	Config
Pt(PhCN)₂Cl₂	(2R,4R)-bis(4R,6R)	24	60	72	16	71/29	40	(S)
Pt(PhCN)₂Cl₂	(2R,4R)-bis(4R,6R)	18	100	56	25	71/29	12	(S)
Rh(acac)(CO)₂	(2R,4R)-bis(4R,6R)	73	24	10	0	87/13	17	(R)
Rh(acac)(CO)₂	(2R,4R)-bis(4R,6R)	5	60	69	0	83/17	15	(R)

CONCLUSION

Optically-active aldehydes are very important as precursors not only for biologi-cally active compounds but also for new materials. Asymmetric hydroformylation is an attractive catalytic approach to the synthesis of a large number of chiral aldehydes. With the platinum precursor (Pt(PhCN)₂Cl₂), anhydrous tin(II) chloride was used as cocatalyst (SnCl₂/Pt = 1), which is essential for catalytic activity. In case of rhodium systems an excess amount (P/Rh = 4) of diphosphite ligand was always added to the catalyst precursor to exclude the formation of HRh(CO)₄, which is an active achiral hydroformylation catalyst.

12.3.2   SYNTHESIS OF (4R,6R)-4,6-DIMETHYL-2-CHLORO-1,3,2-DIOXAPHOSPHORINANE

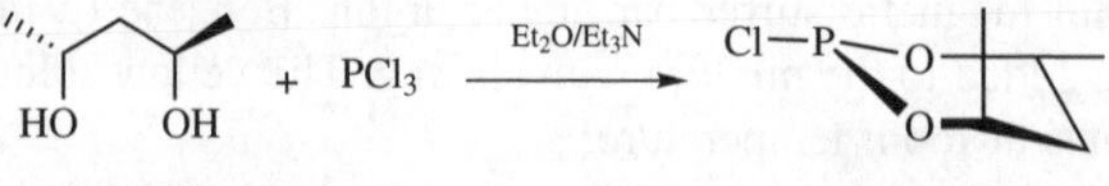

Materials and equipment

- Anhydrous diethyl ether, 450 mL
- Anhydrous toluene, 30 mL
- (2R,4R)-Pentane-2,4-diol, 6.4 g
- Anhydrous triethylamine, 17.7 mL
- Distilled phosphorus trichloride, 5.4 mL

- Two 500-mL Three-necked round-bottomed flasks with magnetic stirrer bar
- Dropping funnel

- Vacuum distillation equipment
- Magnetic stirrer plate
- Inert filter
- Schlenk tube

Procedure

1. The (2R,4R)-pentane-2,4-diol was placed in a 500-mL three-necked round-bottomed flask equipped with magnetic stirrer bar. Diol was azeotropically dried with toluene (3 × 10 mL) and placed under high vacuum to remove residual toluene.
2. Pentanediol was dissolved in diethyl ether (200 mL) under argon and triethylamine (17.7 mL) was added. The mixture was stirred at −20 °C and a solution of phosphorus trichloride (50 mL of ether and 5.4 mL of PCl_3) was added dropwise over one hour. Subsequently, the reaction mixture was stirred for three hours at −20 °C and then overnight at room temperature.
3. The amine hydrochloride which was formed was filtered under an argon atmosphere. The white precipitate was washed several times with diethyl ether (5 × 40 mL). The ether was removed under reduced pressure and the product was purified by vacuum distillation (111 °C / 20 mmHg) to yield (9.1 g) of colourless liquid.

Note: The product is sensitive to air and moisture. Store in Schlenk tube under argon at room temperature.

^{31}P-NMR (121.4 MHz, CDCl$_3$): δ 150.7 ppm

^{1}H NMR (300 MHz, CDCl$_3$): δ 1.35 ppm, (M, CH$_3$ eq), 1.60 ppm (M, CH$_3$ ax), 1.8–2.3 ppm (m CH$_2$), 4.65 ppm (m CH)

^{13}C NMR (75 MHz, CDCl$_3$): δ 22.4 ppm (d, $^3J_{POCC}$ = 1.4 Hz CH$_3$), 38.5 ppm ($^3J_{POCC}$ = 7.7 Hz CH$_2$), 64.6 ppm (d, $^2J_{POC}$ = 2.2 Hz CH), 72.5 ppm $^2J_{POC}$ = 7.3 Hz (CH).

$[\alpha]_D^{20}$ + 125.7° (c 5.425; CHCl$_3$).

12.3.3 SYNTHESIS OF (2R,4R)-2,4-BIS[(4R,6R)-4,6-DIMETHYL-1,3,2-DIOXAPHOSPHORINANE-2-YLOXY]-PENTANE

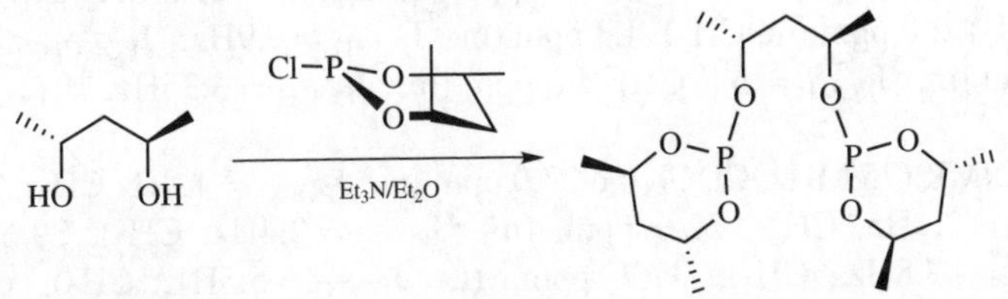

Materials and equipment

- Anhydrous diethyl ether, 300 mL
- Anhydrous toluene, 30 mL

- (2R,4R)-Pentane-2,4-diol, 2.6 g
- Anhydrous triethylamine, 7.7 mL
- (4R,6R)-4,6-dimethyl-2-chloro-1,3,2-dioxaphosphorinane, 8.4 g

- Two 500-mL three-necked round-bottomed flasks with magnetic stirrer bar
- Dropping funnel
- Vacuum distillation equipment
- Magnetic stirrer plate
- Inert filter

Procedure

1. The (2R,4R)-pentane-2,4-diol was azeotropically dried with toluene (3 × 10 mL). Diol was placed under high vacuum to remove residual toluene.
2. (4R,6R)-4,6-dimethyl-2-chloro-1,3,2-dioxaphosphorinane (8.4 g) was placed in a 500-mL three-necked round-bottomed flask equipped with magnetic stirrer bar and dropping funnel. Dioxaphosphorinane was dissolved in diethyl ether (100 mL) under argon. This solution was cooled to 0 °C while stirring and (2R,4R)-pentane-2,4-diol (2.6 g) and triethylamine (7.7 mL) in diethyl ether (100 mL) were added dropwise over one hour. The resulting suspension was stirred for a further one hour at 0 °C. Subsequently the reaction mixture was stirred overnight at room temperature.
3. The amine-salt was then filtered off under an argon atmosphere. The white precipitate was washed several times with diethyl ether (5 × 20 mL). The solvent was removed under reduced pressure. The crude product was purified by vacuum distillation (150 °C / 2 mmHg) to yield (8.8 g) of colourless liquid.

Note: The product is sensitive to air and moisture. Stored under argon at room temperature.

Azeotropically drying of the (2R,4R)-pentane-2,4-diol is necessary to prevent formation of phosphonic-acid derivatives.

^{31}P-NMR (121.4 MHz, CDCl$_3$): δ 132.3 ppm

^{1}H NMR (300 MHz, CDCl$_3$): δ 1.23 ppm (d, $^3J_{HCCH} = 6Hz$, CH$_3$ (eq)), 1.25 ppm (d, $^3J_{HCCH} = 6.4Hz$, CH$_3$), 1.45 ppm (d, $^3J_{HCCH} = 6Hz$ CH$_3$(ax.)), 1.8 ppm (m, CH$_2$), 2.0 ppm (m, CH$_2$), 4.3 ppm (m, $^3J_{POCH} = 9.9Hz$, $^4J_{POCCH} < 1Hz$ CH), 4.4 ppm (m, $^3J_{POCH} = $ n.r, CH), 4.6 ppm (m, $^3J_{POCH} = 3.35Hz$, $^4J_{POCCH} < 1Hz$, CHeq),

^{13}C NMR (75 MHz, CDCl$_3$): δ 23.0 ppm (m, $^3J_{POCC} = 1.8Hz$ CH$_3$), 23.2 ppm, $^3J_{POCC} = 2.4Hz$, CH$_3$), 22.6 ppm, (m, $^3J_{POCC} = 2.4Hz$, CH$_3$), 39.6 pm, (m, $^3J_{POCC} = 7.8Hz$, CH$_2$), 47.7 ppm, (t, $^3J_{POCC} = 5.5Hz$, CH$_2$), 60.9 ppm $^2J_{POC} = 1.6$ Hz, (CH), 67.3 ppm $^2J_{POC} = 21.0Hz$, (CH), 68.1 ppm, $^2J_{POC} = 6.0Hz$, (CH).

$[\alpha]_D^{20}$ 71.3° (c 1.15, CH$_2$Cl$_2$),

MS m/z 368 (0.29%, M^+); 367 (0.39%, $M^+ - 1$); 299 (13.8%, M^+-C_5H_{10}); 267 (10.2%, M^+-$C_5H_{10}O_2^+$); 133 (76.2%, $C_5H_{10}O_2P^+$); 69 (100%, C_5H_{10}).

12.3.4 DETERMINATAION OF OPTICAL PURITY: SYNTHESIS OF MIXTURE OF 2-PHENYLPROPIONIC ACID AND 3-PHENYLPROPIONIC ACID

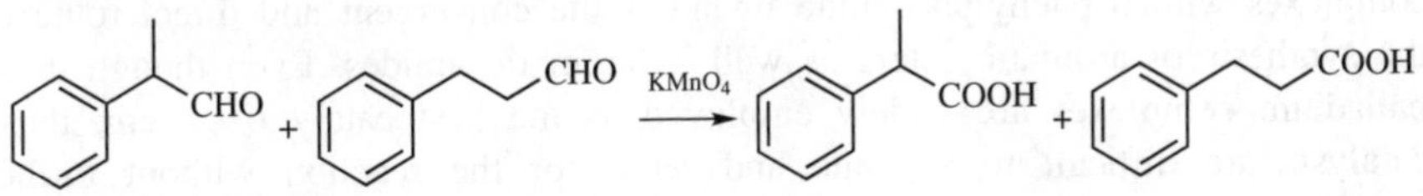

Materials and equipment

- Distilled products (mixture of 2-phenylpropanal and 3-phenylpropanal), 245 mg
- Magnesium sulfate, 350 mg
- Acetone, 50 mL
- Potassium permanganate, 331 mg
- Dichloromethane, 60 mL
- Water, 200 mL
- 1M Hydrochloric acid

- 250-mL Round-bottomed flask with magnetic stirrer bar
- Büchner funnel
- Büchner flask
- Separating funnel, 250 mL
- Two Erlenmeyer flasks, 250 mL
- Magnetic stirrer plate

Procedure

1. To a stirred mixture of aldehydes (245 mg) and magnesium sulphate (350 mg) in acetone (50 mL) potassium permanganate (331 mg) was added over two hours. The mixture was stirred at room temperature overnight.
2. The solvent was evaporated under reduced pressure and the solid residue was treated with 3×50 mL hot water and filtered. The cold aqueous solution was acidified with hydrochloric acid to pH 1, and extracted with dichloromethane. The organic layer was dried over magnesium sulfate. Removal of the solvent under reduced pressure gave 220 mg of the product as colourless liquid.

   Gas chromatographic analysis were run on a Hewlett Packard 5830A equipped with a flame ionisation detector (β-DEX, 30 m, id 0.25 mm)

12.4 PHOSPHINE-FREE DIMERIC PALLADIUM (II) COMPLEX FOR THE CARBONYLATION OF ARYL IODIDES

C. RAMESH, Y. KUBOTA and Y. SUGI*

Department of Materials Science and Technology, Faculty of Engineering, Gifu University, Gifu 501–1193, Japan

The carbonylation of aryl halides with alcohols and amines catalysed by palladium complexes with triphenylphosphine ligand is the convergent and direct route to the synthesis of aromatic esters as well as aromatic amides. Even though these palladium complexes are widely employed as the best catalytic system, those catalysts are difficult to separate and reuse for the reaction without further processing. The major drawbacks are oxidation of triphenylphosphine to phosphine oxide, reduction of palladium complex to metal and termination of the catalytic cycle. The phosphine-free, thermally stable and air resistant catalyst (1) containing a carbon-palladium covalent bond (Figure 12.3) has been found to be a highly selective and efficient catalyst for the carbonylation of aryl iodides.[1]

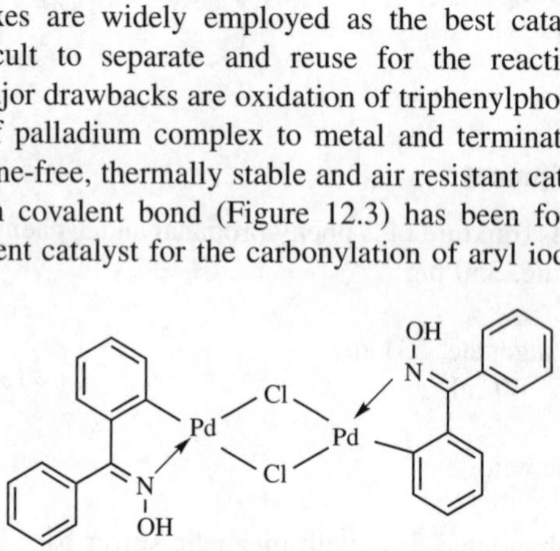

Figure 12.3 Dimeric oximepalladacycle.

12.4.1 SYNTHESIS OF THE DIMERIC OXIMEPALLADACYCLE (1)

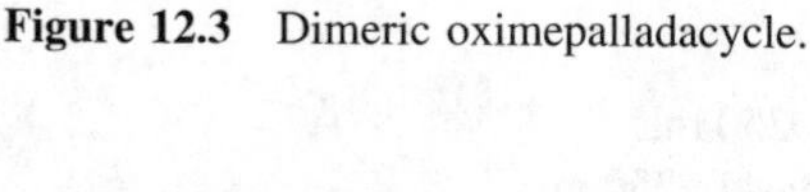

Materials and equipment

- Palladium chloride(98%), 885 mg, 5.0 mmol
- Lithium chloride, 430 mg, 10.0 mmol
- Benzophenone oxime, 980 mg, 5.0 mmol
- Sodium acetate, 410 mg, 5.0 mmol
- Dry methanol, 15 mL
- Distilled water, 250 mL

- One 50-mL round-bottomed flask
- Magnetic stirrer bar
- One glass sintered funnel, diameter 3.5 cm

Procedure

1. A 50-mL round-bottomed flask, equipped with a magnetic stirrer bar and an argon balloon, was charged with a mixture of palladium chloride (885 mg, 5.0 mmol) and lithium chloride (430 mg, 10.0 mmol) in dry methanol (7 mL) and the reaction was allowed to stir for 1 hour at room temperature.
2. To the above mixture, a solution of a mixture of benzophenoneoxime (980 mg, 5.0 mmol) and sodium acetate (410 mg, 5.0 mmol) in dry methanol (8 mL) was added, and the whole mixture was allowed to stir overnight.
3. The mixture was diluted with 250 mL of distilled water, and the pale yellow precipitate which was formed was filtered and dried under vacuum for 3 hours at room temperature.

12.4.2 SYNTHESIS OF PHENYL BIPHENYL-4-CARBOXYLATE

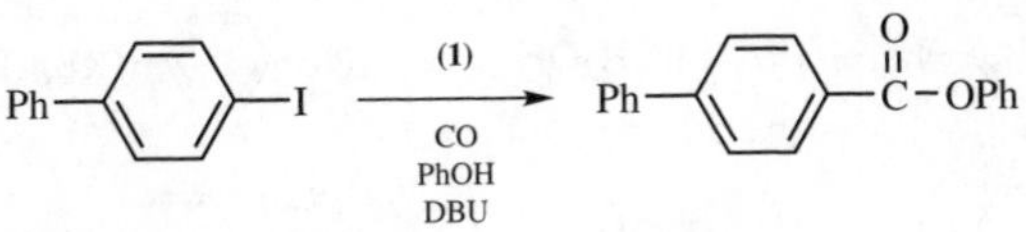

Materials and equipment

- 4-Iodobiphenyl, 583 mg, 2.5 mmol
- Phenol, 282 mg, 3.0 mmol
- 1,8-Diazabicyclo[5.4.0]undec-7-ene (DBU) 456 mg (3.0 mmol)
- Benzene 5.0 mL
- Palladium catalyst (1) 6.0 mg (0.009 mmol)
- Carbon monoxide 0.5 MPa
- Hexane (for purification) ca 250 mL

- One 50-mL stainless steel autoclave (50 mL)
- One oil bath
- Magnetic stirrer bar
- One stethoscope
- Silica gel (particle size 63–200 μm; 70–230 mesh)
- TLC plates, SIL G-60 UV$_{254}$
- Rotary evaporator

Procedure

1. A clean and dry 50 mL autoclave was charged with 4-iodobiphenyl (583 mg, 2.5 mmol), phenol (282 mg, 3.0 mmol), DBU (456 mg, 3.0 mmol), catalyst (1) (6.0 mg, 0.009 mmol) and benzene (5 mL), the autoclave was flushed with nitrogen three times and flushed with carbon monoxide two times, then pressurised with carbon monoxide(0.5 MPa).

    Warning: extreme care should be taken (e.g. operating under excellently ventilated conditions) when handling carbon monoxide throughout the reaction run!

Table 12.3 Carbonylation of aryl iodides catalysed by palladium catalyst (**1**).

Entry	Aryl Iodide	Alcohol	Product	Yield(%)
1	Ph–I	EtOH	Ph–COOEt	95
2	Ph–I	PhCH$_2$OH	Ph–COOCH$_2$Ph	91
3	Ph–I	PhOH	Ph–COOPh	91
4	Ph–C$_6$H$_4$–I	EtOH	Ph–C$_6$H$_4$–COOEt	90
5	Ph–C$_6$H$_4$–I	PhCH$_2$OH	Ph–C$_6$H$_4$–COOCH$_2$Ph	85
6	Ph–C$_6$H$_4$–I	PhOH	Ph–C$_6$H$_4$–COOPh	87
7	MeO–C$_6$H$_4$–I	EtOH	MeO–C$_6$H$_4$–COOEt	83
8	MeO–C$_6$H$_4$–I	PhCH$_2$OH	MeO–C$_6$H$_4$–COOCH$_2$Ph	90
9	MeO–C$_6$H$_4$–I	PhOH	MeO–C$_6$H$_4$–COOPh	92
10	Br–C$_6$H$_4$–I	PhOH	Br–C$_6$H$_4$–COOPh	86
11	EtOOC–C$_6$H$_4$–I	PhCH$_2$OH	EtOOC–C$_6$H$_4$–COOCH$_2$Ph	83
12	1-iodonaphthalene	EtOH	1-(COOEt)naphthalene	76
13	I–C$_6$H$_4$–C$_6$H$_4$–I	EtOH	EtOOC–C$_6$H$_4$–C$_6$H$_4$–COOEt	80
14	I–(dihydrophenanthrene)–I	PhOH	PhOOC–(dihydrophenanthrene)–COOPh	81

2. The autoclave was placed in preheated oil bath at 120 °C and allowed to stir for 3 hours; the stirring was monitored by stethoscope.
3. After 3 hours, the autoclave was cooled to room temperature and the excess carbon monoxide was purged. The reaction mixture was concentrated and purified by silica gel column chromatography (eluent: hexane–benzene) to give phenyl biphenyl-4-carboxylate (599 mg, 87%), which was spectroscopically pure.
4. For further purification for analytical purpose, the product was recrystallised from benzene–hexane and obtained as colorless pillars; m.p. 161.0–161.5 °C.

Anal. Calcd for $C_{19}H_{14}O_2$: C, 83.19; H, 5.14. Found: C, 83.44; H, 5.02.

MS m/z: 274 (M^+), 181 (M^+–OPh).

IR (KBr): 1730 cm^{-1}.

^{1}H NMR (CDCl$_3$) δ: 7.22–7.32 (3H, m), 7.37–7.53 (5H, m), 7.63–7.68 (2H, m), 7.73 (2H, dt, $J = 8.4$, 1.7 Hz), 8.27 (2H, dt, $J = 8.4$ Hz, 1.7 Hz).

^{13}C NMR (CDCl$_3$) δ: 121.72 (2CH), 125.86(CH), 127.21 (2CH), 127.31 (2CH), 128.28 (C), 128.28 (CH), 128.97 (2CH), 129.48 (2CH), 130.69 (2CH), 139.86 (C), 146.31 (C), 150.99 (C), 165.05 (C=O).

CONCLUSION

The complex (1) proved to be an efficient, stable phosphine-free palladium catalyst for the carbonylation of various aryl iodides with different aliphatic alcohols as well as less reactive phenols to give the corresponding esters in excellent yield as shown in Table 12.3.

REFERENCE

1. Ramesh, C., Kubota, Y., Miwa, M. and Sugi, Y., *Synthesis* **2002**, 2171.

12.5 CARBOXYLATION OF PYRROLE TO PYRROLE-2-CARBOXYLATE BY CELLS OF BACILLUS MEGATERIUM IN SUPERCRITICAL CARBON DIOXIDE

TOMOKO MATSUDA,*[a] TADAO HARADA,[a] TORU NAGASAWA[b] and KAORU NAKAMURA[c]

[a]*Department of Materials Chemistry, Faculty of Science and Technology, Ryukoku University, Otsu, Shiga 520–2194, Japan*
[b]*Department of Biomolecular Science, Gifu University, 1–1 Yanagido, Gifu 501–1193, Japan*
[c]*Institute for Chemical Research, Kyoto University, Uji, Kyoto 611–0011, Japan*

Carboxylation of organic molecules in supercritical carbon dioxide has attracted increasing attention as an environmentally benign synthetic method. A decarboxylase from *Bacillus megaterium* has been found to catalyse carbon dioxide fixation reactions on organic molecules.[1–4] Very recently, using this biocatalyst, the carboxylation of pyrrole was conducted in supercritical carbon dioxide, affording pyrrole-2-carboxylate efficiently.[5]

12.5.1 CONSTRUCTION OF SUPERCRITICAL CARBON DIOXIDE REACTION SYSTEM

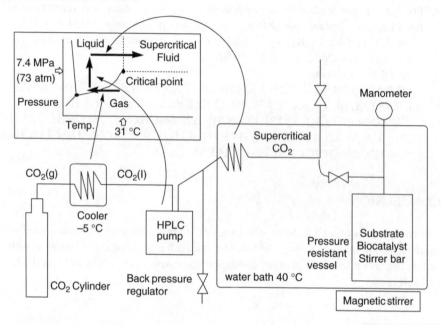

Equipment

- Carbon dioxide gas cylinder
- Cooler
- HPLC pump, Jasco PU-1580
- Stainless steel pressure-resistant vessel, 10 mL, 20 MPa
- Manometer, 15 MPa
- Water bath, 40 °C
- Magnetic stirrer plate
- Magnetic stirrer bar
- 1/16 stainless steel pipe and connectors
- Back pressure regulator
- Stop valve, 40 MPa

Procedure

1. A stainless steel pressure-resistant vessel (Taiatsu Techno, Co., Osaka, TVS-N2 type, 10 mL) was equipped with a stop valve (Whitey Co. SS3NBS4G),

manometer (Taiatsu Techno, Co., Osaka, 15 MPa) and placed in water bath on a magnetic stirrer plate.

2. The carbon dioxide gas cylinder was connected to a cooler ($-5\,°C$), HPLC pump (Jasco PU-1580 pump) and the pressure resistant vesssel as shown.

High-pressure carbon monoxide (10 MPa) is dangerous and the equipment must be constructed with special care.

12.5.2   CARBOXYLATION OF PYRROLE TO PYRROLE-2-CARBOXYLATE

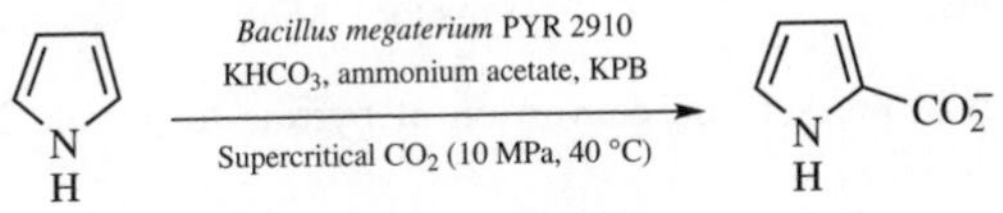

Materials

- 0.40 M pyrrole, 0.50 mL, 0.2 mmol
- 0.40 M potassium phosphate buffer, pH 5.5, 0.50 mL
- 0.56 M ammonium acetate, 0.50 mL
- *Bacillus megaterium* PYR 2910 cells, $OD_{610} = 32$, 0.50 mL
- Potassium hydrogen carbonate, 0.60g
- 0.5N sodium hydroxide, 0.5 mL

Procedure

1. *B. megaterium* PYR 2910 was grown and stored at $-20\,°C$ as previously described.[2]
2. To a stainless steel pressure-resistant vessel containing a magnetic stirrer bar, pyrrole (0.40 M, 0.50 mL), potassium phosphate buffer (pH 5.5, 0.40 M, 0.50 mL), ammonium acetate (0.56 M, 0.50 mL), the cells ($OD_{610} = 32$, 0.50 mL) and potassium hydrogen carbonate (0.60g) were added, and the vessel was sealed.

Note: Contact of the cells with a high concentration of pyrrole solution must be avoided.

3. The vessel was warmed to $40\,°C$, and carbon dioxide, preheated to $40\,°C$, was introduced to a final pressure of 10 MPa.
4. The mixture was stirred at $40\,°C$ for 3 hours.
5. Carbon dioxide was liquefied at $-10\,°C$, and then the gas pressure was released slowly.
6. The reaction was stopped by adding 0.5N sodium hydroxide (0.5 mL). The resulting reaction mixture was filtered, and chemical yield was determined by HPLC analysis as described previously.[2]

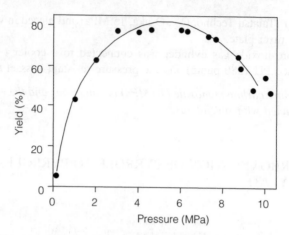

Figure 12.4 Effect of pressure on conversion of pyrrole to pyrrole-2-carboxylate by the cells.

7. The effect of pressure on the yield was also investigated and the maximum yield was obtained when the pressure was between 4 and 7 MPa, just below the critical pressure as shown in Figure 12.4.

CONCLUSION

A biocatalytic carbon dioxide fixation reaction was conducted in supercritical carbon dioxide for the first time. It will play a significant role both in identifying enzyme species suitable for biocatalysis in supercritical carbon dioxide and in developing synthetic methods using carbon dioxide.

REFERENCES

1. Wieser, M., Yoshida, T. and Nagasawa, T. *Tetrahedron Lett.* **1998**, *39*, 4309.
2. Wieser, M., Fujii, N., Yoshida, T. and Nagasawa, T. *Eur. J. Biochem.* **1998**, *257*, 495.
3. Yoshida T. and Nagasawa, T. *J. Biosci. Bioeng.* **2000**, *89*, 111.
4. Wieser, M., Yoshida T. and Nagasawa, T. *J. Mol. Catal. B: Enzymatic*, **2001**, *11*, 179.
5. Matsuda, T., Ohashi, Y., Harada, T., Yanagihara, R., Nagasawa, T. and Nakamura; K. *Chem. Commun.* **2001**, 2194.

Index

Catalysts for Fine Chemical Synthesis, Vol. 3, Metal Catalysed Carbon-Carbon Bond-Forming Reactions
Edited by S. M. Roberts, J. Xiao, J. Whittall, and T. Pickett
© 2004 John Wiley & Sons, Ltd   ISBN: 0-470-86199-1

With thanks to W. Farrington for the creation of this index.